FLOOD HAZARDS

Impacts and Responses
for the Built Environment

FLOOD HAZARDS

Impacts and Responses for the Built Environment

Edited by

Jessica Lamond   Colin Booth
Felix Hammond   David Proverbs

CRC Press
Taylor & Francis Group
Boca Raton   London   New York

CRC Press is an imprint of the
Taylor & Francis Group, an **informa** business

CRC Press
Taylor & Francis Group
6000 Broken Sound Parkway NW, Suite 300
Boca Raton, FL 33487-2742

First issued in paperback 2017

© 2012 by Taylor & Francis Group, LLC
CRC Press is an imprint of Taylor & Francis Group, an Informa business

No claim to original U.S. Government works

ISBN 13: 978-1-138-11825-6 (pbk)
ISBN 13: 978-1-4398-2625-6 (hbk)

Visit the Taylor & Francis Web site at
http://www.taylorandfrancis.com

and the CRC Press Web site at
http://www.crcpress.com

Contents

SECTION I   *Impacts, Preparedness, and Emergency Response*

SECTION II   Recovery, Repair, and Reconstruction

SECTION III   Mitigation and Adaptation to Flood Risk

SECTION IV   The Community Perspective

Aaron Mullins and Robby Soetanto

*Felix N. Hammond, Colin A. Booth, Jessica E. Lamond, and
David G. Proverbs*

Preface

As recent flooding events show worldwide, the impact of flooding on the built environment can cause widespread chaos. These flood events form part of a wider pattern of increasing flood frequency coupled with increased vulnerability of the built environment to flood hazard. Arguably, in the developed world there exists the technology to handle the vast majority of flood events, either through built-in resilience, prevention of small-scale regular flooding, or prediction and evacuation during large-scale disasters. However, events in the United Kingdom (UK) during the summers of 2007 and 2009 have revealed that, while agencies are well prepared and briefed for the majority of circumstances, flood victims remain unconvinced that the best actions have been taken. The debate surrounding the role of government, insurer, property professional, NGO (nongovernmental organization), coastal engineer, and flood victim has been played out in the media.

The idea for this book stemmed from the synergy developed within the EPSRC (Engineering and Physical Sciences Research Council)-funded international network of experts in flood repair (www.floodrepair.net), which is led by the University of Wolverhampton. This unique network of academics, property stakeholders, and restoration specialists, together with the involvement of international experts, provided a forum for exchange of ideas that has proved highly valuable. It emerged that flood management is constantly evolving, and valuable experience has been gained in the emergency management, recovery, and restoration processes in recent years. Some of this experience has yet to be captured in the literature. The aim of this book is to disseminate these discussions to a wider audience. In this book, practitioners, researchers, restorers, insurers, and policy makers discuss their perspectives on the wide issues surrounding flood risk management in one unique volume.

About the Editors

Jessica Lamond, B.Sc. (Honors), M.Sc., Ph.D., is a research fellow in the Construction and Infrastructure Department at the University of Wolverhampton, United Kingdom. Her areas of research specialism are in the field of flood risk management and property valuation studies. She has conducted research projects for research councils, government departments, the Environment Agency, private industry, and the RICS (Royal Institution of Chartered Surveyors), and has published twelve peer-reviewed journal and conference articles in the field. Recent projects include a strategic review of the impact of climate change on the risk of flooding for individual developments, guidance for the drying of buildings following flooding, and availability and cost of property-level flood risk information. She has a particular interest in structural and nonstructural flood resilience, flood insurance, the long-term financial impacts of flooding, information asymmetry in property valuation, the barriers and incentives for adaptation of the built environment to meet the challenges of climate change, and the measurement of environmental impacts within the built environment.

Colin Booth, B.Sc., M.Sc., Ph.D., PGCE, is a senior lecturer in the Environmental Engineering Department at the University of Wolverhampton, United Kingdom, with a background in geology and environmental modeling. Booth has an international research profile and, as such, he has authored or coauthored approximately 80 scientific publications (articles and chapters) in international peer-reviewed journals and books. The themes of these works are reflected in those areas of his research interests, which include environmental magnetism and urban pollution, water engineering and resource management, sustainable development and waste management, soil erosion and conservation management, and coastal and estuarine engineering.

Felix Hammond, B.Sc., PGCE, MBA, LLM, Ph.D., MGIS, FHEA, is a senior lecturer and a member of the Surveying and Property Research Group at the University of Wolverhampton, United Kingdom. He is a real estate economics and finance specialist with considerable experience in real estate policy and development economics–related research and analysis. Hammond has published widely in these specialized areas. He has served on various international development projects that impinge on land and real estate dating back to 1995. An example of such is the Ghana Environmental Resource Management Project (GERMP), which closed in (1995–1998) under the auspices of the World Bank and the Danish International Development Association (DANIDA). Hammond served as a real estate specialist on the 15-year, World Bank–led, $US55.0 million Ghana Land Administration Project, which is aimed at reorganizing the land sector of that country as a way of facilitating poverty reduction.

David Proverbs, Professor of Construction Management, B.Sc. (Honors), PG.Cert. Ed., Ph.D., FCIOB, FHEA, is presently Head of the Construction and Property Department at the University of the West of England. He is vice-chair of the Council of Heads of the Built Environment (CHOBE) in the United Kingdom. He has undertaken numerous research projects, both for the research councils (EPSRC, NERC) and various public (DEFRA, EA, CLG) and private (Lloyds TSB, RICS) sponsors. Areas of research specialism include a number of construction management themes, including international benchmarking, construction productivity, contractor performance, and client satisfaction issues. More recently his research efforts have focused on urban flood management issues, including damage assessment, flood repair, and flood resilience. The products of his research activities have been disseminated through more than 200 refereed academic journal and conference papers; textbooks, research seminars, and workshops; articles and features in the construction press; and by the publication of numerous reports and working papers. He is Editor of the *International Journal of Sustainable Development and Planning* as well as a member of the editorial board for the *CIOB Construction Annual Review*, *Disaster Prevention and Management*, and *Structural Survey* journals.

List of Contributors

Richard Ayton-Robinson
Cunningham Lindsey
Whiteley, United Kingdom

Robert Barker
Baca Architects Ltd
London, United Kingdom

Derek L. Bell
Barnsley Metropolitan Borough
    Council
Barnsley, South Yorkshire, United
    Kingdom

Konrad Bogner
Joint Research Centre
Institute for Environment and
    Sustainability
European Commission
Ispra, Italy

Anthony Boobier
Maidstone, Kent, United Kingdom

Colin Booth
School of Engineering and the Built
    Environment
University of Wolverhampton
Wolverhampton, West Midlands, United
    Kingdom

Peter Burek
Joint Research Centre
Institute for Environment and
    Sustainability
European Commission
Ispra, Italy

Harriet Caldin
Health Protection Agency
Centre for Radiation, Chemical &
    Environmental Hazards
London, United Kingdom

Holger Cammerer
Institute of Geography
University of Innsbruck
Innsbruck, Austria

Susanne Charlesworth
Department of Geography, Environment
    and Disaster Management
Coventry University
Coventry, West Midlands, United
    Kingdom

David C. Crichton
Crichton Associates
Perth, United Kingdom

Mary Dhonau
The National Flood Forum
Bewdley, Worcestershire, United
    Kingdom

Katharine Evans
National Office of Environment Agency
Leeds, United Kingdom

William John Finlinson
AMEC ENTEC
Shrewsbury, United Kingdom

Lee French
Cundall Johnston and Partners LLP
Ashington, Northumberland, United
    Kingdom

Felix Hammond
School of Engineering and the Built
  Environment
Wolverhampton University
Wolverhampton, West Midlands, United
  Kingdom

Timothy Martin Harries
King's College
London, United Kingdom

Bingunath Ingirige
School of the Built Environment
University of Salford
Salford, United Kingdom

Michael Raymond Johnson (retired)
Neant sur Yvel, France

Benjamin Rhys Kidd
CIRIA
London, United Kingdom

William John Lakin
WJL Associates
Stafford, Staffordshire, United
  Kingdom

Jessica Lamond
School of Engineering and the Built
  Environment
Wolverhampton University
Wolverhampton, West Midlands, United
  Kingdom

Kenneth Ian Manktelow
School of Applied Sciences
Wolverhampton University
Wolverhampton, West Midlands, United
  Kingdom

William Peter Medd
Lancaster Environment Centre
Lancaster University
Lancaster, United Kingdom

Aaron Mullins
Coventry University
Coventry, United Kingdom

Virginia Murray
Health Protection Agency
Radiation, Chem & Env. Haz.
London, United Kingdom

Florian Pappenberger
European Centre for Medium-Range
  Weather Forecasts
Reading, United Kingdom

Clemens Pfurtscheller
Institute of Mountain Research: Man
  and Environment
Austrian Academy of Sciences
Innsbruck, Austria

David Proverbs
Department of Humanities, Language
  & Social Sciences
University of Wolverhampton
Wolverhampton, West Midlands, United
  Kingdom

Ad de Roo
Joint Research Centre
Institute for Environment and
  Sustainability
European Commission
Ispra, Italy

Carly Rose
University of Wolverhampton
Wolverhampton, West Midlands, United
  Kingdom

Peter Salamon
Joint Research Centre
Institute for Environment and
  Sustainability
European Commission
Ispra, Italy

Victor Samwinga
University of Northumbria
School of Built & Natural Environment
Newcastle Upon Tyne, United Kingdom

Robby Soetanto
Department of Built Environment
University of Coventry
Coventry, United Kingdom

Swenja Surminski
Association of British Insurers
London, United Kingdom

Annegret Thieken
Institute of Geography
University of Innsbruck
Innsbruck, Austria

Jutta Thielen-del Pozo
Joint Research Centre
Institute for Environment and
    Sustainability
European Commission
Ispra, Italy

Frank Warwick
Department of Geography, Environment
    and Disaster Management
Coventry University
Coventry, West Midlands, United
    Kingdom

Gayan Wedawatta
School of the Built Environment
University of Salford
Salford, United Kingdom

Rebecca Kate Whittle
Lancaster Environment Centre
Lancaster University
Lancaster, United Kingdom

Roger S. Woodhead
Rameses Associates Ltd
Stone, Staffordshire, United Kingdom

1 Flooding in the Built Environment

Changing Risk and an Overview of Impacts

Jessica E. Lamond, David G. Proverbs,
Colin A. Booth, and Felix N. Hammond

CONTENTS

1.1   INTRODUCTION

Humankind's need and desire to control natural forces to preserve life and improve their lifestyles has a long history. One of the most dangerous and frequent challenges to human settlements comes from the flow of water into areas that are not designed for such inundations—in other words, flooding. Notable examples of recent floods include the 2004 Tsunami flooding in Asia, the 2005 flooding of New Orleans by Hurricane Katrina, the Summer 2007 flooding across the United Kingdom (UK), and the 2010 Pakistan flood disaster affecting an estimated 20 million people. Summaries of disaster statistics consistently note that floods are the most frequently occurring natural disaster and that they affect more people than any other single disaster type (Vos et al., 2010; Scheuren et al., 2008). Over the thirty years leading to 2006, floods killed more than 200,000 people directly (Guha-Sapir, 2006) and may have indirectly contributed to millions of illness-related deaths. In 2009, 180 hydrological disasters affected 57.3 million victims worldwide; and despite humankind's best efforts, 3,500 fatalities were attributed to flooding (Vos et al., 2010).

The impacts of these events encompass tragic loss of life, damage to built and natural environments, and massive disruption to the lives of affected populations in the short term. In the longer term, the recovery and post-recovery phases can also cause distress, disruption, health problems, and financial hardship lasting many years. While meteorological events cannot be changed (at least in the short term), the manifestation of a flood event as a result of weather extremes is to some extent controllable. The design and operation of the built environment can make a large contribution to mitigating the effect of weather; indeed, it is one of its prime purposes. The driver for increasing recent interest in adaptation of the built environment has been the extreme events, described above, and commentators have observed that the increased impacts of these events have multiple causes related to increased flood vulnerability and exposure, as well as to the higher frequency of events (Clark et al., 2002). In this book the main focus is on the exposure and vulnerability of the built environment to flooding and possible responses that will mitigate the impact of such events. This chapter is designed to introduce the wider flooding context, including the uncertainties inherent in climate change predictions, and to set out the major impacts within the built environment.

1.2   CHANGING FLOOD RISK

Flood risk is commonly defined as the function of a flood hazard on an exposed receptor that has a certain vulnerability to the hazard. Therefore, a change in risk from flooding can derive from changes in the hazard (such as increased rainfall), increased exposure (such as new development in the floodplain), or alterations in vulnerability (such as the tendency to lower building thresholds for disability access).

1.2.1   THE CHANGING FLOOD HAZARD

Flooding can occur from many sources, including from the sea (coastal and estuarine flooding), from watercourses (fluvial flooding), from overland flow of water that has not reached a natural drainage channel (pluvial flooding), from rising groundwater, and from the failure of artificial water systems (Lancaster, Preene, and Marshal, 2004).

Predictions of climate change perturbations to the water environment that may result in increased flood risks include warming seas, changing patterns of precipitation, and rising sea levels (UNFPA, 2007). Warming seas can lead to sea level rise and increased storminess. Increased precipitation, or more intense rainfall patterns, can lead to increased pluvial flooding, and sea level rise can lead to coastal erosion and coastal flooding. It is therefore possible that all kinds of flood events will increase, and the type of increased flooding will depend on local circumstances. In the United Kingdom, for example, the primary cause for concern would appear to be warmer, wetter winters leading to higher prior wetness and increased flooding, coupled with more intense summer rainfall causing more summer pluvial flooding. However, in more northern areas, the problems of more rapid snowmelt may lead to greater concern about fluvial flooding.

Current meteorological models are not able to predict these changes with any degree of precision or certainty, as they generally depend on assumptions about world emissions of greenhouse gases and mitigation responses. Different world scenarios imply climate change predictions, which also vary widely. For example, under the four different climate scenarios of the Intergovernmental Panel for Climate Change (IPCC), UK winter rainfall is predicted to experience changes from 0% to 25% by the 2050s and from 10% to 40% by the 2080s (Evans et al., 2008). Naturally, these predictions of increased flooding are heavily influenced by climate predictions, and the range of possible flooding futures is correspondingly broad. For example, the "expected annual damages" for the United Kingdom vary between £1 and 20 billion at the extremes (Wheater and Evans, 2009). If the worst-case scenario is realized, then the required scale of adaptation to future flood risk will require action on behalf of governments from all over the world. However, under any future scenario, the need to pursue adaptation within built environments will continue to challenge national and local governance.

1.2.2 CHANGING EXPOSURE AND VULNERABILITY

The close link between rivers and coasts and human prosperity has led to human settlement patterns that leave many major cities and other urban areas exposed to regular or occasional flooding. Recent population growth has meant that urban areas have expanded massively. Between 1950 and 2000, the urban population almost quadrupled to just less than 3 billion (Cohen, 2006). Over half of the world's population is now resident in urban settlements; by 2025 some estimate that two thirds of the population will live in cities and towns (i.e., more than 4 billion individuals), with 90% of these in developing countries (UNFPA, 2007). In the United Kingdom in 2001, 8 out of 10 people lived in an urban area (Office of National Statistics, 2005).

Urban development has frequently proceeded without consideration of flood risk, both in the United Kingdom and worldwide (Smith and Tobin, 1979; Lancaster, Preene, and Marshal, 2004; Crichton, 2005; Tucci, 2007). This has resulted in an increase in the population and associated buildings and infrastructure at risk for flooding (EA, 2009). Increased wealth has also led to an increase in the value of property exposed to flood hazard (Chagnon et al., 2000). As a result, the insured losses from flood events continue to grow and will do so for the foreseeable future even without the predicted increase in flood hazard (Pielke, 2006).

1.3 IMPACTS FROM FLOODING WITHIN THE BUILT ENVIRONMENT

Flooding can cause damage in several ways within an urban setting. There is a threat to personal safety given the ingress of water to normally dry areas, where escape from buildings via boats, helicopters, or other emergency vehicles is often necessary. In a high-velocity flood, people may be swept away or drowned before emergency services can reach them. Physical damage to buildings and their contents is another primary impact. As an illustration, insured losses incurred during the 2007 summer flooding in the United Kingdom were estimated at £3 billion (ABI, 2008). Damage to buildings varies greatly, depending on factors such as the velocity, depth, and duration of flooding (Kelman and Spence, 2004; Proverbs and Soetanto, 2004). Repairing that damage to habitable standards is a major undertaking that can take longer than 18 months.

Floodwaters also impede transport, covering roads during the flood and causing long-term and expensive damage. This causes disruption to commercial and domestic life, as was seen in the Cumbrian floods in 2009, where the destruction of a bridge tragically caused one death and effectively divided a town (Workington) and necessitated a 50-kilometer (30-mile) detour to get supplies and emergency services to residents (BBC News, 2009; Whipple and Brown, 2009) and later to get children to school. Essential services, such as electricity and water supplies, can also be cut off, causing discomfort in addition to health and safety hazards. In the years following a flood, there are many longer-term, more intangible impacts of flooding such as financial hardship for flood victims and stress-related problems.

1.4   THE ROLE OF FLOOD DEFENSE AND FLOOD MANAGEMENT

The purpose of flood defense is to reduce the risk of flooding, to protect property, and to safeguard life (NAO, 2001). It may seem that the natural and preferred response to increasing flood risk should be to improve and increase the structural flood barriers protecting valuable property. During the 2000 flood event that affected large parts of the country from Lewes (United Kingdom) in the south to the northern borders, although an estimated 10,000 properties were flooded, 280,000 properties were deemed to be protected against flooding by existing defenses (EA, 2001). Flood defenses can be permanent (for example, levees or flood walls) but they can also be temporary barriers, erected when flooding is imminent.

It is clear that defenses may be overtopped by an extreme event exceeding the design specification of the defense (Fleming, 2001). While it might be thought that the construction of flood defenses would always reduce the losses due to flooding, Smith and Tobin (1979) point out that this is not always the case. Often the construction of defenses allows development in what are now regarded as safe areas. The result of a catastrophic flood will consequently affect far more property and can more than offset the gains made by protecting the original settlement. This effect, known as the *levee effect*, means that flood defenses reduce the costs due to regular anticipated flooding but increase the flood loss burden overall unless development is prevented behind the flood defense. If defenses are built and poorly maintained, as has been suggested to be the case in the United Kingdom (Clark et al., 2002), the impact of the levee effect is further heightened. Thus, there is an increasing trend toward flood management rather than flood defense, which in the United Kingdom is encapsulated by the phrase "making space for water" (Rooke, 2007) and which relies on a wider range of responses to flood risk.

Within the United Kingdom, the Environment Agency is the primary body responsible for the construction and maintenance of structural flood defenses. A large part of this agency's budget is devoted to the maintenance of existing defenses and the improvement or construction of new ones (NAO, 2007). The construction of new flood defense structures now results from lengthy consultation and cost–benefit analysis of the consequences to the flood-affected community and beyond as laid out in the multicolored manual and recent Defra (Department for Environment, Food and Rural Affairs, United Kingdom) guidelines (Penning-Rowsell, Chatterton, and Wilson, 2005; EA, 2010). However, many of the chapters in this book are directed more toward other responses that come under the wider flood risk management agenda.

1.5 STRUCTURE OF THE BOOK

The range of authors who have contributed to this collection of chapters forms a unique diversity of viewpoints and experiences on the topic of flood impacts and responses. Apart from leading academics in the United Kingdom and Europe, there are also chapters written by flood practitioners, damage repair specialists, loss adjusters, environmental consultants, architects, emergency planners, insurers, government advisers, and policy makers. Organizations such as the National Flood Forum (NFF), the Health Protection Agency, the Environment Agency (EA), communities and local governments, the Construction Industry Research and Information Association (CIRIA), and the Association of British Insurers demonstrate the wealth of perspectives surrounding this critical issue.

The chapters in this book are grouped loosely to correspond to the phases of the disaster management cycle. However, in the holistic view of flood risk management adopted by many practitioners today, the boundary between these stages is blurred, with reconstruction and adaptation, for example, going hand in hand. Disasters are immediately preceded by emergency preparation and involve instant impacts and a period of emergency response. These are covered in Section 1, which looks at the different types of impacts and at the topics of forecasting and emergency warning. Section 2 explores the recovery, repair, and reconstruction phase and describes the process and good practice in recovery of flood-damaged property from the perspectives of the insurance industry, restorers, and loss adjusters. Chapters on mitigation and adaptation responses flow throughout Sections 3 and 4. Section 3 concentrates on the regulation and design of the built environment toward risk reduction, including chapters on business continuity, land use planning, property-level and infrastructure protection, and urban drainage. The community response to flooding is the focus of Section 4, shedding light on the experience of flood-affected families and the barriers and enablers that may change behaviors toward increased flood resilience in the broadest sense.

REFERENCES

ABI (Association of British Insurers) (2008). *The Summer Floods 2007: One Year on and Beyond, 2008*. London: ABI.

BBC News (2009). Rail Station for Flood Hit Town [online]. Available at http://news.bbc.co.uk/1/hi/england/cumbria/8375981.stm [Accessed October 2010].

Chagnon, S. A., Pielke, R. A., Chagnon, D., Sylves, R. T., and Pulwarty, R. (2000). Human factors explain the increased losses from weather and climate extremes. *Bulletin of the American Meteorological Society*, 81, 437–442.

Clark, M. J., Priest, S. J., Treby, E. J., and Crichton, D. (2002). Insurance and UK Floods: A Strategic Reassessment, Report of the TSUNAMI Project. Universities of Southampton, Bournemouth, and Middlesex.

Crichton, D. (2005). Flood Risk and Insurance in England and Wales: Are There Lessons to be Learned from Scotland? Technical report. London: Benfield Hazard Research Centre, University College London.

EA (Environment Agency) (2001). Lessons Learned: Autumn 2000 Floods. Bristol: Environment Agency.

EA (Environment Agency) (2009). Flooding in England: A National Assessment of Flood Risk. Bristol: Environment Agency.

EA (Environment Agency) (2010). Flood and Coastal Erosion Risk, Management. Appraisal Guidance. Bristol: Environment Agency.

Evans, E. P., Simm, J. D., Thorne, C. R., et al. (2008). An Update of the Foresight Future Flooding 2004 Qualitative Risk Analysis. London: Cabinet Office.

Fleming, G. (2001). Learning to Live with Rivers. London: The Institution of Civil Engineers.

Guha-Sapir, D. (2006). Climate change and human dimension: Health impacts of floods. *International Workshop on Climate Change Impacts on the Water Cycle, Resources and Quality.* Brussels.

Kelman, I., and Spence, R. (2004). An overview of flood actions on buildings. *Engineering Geology,* 73, 297–309.

Lancaster, J., Preene, M., and Marshal, C. (2004). Development and Flood Risk— Guidance for the Construction Industry. London: Construction Industry Research and Information Association

NAO (National Audit Office) (2001). Inland Flood Defence, Report by the Controller and Auditor General. London: National Audit Office.

NAO (National Audit Office) (2007). Environment Agency Building and Maintaining River and Coastal flood Defences in England. London: National Audit Office.

Office of National Statistics (2005). Focus on People and Migration. Office of National Statistics.

Penning-Rowsell, E. C., Chatterton, J., and Wilson, T. (2005). *The Benefits of Flood and Coastal Risk Management, a Handbook of Assessment Techniques.* London: University Press.

Pielke, R. A. (2006). White paper prepared for the workshop. Climate Change and Disaster Losses Workshop, Understanding and Attributing Trends and Projections. Hohenkammer, Germany.

Proverbs, D., and Soetanto, R. (2004). *Flood Damaged Property: A Guide to Repair.* Blackwell: London.

Rooke, D. (2007). The summer of storm. *Water and Environment Magazine,* 10, 8–9.

Scheuren, J.-M., le Polain de Waroux, O., Below, R., Guha-Sapir, D., and Ponserre, S. (2008). Annual disaster statistical review: The numbers and trends 2007. Brussels: Center for Research on the Epidemiology of Disaster (CRED).

Smith, K., and Tobin, G. A. (1979). *Human Adjustment to the Flood Hazard,* London: Longman.

Tucci, C. E. M. (2007). Urban Flood Management. WMO/ Capnet.

UNFPA (United Nations Environment Programme) (2007). *Global Environment Outlook (GEO-4): Environment for Development.* Valletta: United Nations Environment Programme.

Vos, F., Rodriguez, J., Below, R., and Guha-Sapir, D. (2010). Annual disaster statistical review: The numbers and trends 2009. Brussels: Center for Research on the Epidemiology of Disaster (CRED).

Wheater, H., and Evans, E. (2009). Land use, water management and future flood risk. *Land Use Policy,* 26, S251–S264.

Whipple, T., and Brown, D. (2009). Flood town of Workington is cut in two as bridges collapse. *The Times.* November 22, 2009 edition. London.

Section I

Impacts, Preparedness, and Emergency Response

2 State of the Art of Flood Forecasting

From Deterministic to Probabilistic Approaches

Jutta Thielen, Florian Pappenberger, Peter Salamon,
Konrad Bogner, Peter Burek, and Ad de Roo

CONTENTS

2.1 THE FIRST STEP: ACCEPTING UNCERTAINTY IN THE FLOOD FORECASTING CHAIN

Flood forecasting systems form a key part of preparedness strategies for disastrous floods and provide hydrological services, civil protection authorities, and the public with information on upcoming events. Provided the warning lead time is sufficiently long, adequate preparatory actions can be taken to efficiently reduce the impacts of the flooding (Penning-Rowsell et al., 2000; de Roo et al., 2003).

In 1674, Pierre Perrault had established a quantitative relationship between rainfall and flow for the river Seine (Perrault, 1674), effectively allowing real-time forecasting, but only the development of technology, computers, and numerical models starting in the 1960s made quantitative flood forecasts possible as we know it today.

Increasingly powerful computing systems, data storage capacities, and remote sensing technology have led to enhanced observational data collection systems, high-resolution spatial data sets over land surfaces and the oceans, and complex mathematical models, thereby furthering an understanding of the complex physical hydrometeorological processes in a river basin.

Research has shown that the combination of the particular rainfall climatology in space and time and the manifold and interactive processes at the surface and in the soil results in such highly nonlinear hydrologic responses that they become characteristic to individual catchments (Obled, Wendling, Beven, 1994; Arnaud et al., 2002; Smith et al., 2004; Segond, 2006; Smith, Beven, and Tawn, 2008). Therefore, the design of the best flood forecasting system may differ from catchment to catchment. Such a system needs to balance the availability and quality of data on the one hand and the computational representation of the processes in the atmosphere, surface, soil, and channels contributing to flooding on the other. Furthermore, it needs to respect the particular demands of the end user, as decision makers have different priorities. For example, urban areas require a significantly different management approach than reservoir operations.

Despite the differences in concept and data needs, there is one underlying issue that spans across all systems. There has been an increasing awareness and acceptance that uncertainty is a fundamental issue of flood forecasting and must be dealt with at the different spatial and temporal scales as well as at different stages of the flood generating processes (Cloke et al., 2009). The main sources of uncertainties arise either from input data (i.e., physical measurement errors, the difference in spatio-temporal scale between model and measurements, and meteorological forecasts) or from the model itself through the mathematical simplification and parameterization of the different physical processes contributing to runoff (Thielen et al., 2008).

Today, operational flood forecasting centers are changing increasingly from single deterministic forecasts to probabilistic forecasts with various representations of the different contributions of uncertainty. The move toward these so-called hydrological ensemble prediction systems (HEPS) in flood forecasting represents the state of the art in forecasting science (Schaake et al., 2006; Thielen et al., 2008), following up on the success of the use of ensembles for weather forecasting (Buizza et al., 2005) and paralleling the move toward ensemble forecasting in other related disciplines such as climate change prediction (Collins and Knight, 2007).

The advantage of using ensembles in flood forecasting for increasing the warning time and for producing reliable warning information has been demonstrated in numerous research studies (Bartholmes and Todini, 2005; Roulin, 2007; Bogner and Kalas, 2008, Kalas et al, 2008; Pappenberger et al., 2008; Thielen, Bogner et al., 2009). The use of HEPS has been internationally fostered by initiatives such as The Hydrologic Ensemble Prediction Experiment (HEPEX), created with the aim of investigating how best to produce, communicate, and use hydrologic ensemble forecasts in hydrological short-, medium-, and long-term prediction of hydrological processes (Schaake et al., 2006; Thielen et al., 2008). However, despite the demonstrated advantages, worldwide the incorporation of HEPS in operational flood forecasting is still limited (Cloke and Pappenberger, 2009). In Europe, the European Flood Alert System (EFAS) has provided medium-range probabilistic flood forecasting information for large trans-national river basins since 2005 (Bartholmes et al., 2009; Thielen, Bartholmes et al., 2009). Also, the national flood forecasting centers of Sweden, Finland, and the Netherlands have already implemented HEPS in their operational forecasting chain; and, for example, in France, Germany, Czech Republic, and Hungary, hybrids or experimental chains have been installed (Cloke et al., 2009). The applicability of HEPS for smaller river basins was tested in MAP D-PHASE, an acronym for Demonstration of Probabilistic Hydrological and Atmospheric Simulation of Flood Events in the Alpine Region, which was launched in 2005 as a Forecast Demonstration Project of the World Weather Research Programme of the WMO (World Meteorological Organization), and entered a pre-operational and still-active testing phase in 2007 (Zappa et al., 2008). Examples outside Europe include the system of Hopson and Webster (2008, 2010), who develop and run an operational ensemble flood forecasting system for Bangladesh.

Perhaps one reason for the slow transition from deterministic to probabilistic systems is the radically different way of thinking, communication, and decision making required. In a deterministic framework, decisions can always be related clearly to a single point in time and space. In the case of probabilistic forecasts, this is no longer true as other factors such as cost–loss estimations or long-term rentability and reliability considerations will play a role in decision making. Therefore, moving from deterministic to probabilistic forecasting systems requires training of staff to understand the sources of uncertainty and how they propagate through a highly nonlinear system, ways of visualizing multiple forecasts without losing focus, and guidance on how to communicate this information to different end users and decision makers in particular to obtain the best forecast information. It is an illusion that one can make reliable and accurate flood forecasts by relying solely on deterministic predictions without considering uncertainty (Pappenberger and Beven, 2006). Even with increasing technology and knowledge, uncertainties will remain—and ignoring them does not result in their disappearance!

This chapter discusses flood forecasting with regard to the major sources of uncertainty in the forecasting chain and how these have shaped state-of-the-art probabilistic flood forecasting frameworks today. The subsequent Section 2.2 deals with uncertainties in meteorological inputs, one of the driving forces of flood forecasting. Emphasis is placed on ensemble prediction systems and how they can be corrected to reduce bias. Section 2.3 discusses the uncertainties present in the hydrological model and also a method of post-processing the hydrological output that has recently been implemented in the EFAS. Section 2.4 describes how these aspects have been incorporated in the EFAS. Conclusions drawn and the outlook for the future conclude the chapter.

2.2 DEALING WITH UNCERTAINTY IN METEOROLOGICAL INPUTS FOR FLOOD FORECASTING

Operational hydrological forecasting systems that do not only predict downstream water levels (discharge) by routing typically utilize weather and radar observations and numerical weather predictions. Weather observations are used to determine the initial conditions at the beginning of the flood forecast. Additionally, they allow the performance of short-term predictions on the order of the response time of the catchment. For any longer-term predictions, numerical weather prediction (NWP) products are needed.

2.2.1 Meteorological Observations

For observation of rainfall, the weather services rely primarily on weather radars and rain gauges. Gauge data measure the rainfall depth at a point but have the disadvantage that they are collected over an unevenly spaced network, often not covering hydrologically vital areas such as steep slopes, and need further processing to obtain spatial fields (e.g., through interpolation). Numerous interpolation methods have been developed and documented in the literature. However, whatever the method, a considerable degree of uncertainty remains in the determination of the total rainfall volume and will remain in the foreseeable future. Despite the large quantity of research in this area, the significance of various sources of input uncertainty on runoff estimation is complex, sometimes shows contradictory results, and is largely catchment dependent (Michaud and Sorooshian, 1994a,b; Segond, 2006). Known effects on stream flow include change in runoff volume, peak flows, and hydrograph shifts (Arnaud et al., 2002; Smith et al., 2004). The best way of dealing with gauge data in a forecasting system will depend on the spatial density of the data and the requirements in terms of processing speed.

In contrast to gauge networks, weather radars produce estimations of spatial fields of rainfall (Battan, 1973; Collier, 1989; Sauvageot, 1991; Doviak and Zrnic, 1993; Rinhart, 1997). Although the qualitative information of the spatial and temporal evolution of the precipitation fields from radar was used quite early on by meteorologists to improve weather prediction, its operational use in hydrologic applications started only in the 1990s (Krajewski and Smith, 2002). This was due to the high degree of uncertainty in quantitative rainfall estimation, in particular when dealing with severe events that are of particular interest in flood forecasting. A promising approach to handle the uncertainty in radar-derived rainfall fields is through the generation of radar ensembles by combining stochastic simulation and detailed knowledge of the radar error structure (for recent research on this issue, see, for example, Germann et al., 2006a,b, 2009; Rossa, Cenzon, and Monai, 2010). The potential benefit of radar ensembles for increasing the warning time for flash floods was demonstrated by Zappa et al. (2008).

Gauges and weather radars have been complemented by special-purpose rain gauges, disdrometer networks (instruments measuring drop size distribution and velocity of falling hydrometeors), and lidars (Ciach and Krajewski, 1999; Sieck, Burges, and Steiner, 2006; Lewandowski et al., 2009) for capturing small-scale rainfall phenomena, while for continental or global applications, satellite remote sensing technology is increasingly used to derive rainfall information in areas with sparse gauge networks (Amitai et al., 2006; Hughes, 2006; Kidder and Jones, 2007; Levizzani et al., 2007; Liao and Meneghini, 2009; Li and Shao, 2010). However, there is general consensus that the best way forward to obtain an acceptable estimate of the "true" rainfall field is through the combination—or blending—of information from different sources and types of measuring devices (Todini, 2001; Sinclair and Pegram, 2005; Velasco-Forero et al., 2009; Li and Shao, 2010).

2.2.2 Weather Forecasting Data

A dramatic performance increase in NWP since its inception in the 1950s was fostered by high-density data collection through satellite—in particular, over the oceans—novel data assimilation techniques, and an increase in computing power allowing the simulation of the physical processes at much finer grid resolutions and a higher number of vertical levels. NWP models run at spatial grid resolutions of a few tenths of kilometers globally (e.g., the European Centre for Medium-Range Weather Forecasting [ECMWF] high-resolution deterministic forecast now has a horizontal grid spacing of about 16 kilometers) and on kilometer scale for limited areas, so-called limited area model simulations.

However, despite the general improvements in weather prediction, the skill in predicting precipitation has remained low over the past decade (Hamill, Hagedorn, and Whitaker, 2008). McBride and Ebert (2000) have shown that the occurrence of rain is much better predicted than the magnitude and location of the peak values, which are particularly important for hydrological applications. Furthermore, error and uncertainty significantly increase at higher lead times. Consequently, hydrological services rather relied on observations and short-time NWP products instead of using longer-term forecasts that would have rendered the results unreliable and therefore not useful for decision making.

Lately, however, the hydrological community has been looking increasingly at the use of ensemble prediction systems (EPS) instead of single (deterministic) weather forecasts to increase flood warning times beyond 48 hours. Since their operational launch in 1992 by the U.S. National Meteorological Centre (NMC) and the ECMWF (Palmer et al., 1993; Tracton and Kalnay, 1993), EPS have become an integral part of operational weather forecasts over the intervening years (Molteni et al., 1996; Buizza et al., 1999, 2005, 2007; Palmer and Hagedorn, 2006). EPS account for the sensitivity of the nonlinear set of equations to the initial conditions as well as errors introduced through imperfections in the model, such as parameterizations or the finite grid spacing. Ensemble forecasts can be generated by the same model with varying initial conditions and parameters (so-called stochastic physics) or by combining results from different NWP models (so-called poor-man's ensembles). To date, EPS provide valuable information on the predictability of the weather and uncertainty in the model solution for lead times up to two weeks, which is considered well outside the range of predictability for deterministic models.

Unfortunately, although EPS allow some quantification of the magnitude of the uncertainty in comparison to deterministic forecasts, they do not provide a ready-made solution to the problem of skillful flood forecasting—in fact, they have opened the door to entire new fields of research, ranging from the quantification of probabilities to the evaluation of probabilistic forecasts and the communication of probabilistic results to decision makers. One major issue with using NWPs, or EPS, is that they are often unsatisfactory in direct comparison to observations and, in particular, to point observations or catchment averages. They usually exhibit two main types of errors (bias and under- or over-dispersive). Figure 2.1 illustrates these two main error sources.

The ensemble spread relates to the probability distribution of an ensemble. In a perfect EPS, which fully accounts for all sources of uncertainty, the measurement (or

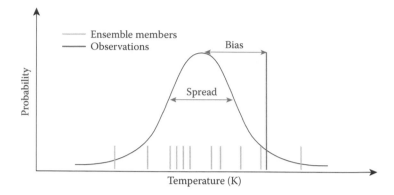

FIGURE 2.1   Illustration of ensemble spread or dispersion and forecast bias or error.

verifying truth) should be indistinguishable from the members of the forecast ensemble (Anderson, 1996; Hamill, 2001). Thus, in a reliable system, the distribution of the spread replicates the distribution of the observations. Currently, many ensemble forecasting systems have a spread that is too low at the beginning of the forecast time (Buizza et al., 1999, 2005). It is important to point out that a low ensemble spread does not necessarily imply high confidence in the ensemble mean. Any relationship between spread and skill (e.g., model quality) depends on the modeled region, the forecasted variable, the atmospheric model in use, and the application (Grimit and Mass, 2007).

In addition, many forecast models exhibit biases or systematic errors. These errors can, for example, be a result of a coarse grid resolution (other reasons may be possible, such as an incorrect simulated snow cover leading to a temperature bias). For example, in Figure 2.2, the representation of the topography of a model is displayed in connection with the real topography. In addition, in this figure a measurement station is illustrated. A temperature forecast for the model would be based on the topography of the model and thus have a systematic offset from the measurement station.

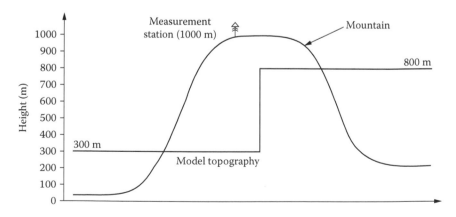

FIGURE 2.2   Comparison of model topography and "real" topography. Model forecasts are usually computed on a grid that averages the real topography.

Traditionally, a hydrological forecast is either based on a finer grid than the meteorological model or lumped, and as such is not directly compatible with forecasted variables. It has been shown that for many locations, forecasts can be improved using post-processing/pre-processing techniques (e.g., Hagedorn, Hamill, and Whitaker, 2008; Hammill, Hagedorn, and Whitaker, 2008).

2.2.3 CORRECTING ENSEMBLE PREDICTION DATA FOR FLOOD FORECASTING

The need for adapting EPS to hydrological applications was recognized as a key part of the international HEPEX (Schaake et al., 2006) and the Thorpex-Hepex Ensemble Prediction Experiment (THEPS) (Cloke and Pappenberger, 2009).

There is a wide range of methods applied to post-process ensemble forecasts, but traditionally they have been developed for meteorological applications and are not always applicable for hydrological applications. Hydrological applications require that the correction methods conserve spatial and temporal correlations (e.g., precipitation patterns), are physically coherent, maintain correlations between variables, and adequately preserve extreme events.

Only a limited number of inter-comparison studies to establish the best correction methods are available (e.g., by Wilks, 2006; Figure 2.2). Wilks (2006) concluded that the most promising approaches were logistic regression, nonhomogeneous Gaussian regression, and ensemble dressing. An application of these methods with realistic data sets yielded that logistic regressions and nonhomogeneous Gaussian regressions are generally preferred for daily temperature, and for medium-range (6- to 10- and 8- to 14-day) temperature and precipitation forecasts (Wilks and Hamill, 2007).

Four methods for the calibration of ensemble forecasts of precipitation and temperature as input to a hydrological model for flood forecasting have been tested (Pappenberger, Thielen-del Pozo, and del Medico, 2010). The four methods range from very simple to more complex approaches: (1) bias correction, (2) local quantile-to-quantile transformation (LQQT), (3) nonhomogeneous Gaussian regression (NGR), and (4) Bayesian processor of forecasts (BPF). They found that simple bias correction underperforms significantly in comparison with the other methods. Moreover, temperature is significantly easier to correct than precipitation. The NGR method outperforms all other methods at the temperature variable example, closely followed by the BPF and LQQT methods. For precipitation, the BPF method is better than the NGR method and nearly equally as good as the LQQT method. To keep the physical coherence between the variables, it is necessary to select a single method of correction for both temperature and precipitation. Because both the BPF and NGR methods produce spatially inconsistent fields, they must be rejected despite their good scores. Therefore, the LQQT method is recommended for use for data correction in hydrological modeling.

2.3   DEALING WITH UNCERTAINTY IN HYDROLOGICAL MODELING: A CHALLENGE

Hydrological research has been dealing for decades with the issue of how to account for the uncertainties in hydrological modeling (e.g., Beven and Binley, 1992; Molteni et al., 1996; Zhang and Lindstrom, 1997; Beven and Freer, 2001; Guinot and Gourbesville, 2003; Pappenberger and Beven, 2006).

Hydrological models are a mathematical combination of simplifications and suppositions of the complex processes occurring in reality that try to predict the behavior of the natural system to user-defined inputs. The simplified conceptualization of those complex natural processes inevitably leads to errors in the output of the numerical model. An additional difficulty is that the meteorological forcing required by the model already has an associated uncertainty, as described in the previous section. Hence, the model output is influenced by a mixture of errors stemming from the different inputs, initial conditions, and the hydrological model itself.

To adapt the model output to the measured discharge (being itself not free of uncertainties), a calibration is usually performed on a set of parameters that represents conceptual aggregates of spatial and temporal properties of the reality. A large variety of methods exists for this purpose (e.g., Duan, Gupta, and Sorooshian, 1992; Vrugt et al., 2003), ranging from simple least-squares approaches to complex stochastic optimization taking into account parameter uncertainty. Nonetheless, even the application of the most sophisticated calibration routines and the use of physically distributed models with high resolution and a high number of parameters may reduce the model uncertainty but will never eliminate it.

Furthermore, accounting for uncertainty in model parameters only is not sufficient to describe the total predictive uncertainty of the forecasting system. Thus, different strategies and methods have been developed in recent years to address the various sources of uncertainty.

To quantify the input errors during calibration, multiplicative error models can be applied by conceptualizing the error in meteorological observations (Kavetski, Kuczera, and Franks, 2002; Vrugt et al., 2008). The multipliers can then be calibrated for using, for example, Markov chain Monte Carlo sampling (Vrugt et al., 2008), Bayes theorem (Kavetski, Kuczera, and Franks, 2006; Salamon and Feyen, 2009), or a generalized Bayesian framework such as the generalized likelihood uncertainty estimation (GLUE) methodology introduced by Beven and Binley (1992). The resulting posterior distributions of the forcing data error provide an important diagnostic tool to quantify the magnitude and bias of the forcing data error.

It is also important to acknowledge the uncertainty due to the model structure, which requires using multiple model structures and combining them. One combination approach is achieved using so-called Bayesian model averaging (Ajami, Dunn, and Sorooshian, 2007; Vrugt and Robinson, 2007). However, the Bayesian model averaging paradigm is not directly applicable to flood forecasting systems. As mentioned above, it is more appropriate to think of hydrological models as approximations and that the "true" model does not exist in reality, whereas the goal of Bayesian model averaging is to find the true model out of a class of models under consideration.

Having demonstrated the importance of identifying the different contributions of uncertainty in a system, these can lead to such a spread in the final output that the results become, in fact, meaningless for decision making. Therefore, although there is a need to quantify all uncertainties in the system, there is a practical need to reduce them to an acceptable level (see also the discussion in Todini [2007] on this subject). Therefore, various post-processing methods have been developed in an effort to minimize the error between predicted (forecasted) and observed runoff and in view of estimating the total predictive uncertainty (Krzysztofowicz, 1999). Data assimilation and error correction are usually performed at points in the river network where observed river discharge data are available, and probabilistic forecasts can be produced through the integration of hydrological and meteorological uncertainties.

To minimize these errors, the operational model predictions must be put in better compliance with the current, latest available observations prior to using them in forecasting mode. In hydrology, this kind of error correction has also been termed *updating*. If the forecasting system shows some biases, such as systematically underestimating the peak, then the procedure of minimizing the difference between the observed and forecasted runoff is called bias correction. Especially for removing the biases from EPS, various methods have been developed, ranging from parametric (Wood and Schaake, 2008) to nonparametric approaches (Brown and Seo, 2010). O'Connell and Clarke (1981) and Refsgaard (1997) reported on the various methodologies used for model updating. It is important to note that the updating of the model output according to the latest available observed runoff values is most crucial for the accuracy of the flood forecasting system to provide end users with detailed information about the quality, reliability, and sharpness of the forecast.

Previous sections in this chapter have discussed the different sources of uncertainty. The Bayesian uncertainty post-processor (BUPP) is an excellent method to estimate the full predictive uncertainty, which was developed by Krzysztofowicz (1999) and is divided into a so-called hydrological uncertainty processor (HUP), capturing all model uncertainties arising from imperfections of the model, incorrect estimates of the parameters, and measurement errors; the input processor taking into account the meteorological forecast uncertainty forcing the hydrological model; and finally the integrator, which combines the HUP and the input processor optimally. According to this methodology, the HUP will be applied to the normal quantile transformed (Kelly and Krzysztofowicz, 1997) and possibly error-corrected discharge series at first in order to derive the predictive conditional distribution under the hypothesis that there is no input uncertainty. The uncertainty of the forecasted meteorological input is then derived from the combination of deterministic weather forecasts and EPS, and the input processor maps this input uncertainty into the output uncertainty under the hypothesis that there is no hydrological uncertainty (see Figure 2.3).

2.4 THE EUROPEAN FLOOD ALERT SYSTEM: AN EXAMPLE OF A PROBABILISTIC FLOOD FORECASTING SYSTEM

In the previous section, the theoretical outline for an operational medium-range flood prediction system was presented. The EFAS is a practical application of such a

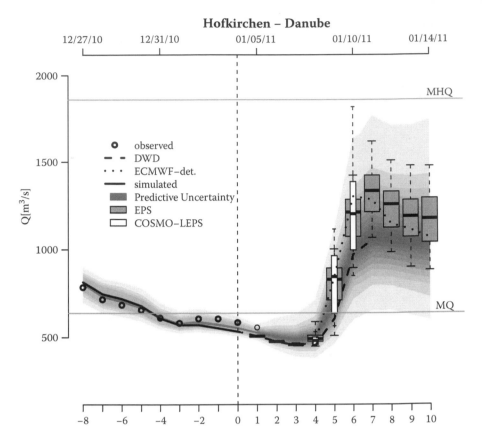

FIGURE 2.3 Example of an operational forecast for the Danube River at the gauging station Hofkirchen (Germany), showing the discharge forecasts for the next ten days starting from the 5th of January. Forecasted discharge values are corrected using the method based on wavelet transformation and VARX (Vector AutoRegressive model with eXogeneous input) and the derived predictive uncertainty is shown in gray shades.

forecasting chain and was launched in 2003 with the aim of increasing flood warning time for transnational riverine floods in Europe. It has been quasi-operational since 2005 and disseminates early flood warning information to the national hydrological services in Europe (Bartholmes et al., 2009; Ramos, Bartholmes, and Thielen-del Pozo, 2007; Thielen, Bartholmes et al., 2009). Because it is set up for the whole of Europe, it captures a higher number of events over a wider range of climatological regions than a local probabilistic flood forecasting system could do.

EFAS forecasts are based on two deterministic, medium-range forecasts from different weather services (and thus different models) and on two sets of EPS, of which one covers the medium range up to 15 days (with a spatial resolution of approximately 30 kilometers and 51 members) and the other is a limited area model EPS with a shorter range of up to 5 days (with a spatial resolution of 7 kilometers and 16 members). The reason for the shorter term EPS is to enhance the spread of

EPS within the first few days and to have a finer grid, in particular for mountainous areas. This allows for better identification of the location of the floods within the river basin (Thielen, Bartholmes et al., 2009). In a case study, it has been demonstrated that using the eight global medium-range EPS available worldwide can provide a higher reliability for the results (Pappenberger and Buizza, 2009) but it is computationally intensive.

At present, each EPS member is cascaded through the system as a deterministic forecast, and only at the end are all forecasts combined to probabilistic products (Bogner and Kalas, 2008; Bartholmes et al., 2009). Again, for computational reasons, parameter uncertainty and its impact on the flood forecasting performance has been tested offline (Salamon and Feyen, 2009) and will possibly be incorporated in the forecasting chain once it is fully operational. Pre-operationally already included is, however, the estimation of uncertainty in the post-processing as well as a correction of the output as described in the previous section.

Apart from this quantitative information provided at those locations where the forecasts can be post-processed, EFAS products are evaluated and visualized as exceedance of critical levels only—both in forms of maps and time series (Ramos, Bartholmes, and Thielen-del Pozo, 2007). Limiting the information to exceedances of critical thresholds reduced not only the wealth of information—information only produced when critical levels are exceeded—but also the uncertainty in the results—information reduced to which critical level has been exceeded and not to what extent.

Finally, probabilistic forecasting results cannot be reduced to simple hit/false alarm information, as is the case in deterministic systems. Probabilistic results need to be evaluated statistically over a long time period; for example, if there is a 25% chance of an event being forecast, then this event should take place once out of four similar forecasts. Because probabilities are mostly low, in particular for severe events, this means that a lot of data and long time series should be evaluated to find out whether or not a system is skillful (Bartholmes et al., 2009). For EFAS, a 10-year analysis has been made, showing that the skill of the system has been steadily increasing over the past decade (Pappenberger, Thielen-del Pozo, and del Medico, 2010).

2.5 SUMMARY AND CONCLUSION

In this chapter the state of the art of flood forecasting was discussed with regard to the treatment of uncertainties that arise from observational and forecasting data, as well as hydrological modeling. With improving numerical weather prediction products at spatial and temporal resolutions that approach hydrological relevant units, flood forecasting systems increasingly make use of NWP products at the short and medium range. Using meteorological EPS products as inputs into flood forecasting, reliable flood warnings can be issued at longer lead times, but at the cost of having to deal with the uncertainties that are introduced into the system at different stages and then cascade through the system.

Novel methods to correct bias in the input data were discussed, as were possibilities to account for uncertainties in the input data in the calibration of the hydrological model in addition to the consideration of parameter uncertainty. The need for output post-processing to reduce the total predictive uncertainty to a level that remains meaningful for decision makers was discussed. Novel post-processing methods explored in an operational context were presented.

Finally, an overview of flood forecasting centers using HEPS in an operational context was given and illustrated in the example of the EFAS.

ACKNOWLEDGMENTS

The authors wish to acknowledge the support of their colleagues in the FLOODS action and the EFAS team in particular for their daily support in research and keeping up with the state-of-the-art of flood forecasting systems. They would also like to acknowledge financial support from DG RTD projects IMPRINTS and KULTURisk as well as GMES-funded projects PREVIEW and SAFER for contributing to the underlying research.

REFERENCES

Ajami, N. K., Duan, Q., and Sorooshian, S. (2007). An integrated hydrologic Bayesian multimodel combination framework: Confronting input, parameter, and model structural uncertainty in hydrologic prediction. *Water Resources Research*, 43, W01403. doi:10.1029/2005WR004745

Amitai, E., Marks, D. A., Wolff, D. B., Silberstein, D. S., Fisher, B. L., and Pippitt, J. L. (2006). Evaluation of radar rainfall products: Lessons learned from the NASA TRMM validation program in Florida. *Journal of Atmospheric and Oceanic Technology*, 23(11), 1492–1505.

Anderson, J. L. (1996). A method for producing and evaluating probabilistic forecasts from ensemble model integrations. *Journal of Climate*, 9, 1518–1530.

Arnaud, P., Bouvier, C., Cisneros, L., and Dominguez, R. (2002). Influence of rainfall spatial variability on flood prediction. *Journal of Hydrology*, 260(1-4), 216–230.

Bartholmes, J. C., and Todini, E. (2005). Coupling meteorological and hydrological models for flood forecasting. *Hydrology and Earth System Sciences*, 9, 333–346.

Bartholmes, J. C., Thielen, J., Ramos, M.H., and Gentilini, S. (2009). The European Flood Alert System EFAS. 2. Statistical skill assessment of probabilistic and deterministic operational forecasts. *Hydrologyand Earth System Sciences*, 13, 141–153.

Battan, L. J. (1973). *Radar Observation of the Atmosphere*. Chicago: The University of Chicago Press, 1973.

Beven, K. J, and Binley, A. M. (1992). The future of distributed models: Model calibration and uncertainty prediction. *Hydrological Processes*, 6, 279–298.

Beven, K. J., and Freer, J. (2001). Equifinality, data assimilation, and uncertainty estimation in mechanistic modelling of complex environmental systems. *Journal of Hydrology*, 249, 11–29.

Bogner, K., and Kalas, M. (2008). Error-correction methods and evaluation of an ensemble based hydrological forecasting system for the Upper Danube catchment. *Atmospheric Science Letters*, 9(2), 95–102.

Brown, J. D., and Seo, D. J. (2010). A nonparametric post-processor for bias correction of hydrometeorological and hydrologic ensemble forecasts. *Journal of Hydrometeorology*, 11, 642–665.

Buizza, R., Asensio, H., Balint, G., Bartholmes, J., Bliefernicht, J., Bogner, K., Chavaux, F., de Roo, A., Donnadille, J., Ducrocq, V., Edlund, C., Kotroni, V., Krahe, P., Kunz, M., Lacire, K., Lelay, M., Marsigli, C., Milelli, M., Montani, A., Pappenberger, F., Rabufetti, D., Ramos, M.-H., Ritter, B., Schipper, J. W., Steiner, P., Thielen-del Pozo, J., and Vincendon, B. (2007). EURORISK/PREVIEW report on the technical quality, functional quality and forecast value of meteorological and hydrological forecasts. ECMWF Research Department Technical Memorandum No. 516, Shinfield Park, Reading, UK: ECMWF.

Buizza, R., Hollingsworth, A., Lalaurette, F., and Ghelli, A. (1999). Probabilistic predictions of precipitation using the ECMWF ensemble prediction system. *Weather and Forecasting*, 14, 168–189.

Buizza, R., Houtekamer, P. L., Toth, Z., Pellerin, G., Wei, M. Z., and Zhu, Y. J. (2005). A comparison of the ECMWF, MSC, and NCEP global ensemble prediction systems. *Monthly Weather Review*, 133(5), 1076–1097.

Ciach, G. J., and Krajewski, W. F. (1999). On the estimation of radar rainfall error variance. *Advances in Water Resources*, 11, 585–595.

Cloke, H. L., and Pappenberger, F. (2009). Ensemble flood forecasting: A review. *Journal of Hydrology*, 375(1-4), 613–626; a continuous update is provided at <http://www.ecmwf.int/staff/florian_pappenberger/heps_review.html>.

Cloke, H. L., Thielen, J., Pappenberger, F., Nobert, S., Balint, G., Edlund, C., Koistinen, A., de Saint-Aubin, C., Sprokkereef, E., Viel, C., Salamon, P., and Buizza, R. (2009). Progress in the implementation of Hydrological Ensemble Prediction Systems (HEPS) in Europe for operational flood forecasting, *ECMWF Newsletter,* 121 (Autumn), 20–24.

Collier, C. G. (1989). *Applications of Weather Radar Systems: A Guide to Uses of Radar Data in Meteorology and Hydrology.* New York: John Wiley & Sons, 1989, 294 pp.

Collins, M., and Knight, S. (2007). Ensembles and probabilities: A new era in the prediction of climate change. *Philosophical Transactions of the Royal Society A*, 1471–2962.

de Roo, A., Gouweleeuw, B., Thielen, J., Bartholmes, J., Bongioannini-Cerlini, P., Todini, E., Bates, P., Horritt, M., Hunter, N., Beven, K.J., Pappenberger, F., Heise, E., Rivin, G., Hills, M., Hollingsworth, A., Holst, B., Kwadijk, J., Reggiani, P., van Dijk, M., Sattler, K., and Sprokkereef, E. (2003). Development of a European flood forecasting system. *International Journal of River Basin Management*, 1, 49–59.

Doviak, R. J., and Zrnic, D. S. (1993). *Doppler Radar and Weather Observations.* San Diego, CA: Academic Press Inc.

Duan, Q., Gupta V. K., and Sorooshian, S. (1992). Effective and efficient global optimization for conceptual rainfall-runoff models. *Water Resources Research*, 28, 1015–1031.

Germann, U., Berenguer, M., Sempere-Torres, D., and Salvadè, D. (2006b). Ensemble radar precipitation estimation — A new topic on the radar horizon. *Proceedings of the 4th European Conference on Radar in Meteorology and Hydrology (ERAD),* Barcelona, September 18–22, 2006, pp. 559–562.

Germann, U., Berenguer, M., Sempere-Torres, D., and Zappa, M. (2009). REAL — Ensemble radar precipitation estimation for hydrology in a mountainous region. *Quarterly Journal Royal Meteorological Society*, 135, 445–456.

Germann, U., Galli, G., Boscacci, M., and Bolliger, M. (2006a). Radar precipitation measurement in a mountainous region. *Quarterly Journal Royal Meteorological Society*, 132, 1669–1692.

Grimit, E. P., and Mass, C. F. (2007). Measuring the ensemble spread–error relationship with a probabilistic approach: Stochastic ensemble results. *Monthly Weather Review*, 135, 203–221.

Guinot, V., and Gourbesville, P. (2003). Calibration of physically based models: Back to basics. *Journal of Hydroinformatics*, 5, 233–244.

Hagedorn, R., Hamill, T. M., and Whitaker, J. S. (2008). Probabilistic forecast calibration using ECMWF and GFS ensemble reforecasts. I. Two-meter temperatures. *Monthly Weather Review*, 136(7), 2608–2619.

Hamill, T. M. (2001). Interpretation of rank histograms for verifying ensemble forecasts. *Monthly Weather Review*, 129, 550–560.

Hamill, T. M., Hagedorn, R., and Whitaker, J. S. (2008). Probabilistic forecast calibration using ECMWF and GFS ensemble reforecasts. II. Precipitation. *Monthly Weather Review,* 136(7), 2620–2632.

Hopson, T., and Webster, P. (2008). Three-tier flood and precipitation forecasting scheme for South-east Asia. Available at <http://cfab2.eas.gatech.edu/>.

Hopson, T., and Webster, P. (2010). A 1–10 day ensemble forecasting scheme for the major river basins of Bangladesh: Forecasting severe floods of 2003–2007. *Journal of Hydrometeorology*, 11, 618–641.

Hughes, D. A. (2006). Comparison of satellite rainfall data with observations from gauging station networks. *Journal of Hydrology*, 327(20), 399–410.

Kalas, M., Ramos, M.-H., Thielen, J., and Babiakova, G. (2008). Evaluation of the medium-range European flood forecasts for the March–April 2006 flood in the Morava River. *Journal of Hydrology and Hydromechanics*, 56, 2.

Kavetski, D., Kuczera, G., and Franks, S. W. (2002). Confronting input uncertainty in environmental modelling, in calibration of watershed models. *Water Science and Application*, 6, 49–68.

Kavetski, D., Kuczera, G., and Franks, S. W. (2006). Bayesian analysis of input uncertainty in hydrological modeling. 1. Theory. *Water Resources Research*, 42(3), W03407, doi:10.1029/2005WR004368.

Kelly, K., and Krzysztofowicz, R. (1997). A bivariate meta-Gaussian density for use in hydrology. *Stochastic Hydrology and Hydraulics*, 11, 17–31.

Kidder, S. Q., and Jones, A. S. (2007). A blended satellite total precipitable water product for operational forecasting. *Journal of Atmospheric Oceanic Technology,* 24, 74–81.

Krajewski, W. F., and Smith, J. A. (2002). Radar hydrology: Rainfall estimation. *Advances in Water Resources,* 25(8–12), 1387–1394.

Krzysztofowicz, R. (1999). Bayesian theory of probabilistic forecasting via deterministic hydrologic model. *Water Resources Research*, 35, 2739–2750.

Levizzani, V., Bauer, P., and Turk, F. Joseph (2007). Measuring precipitation from space, EURAINSAT and the future. *Advances in Global Change Research*, Vol. 28, Springer, the Netherlands, XXVI, p. 722.

Lewandowski, P. A., Eichinger, W. E., Kruger, A., and Krajewski, W. F. (2009). Lidar-based estimation of small-scale rainfall: Empirical evidence. *Journal of Atmospeheric and Oceanic Technology,* 26, 656–664.

Li, M., and Shao, Q. (2010). An improved statistical approach to merge satellite rainfall estimates and raingauge data. *Journal of Hydrology*, 385(1–4), 51–64.

Liao, L., and Meneghini, R. (2009). Validation of TRMM precipitation radar through comparison of its multiyear measurements with ground-based radar. *Journal of Applied Meteorology and Climatology*, 48(4), 804–817.

McBride, J. L., and Ebert, E. E. (2000). Verification of quantitative precipitation forecasts from operational numerical weather prediction models over Australia. *Weather and Forecasting*, 15, 103–121.

Michaud, J., and Sorooshian, S. (1994a). Comparison of simple versus complex distributed runoff models on a midsized semiarid watershed. *Water Resources Research*, 30(3), 593–605.

Michaud, J. D., and Sorooshian, S. (1994b). Effect of rainfall-sampling errors on simulations of desert flash floods. *Water Resources Research*, 30(10), 2765–2775.

Molteni, F., Buizza, R., Palmer, T. N., and Petroliagis, T. (1996). The ECMWF ensemble prediction system: Methodology and validation. *Quarterly Journal of the Royal Meteorological Society,* 122(529), 73–119.

Obled, C., Wendling, J., and Beven, K. (1994). The sensitivity of hydrological models to spatial rainfall patterns—An evaluation using observed data. *Journal of Hydrology*, 159(1–4), 305–333.

O'Connell, P., and Clarke, R. (1981). Adaptive hydrological forecasting—A review. *Hydrological Sciences Bulletin*, 26, 179–205.

Palmer, T. N., Molteni, F., Palmer, T. N., Mureau, R., Buizza, R., Chapelet, P., and Tribbia, J. (1993). Ensemble prediction. ECMWF Seminar Proceedings Validation of Models over Europe: Vol. 1. Shinfield Park, Reading, UK: ECMWF.

Palmer, T., and Hagedorn, R. (Eds.) (2006). *Predictability of Weather and Climate*. Cambridge: Cambridge University Press, 696 pp.

Pappenberger, F., and Beven, K. J. (2006). Ignorance is bliss: Or 7 reasons not to use uncertainty analysis. *Water Resources Research*, 42(5), doi: 10.1029/2005WR004820.

Pappenberger, F., Bartholmes, J., Thielen, J., Cloke, H. L., Buizza, R., and de Roo, A. (2008). New dimensions in early flood warning across the globe using grand-ensemble weather predictions. *Geophysical Research Letters*. 35(10), L10404, doi: 10.1029/2008GL033837.

Pappenberger, F., and Buizza, R. (2009). The skill of ECMWF predictions for hydrological modelling. *Weather and Forecasting*, 24(3), 749–766.

Pappenberger, F., Thielen-del Pozo, J., and del Medico, M. (in press). The impact of weather forecast improvements on large scale hydrology: Analysing a decade of forecasts of the European Flood Alert System. *Hydrological Processes*, doi: 10.1002/hyp.7772

Penning-Rowsell, E. C., Tunstall, S. M., Tapsell, S. M., and Parker, D. J. (2000). The benefits of flood warnings: Real but elusive, and politically significant. *Journal of Water and Environment*, 14 (February), 7–14.

Perrault, P. (1674). *De l'oriegine des fontaines* (Paris, 1674), translated by A. LaRocque as *On the origin of springs*. Translated from the Paris, 1674, edition by Aurele LaRoque. Hafner, New York, 1967, 213 pp.

Ramos, M.-H., Bartholmes, J., and Thielen-del Pozo, J. (2007). Development of decision support products based on ensemble forecasts in the European flood alert system. *Atmospheric Science Letters*, 8(4), 113–119.

Refsgaard, J. C. (1997). Validation and intercomparison of different updating procedures for real-time forecasting. *Nordic Hydrology*, 28, 65–84.

Rinhart, R. (1997). *Radar for Meteorologists*. Nevada, MO: Rinhart Publications.

Rossa, A. M., Cenzon, G., and Monai, M. (2010). Quantitative comparison of radar QPE to rain gauges for the 26 September 2007 Venice Mestre flood. *Natural Hazards and Earth System Science*, 10(2), 371–377.

Roulin, E. (2007). Skill and relative economic value of medium-range hydrological ensemble predictions. *Hydrology and Earth System Sciences,* 11(2), 725–737.

Salamon, P., and Feyen, L. (2009). Assessing parameter, precipitation, and predictive uncertainty in a distributed hydrological model using sequential data assimilation with the particle filter. *Journal of Hydrology,* 376(3-4), 428–442.

Sauvageot, H. (1991). *Radar Meteorology*. London: Artech House, Inc., 315 pp.

Schaake, J., Franz, K., Bradley, A., and Buizza, R. (2006). The Hydrological Ensemble Prediction EXperiment (HEPEX). *Hydrological and Earth System Sciences Discussions*, 3, 3321–3332.

Segond, M.-L. (2006). Stochastic modelling of space-time rainfall and the significance of spatial data for flood runoff generation. Ph.D. thesis. London: Imperial College London, 222 pp.

Sieck, L. C., Burges, S. J., and Steiner, M. (2006). Challenges in obtaining reliable measurements of point rainfall. *Water Resources Research*, 43, W01420, doi: 10.1029/2005WR004519.

Sinclair, S., and Pegram, G. (2005). Combining radar and rain gauge rainfall estimates using conditional merging. *Atmospheric Science Letters*, 6(1), 19–22.

Smith, M. B., Koren, V. I., Zhang, Z., Reed, S. M., Pan, J. J., and Moreda, F. (2004). Runoff response to spatial variability in precipitation: An analysis of observed data. *Journal of Hydrology*, 298(1–4), 267–286.

Smith, P. J., Beven, K. J., and Tawn, J. A. (2008). Detection of structural inadequacy in process-based hydrological models: A particle-filtering approach. *Water Resources Research*, 44, W01410, doi: 10.1029/2006WR005205.

Thielen, J., Bartholmes, J., Ramos, M.-H., and de Roo, A., (2009). The European Flood Alert System. 1. Concept and development. *Hydrology and Earth System Sciences*, 13, 125–140.

Thielen, J., Bogner, K., Pappenberger, F., Kalas, M., del Medico, M., and de Roo, A. (2009). Monthly-, medium-, and short-range flood warning: Testing the limits of predictability. *Meteorological Applications*, 16, 77–90.

Thielen, J., Schaake, J., Hartman, R., and Buizza, R. (2008). Aims, challenges and progress of the Hydrological Ensemble Prediction Experiment (HEPEX) following the third HEPEX workshop held in Stresa 27 to 29 June 2007. *Atmospheric Science Letters*, 9, 29–35.

Todini, E. (2001). Bayesian conditioning of RADAR to rain gauges. *Hydrology and Earth System Sciences*, 5, 225–232.

Todini, E. (2007). Hydrological catchment modelling: Past, present and future. *Hydrology and Earth System Sciences*, 11(1), 468–482.

Tracton, M. S., and Kalnay, E. (1993). Operational ensemble prediction at the National Meteorological Center: Practical aspects. *Weather and Forecasting*, 8, 379–398.

Velasco-Forero, C. A., Sempere-Torres, D., Cassiraga, E. F., and Gómez-Hernández, J. J. (2009). A non-parametric automatic blending methodology to estimate rainfall fields from rain gauge and radar data. *Advances in Water Resources*, 32(7), 986–1002.

Vrugt, J., Gupta, H. V., Bouten, W., and Sorooshian, S. (2003). A shuffled complex evolution metropolis algorithm for optimization and uncertainty assessment of hydrologic model parameters. *Water Resources Research*, 39(8), 1201.

Vrugt, J. A., and Robinson, B. A. (2007). Treatment of uncertainty using ensemble methods: Comparison of sequential data assimilation and Bayesian model averaging. *Water Resources Research*, 43, W01411, doi: 10.1029/2005WR004838.

Vrugt, J. A., Braak, C. J. F., Clark, M. P., Hyman, J. M., and Robinson, B. A. (2008). Treatment of input uncertainty in hydrologic modeling: Doing hydrology backwards with Markov Chain Monte Carlo simulation. *Water Resources Research*, 44, W00B09, doi: 10.1029/2007WR006720.

Wilks, D. S. (2006). Comparison of ensemble-MOS methods in the Lorenz '96 setting. *Meteorological Applications*, 13(3), 243–256.

Wilks, D. S., and Hamill, T. M. (2007). Comparison of ensemble-MOS methods using GFS reforecasts. *Monthly Weather Review*, 135(6), 2379–2390.

Wood, A. W., and Schaake, J. (2008). Correcting errors in streamflow forecast ensemble mean and spread. *Journal of Hydrometeorology*, 9, 132–148.

Zappa, M., Rotach, M.W., Arpagaus, M., Dorninger, M., Hegg, C., Montani, A., Ranzi, R., Ament, F., Germann, U., Grossi, G., Jaun, S., Rossa, A., Vogt, S., Walser, A., Wehrhan, J., and Wunram, C. (2008). MAP D-PHASE: Real-time demonstration of hydrological ensemble prediction systems. *Atmospheric Science Letters*, 2, 80–87.

Zhang, X. N., and Lindstrom, G. (1997). Development of an automatic calibration scheme for the HBV.15 hydrological model. *Hydrological Processes*, 11, 1671–1682.

3 Flood Warning and Incident Management

Katharine Evans

CONTENTS

3.1 INTRODUCTION: WHY HAVE A FLOOD WARNING SERVICE?

Flood warning is an essential tool in the management of floods, providing people with advance notice of flooding in an effort to save lives and help people prepare before it happens. Flood warnings are provided to reduce the impacts of flooding on people's homes and businesses, through such means as moving belongings and furniture to a safer place or putting in place temporary measures to prevent floodwater from entering buildings. These actions contribute to saving money, stress, and time during the recovery period after floods.

Great advances in the fields of flood forecasting and warning have been made over the past decade, bringing improvements to the systems, techniques, and delivery of services. This has resulted in lengthening the forecast lead time, improved accuracy of areas of impact, and coverage of the service to many more communities.

Although many countries provide a flood warning service, the responsibilities and institutional structures present in different countries have led to the development of many different service models, also due to the end customer or outcome sought— for example, to warn the public or to provide forecasts to authorities only. As such, there is no consistent approach but many methodologies and the core principles of what constitutes a good warning service are shared and understood by all.

Benefits can be derived long before any official flood warnings have been issued to the public, through the investment in flood forecasting measures that underpin the ability to warn citizens. Early notification of impending floods enables responders to make preparations to assist the public, manage the event and flood defense structures, and have watercourses checked and blockages removed. Flood forecasts also enable flood defense systems to be erected or barriers closed, such as the Thames Barrier that protects London from tidal surges.

Much analysis has been undertaken into the actual monetary savings made at the household level by investing in flood warning and response activities in the United Kingdom, most recently as part of a review undertaken by the Department for Environment, Food and Rural Affairs (Defra) in July 2006: "Objective 13: The Damage Reducing Effects of Flood Warnings: Results from New Data Collection" (Tunstall, Tapsell, and Fernandez-Bilbao, 2006). This research assessed the average value of savings as £2,373 for those warned versus £1,552 for those not receiving any kind of warning. These figures are based on the following equation:

$$FDA = PFA \times R \times PRA \times PHR \times PHE$$

where FDA = actual flood damage avoided; PFA = potential flood damage avoided (property plus road vehicle damage avoided was specified in 1991 but vehicles were not included in this analysis); R = reliability of the flood warning process (i.e., the proportion of the population at risk that is warned with sufficient lead time to take action); PRA = availability (the proportion of residents or households available to respond to a flood warning); PHR = ability (the proportion of households able to respond to a flood warning); and PHE = effective response (the proportion of households who respond effectively).

Further justifications for investment in flood warning arise through increasing use of temporary flood protection measures and their heavy dependence on reliable flood forecasts. Flooding can be prevented by erecting temporary barriers that are transported to areas with the greatest need.

3.2   HISTORY OF FLOOD WARNING IN ENGLAND AND WALES

Rudimentary warning systems have been in place for many years. In 1968 the Ministry of Agriculture tried to establish a national flood warning system after widespread flooding in the same year. Forecasting of floods on some of the United Kingdom's larger river systems, such as the Severn, has been undertaken for many decades. Coastal flood forecasting was given much greater emphasis after the catastrophic flooding along the East coast in 1953, resulting in the establishment of a new Storm Tide Warning Service that has now developed into the United Kingdom Coastal Monitoring and Forecasting Service (UKCMF). With the introduction of the Environment Act in 1995 and the formation of the Environment Agency (EA) in 1996, flood warning was given even greater emphasis. Responsibility for both forecasting floods and disseminating warnings to people was given to the newly formed body. Prior to 1996 it was the responsibility of the police to disseminate warnings to affected residents.

Widespread flooding at Easter 1998 resulted in the first big push for change. There was a call for a greater effort to be made on flood warning systems and a review of the warning messages used. At this point, a color-coded system of warnings was used to convey the severity and impact of flooding for each color stage. This was poorly understood by those who had little routine exposure to the service.

> Colour-coded warnings appear to be misunderstood by nearly all who receive them. This is because the colours are spontaneously linked with the escalating probability of flooding actually occurring and not with the extent definitions to which the colours relate The interests of the public are not well served by warnings given on a colour-coded basis. (Bye and Horner, 1998)

An action plan was established to review the color-coded warning system and develop a new system that was easier to understand and gave the recipient a clearer set of advice. Coupled with this was the need to review the capacity of the 26 automatic voice messaging (AVM) systems that were used and start work to develop a single national flood warning dissemination system. The Easter floods also prompted the establishment of a sustained and targeted awareness campaign to encourage people to think about their flood risk and take action by signing up for flood warnings. The campaign started in 1999 and was delivered on an annual basis. The campaign approach has now changed from an annual event to a continual effort throughout the year, with an emphasis on local engagement and community participation, rather than through national advertising and widespread direct mail shots.

After 12 months of comprehensive research, the Environment Agency launched its new warning codes on September 12, 2000. Almost immediately they were put to good use in the widespread flooding that occurred in October and November of that year. Although there was much review and analysis of the autumn 2000 floods, it was concluded that the codes had served their purpose and were understood by those who had received them. In terms of the flood warning service, the review concluded that it should be expanded to all those at risk and, as a result, the National Flood Warning Investment Strategy to direct this effort was revised and supported by Defra and Her Majesty's Treasury. This strategy set out a £247-million, 10-year spending plan linked to a comprehensive benefits assessment program that would justify the provision of new forecasting and warning systems and an increased expenditure on detection networks and weather services. The aim of the strategy is to provide flood warning services to 80% of households and businesses at risk of flooding in the extreme flood outline by the end of the 2012/2013 financial year. To ensure that this target is met, the establishment of new services and the enhancement of existing ones are governed through a levels-of-service strategy. The underlying principle of a levels-of-service strategy is that the level of service provided to a community should be proportional to the level of flood risk in that particular community—that is, the higher the risk, the higher the level of service. Where, for flood warning, risk is measured as the product of probability and consequence, probability is the return period and consequence is the number of properties at risk. A service based on risk provides clear and objective justification for prioritizing development of the service, which in turn helps resourcing and programming of service expansion. Although an appropriate assessment for establishing warning services in general, it lacks the ability to pick up characteristics of certain flood types that would give more call for reliable and effective warnings, such as flashy catchments. Further review and consideration of the risk matrix used is currently underway in order to incorporate and assess wider flood hazard factors.

Further flooding has also challenged the flood warning service in England and Wales, and there continues the assessment of the service's performance and value, most notably those floods experienced in summer 2007. Yet again, the understanding of warning codes and the public's experience of the service were called into question. This flood review placed an emphasis on the greater need to personalize the warning service and make warnings as locally relevant and intuitive as possible. The floods also challenged the scope of the Environment Agency's warning service, which currently covers only river and coastal floods, to broaden it out to forecasting flooding from surface and groundwater sources.

[I]nstead of a one-size fits-all approach, the warnings should be tailored to different types of people and places, particularly addressing vulnerabilities, and possibly different types of flooding. (Pitt, 2008)

3.3 WHAT CONSTITUTES A GOOD WARNING SERVICE?

Although there is a tangible product of a flood warning service—that being the warning that is disseminated to the public, media, and response agencies—the service itself actually relies on many component parts. These are mainly regarded as

- *Detection:* This is the ability to detect when periods of potential flooding are likely (e.g., weather radar/forecast services and rain gauges).
- *Forecasting:* This is the ability to interpret the effects of adverse weather on flood risk areas (e.g., pre-prepared scenario predictions and modeling systems).
- *Warning:* This is the interpretation of a forecast into a locally relevant message and the dissemination of such messages through widespread channels (e.g., automatic message dissemination systems, media broadcasts, and representations of warnings over the Internet).
- *Response:* This is the ability of those at risk and those who support them (e.g., the police) to understand and prepare for flooding on receipt of a warning (e.g., awareness campaigns to inform and prepare the public, flood response planning).

See Figure 3.1 for a summary of flood warning service elements in England and Wales.

Failure to develop and deliver any one of the above elements can mean the failure of the whole system and being able to achieve the end outcome, the reduction of the impact of flooding through the preparedness of the public at risk. The elements most often lacking are those that involve the recipient end of the service, namely, the warning and response aspects. "Among both developed and developing nations, the weakest elements concern warning dissemination and preparedness to act" (UN/ISDR, 2006). This was effectively demonstrated during Hurricane Katrina in August

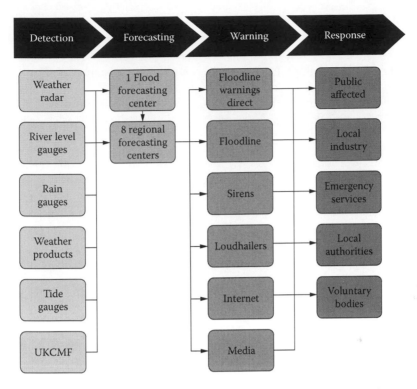

FIGURE 3.1   Summary of flood warning service elements for England and Wales.

2005, where advanced detection and forecasting systems were able to predict the hurricane many days in advance, but the translation of this vital information into local-scale impacts and the systems to warn people about the impending flood were not adequate.

Due to the clarity of responsibilities and the relative ease of establishing them, the technical elements of a warning service (such as detection or forecasting systems), usually see the most investment and effort. This is supported by recent research carried out in England and Wales that concluded:

> A key issue for improving flood warnings is to stop thinking about the warning as an "end." A flood warning should start a chain of actions that will result in people taking action to protect themselves, their family, neighbours or wider community so that there is no threat to life or property. (Fernandez-Bilbao and Twigger-Ross, 2009)

Increased effort in understanding the social context in which warnings are received and actions taken must form part of the development of an effective warning service to capitalize on local knowledge and community networks during a flood and to provide relevant information at the appropriate time for each community.

Flood warnings must make sense contextually and be intuitive to follow. There are many warnings in the public domain that have an influence over people's reactions; confusion among such services can hamper people's ability to interpret and react to flood warnings. For example, in the United Kingdom, the Meteorological Office National Severe Weather Service warnings are very prominent in media weather bulletins and reports. The presence of two warning systems for both weather and flooding in the United Kingdom has caused some confusion among those at risk.

> You look on the Internet and you look on three different Internet browsers. Three different programmes for weather and all have three different reports but same area and you are like which one.... (Pitt, 2008)

For a warning service to be most effective, it must be accurate, reliable, and timely. The area for which warnings are issued must be appropriate to the flood expected, so that the right people are warned. The systems that generate both the forecasts and the warnings must be sound and well practiced. The warning must be provided in a timely manner, not too far in advance that the forecast is not certain or people start to disbelieve the warning, but also not so close to the flooding to be of no use.

3.3.1 Flood Warnings: Names, Stages, or Colors?

There are multiple methods for designating expected states or likelihoods of flooding around the world. However, they commonly fall into two types: either a color-coded system as seen in France, or a name-based system as currently used in the United Kingdom and the United States. There is currently no consensus on the best method for designating states of flooding in an official warning service. However, the most important factor for recipients is having a system that clearly informs them when their property will flood and the severity and likelihood of such flood. Whatever method is chosen, warnings should also be used in a staged way so that people are taken through a coherent process, moving up and back down through various states of alert. This was supported in research conducted into people's interpretation of the Environment Agency's flood warnings codes following the flooding in 2007:

> According to participants, the code names have no relevance to the official definitions. The terms "Watch," "Warning" and "Severe" were misinterpreted by the public. The public perception is that the codes refer to probability and timing of flooding and act as a traffic light system. In their eyes, Watch means monitor, Warning means its coming, and Severe Warning means it is coming very soon. (Ipsos MORI, 2009)

3.3.2 Methods of Warning Dissemination

To be truly effective, warnings must be disseminated through multiple channels and backed up with consistent information in supporting sources, most successfully through local and national media and through the use of informed and trusted community representatives. Unofficial networks of local information occur regardless of whether or not official warning services exist. In receipt of an official warning or any form of flood communication, for example, through the news or by word of mouth, people draw on others' experiences and knowledge in order to assess the situation before deciding how best to react. The most effective warning services should seek to join up both the official and unofficial sources of information to avoid contradictory advice. The ability of local champions to interpret official warning information into a relevant local context can add great benefit to the service and encourage people to make appropriate responses.

The most common method of disseminating warnings used in England and Wales is by a computerized calling system known as Floodline Warnings Direct. The system was delivered in 2006, costing approximately £10 million to develop and install—the equivalent of £2.30 per person living in an area at risk of flooding in England and Wales. Flood warnings are sent to people via phone, fax, text message, e-mail, or pager. People can register up to five contact numbers and do this online or by calling the Floodline, a 24/7 helpline. Recruiting new customers to the system has always been a challenge; barriers to sign up are mainly created by people's disbelief that they are at risk. An average recruitment drive would generally see a customer take-up of approximately 30% of those targeted. Since the flooding in 2007, the Environment Agency has been working with telephone providers to automatically register an additional 400,000 customers to the service, as an "opt-out" rather than "opt-in" approach. This will take the total number of registered properties who will be able to receive flood warnings to just under one million.

Other methods of warning are the use of sirens and megaphones (loudhailers). The effectiveness of these systems has been called into question many times, as they do not penetrate fully into the community and properties at risk, and do not get messages to those away from the area. They work well in rapidly responding catchments, as sirens can be directly linked to telemetry systems and automatically triggered. They are also useful methods for warning transient populations, such as in tourist hotspots where people will not be aware of the risk.

3.3.3 THE FLOOD WARNING MESSAGE

The flood warning message must contain several elements of information in order to be understood and trigger the appropriate response by those receiving them. These elements include

1. Clear initial opening line
2. Specific local context of where it is going to flood
3. Severity of the impact or flood
4. Actions people should take
5. Timing of the expected flood; how long people have to act
6. Language that conveys the appropriate urgency
7. Where to get further information
8. Repetition of the most important points

The most important factor is a clear initial sentence, so that people receiving the message know its purpose.

Well I like the line straightaway "Flooding of homes and businesses expected in your area." I think that goes straight to the [point], you've got the warning straightaway. (Ipsos MORI, May 2009)

Where possible, warning messages should also be personal and use terms such as "*Your* home is at risk from flooding" or "*You* should take action" rather than more formal and impersonal style such as "Homes in the area are at risk from flooding."

Both the warning areas themselves and the descriptions of where and what is going to flood must be as locally specific as possible and use locations commonly known in the community.

> The Review heard that, during the 2007 floods, warnings based on named stretches of watercourse—for example, "between 'x' brook and 'y' stream"—were considered unhelpful, both to emergency responders and the public. Most people do not use watercourses as a reference point and struggle to understand information issued on that basis. (Pitt, 2008)

Is it also important to provide an assessment of what the flood will do: Is it homes and businesses that are expected to flood, or just riverside footpaths? How deep or fast flowing will the water be?

The message should also contain, in a clear informative manner, advice on actions to take. These actions should not contradict each other or this will lead to confusion—for example, "Listen to your radio or look on the Internet for more information" in the same message as "Turn off your electricity." To decide on the best course of actions to take, and to prioritize them, people should be given information about when the flood is expected to arrive or how long they will have to prepare themselves. Coupled with this, the choice of language is important in conveying the urgency of the situation. For earlier, less certain warnings, words such as "Flooding is *possible*" and "You *should* keep an eye on the situation" are appropriate; but when warnings are issued for more certain or hazardous flooding, words such as "Flooding is *expected*" or "You *must* prepare to evacuate" become necessary.

Once all the contextual information is conveyed in a message, it is important to specify where people can obtain further information—for example, the radio stations in their area that will broadcast information or the details of any call center helplines. Invariably with automated or broadcast messages, people will only be able to digest limited amounts of information, as it usually takes several seconds for people to register the purpose of the message and start listening to the content. Therefore, it is important to repeat the most important points of the message or give people the option to repeat the message or be able to hear it again by some other means. However, there is a balance to be achieved in order to achieve the optimum message, a balance between the length of the message versus the inclusion of relevant information. Whereby,

> Participants prefer that warning messages be as short and simple as possible, while conveying the key information of likelihood of flooding, expected time of flooding, and the specific locations which will be affected. All feel the message should be repeatable several times. (Ipsos MORI, 2009)

Message length is an important factor when considering the time taken to disseminate warnings through computerized calling systems, especially for large towns and cities where there is a large population to warn.

3.4   FLOOD INCIDENT MANAGEMENT

As indicated, the response element of the warning service chain is vital in reducing the effects of a flood. Much attention has focused on this in recent years.

3.4.1   FLOOD RESPONSE PLANNING IN THE UNITED KINGDOM

Flood response planning in the United Kingdom has recently made a significant move forward after the introduction of the Civil Contingencies Act in 2004 ("the Act"). The Act established the legal framework for emergency preparedness at the local level in the United Kingdom, creating Regional and Local Resilience Forums to carry out contingency planning. The Act recognizes authorities as either Category 1 or Category 2 responders, which determines their role in planning for and responding to emergencies. Category 1 responders are deemed as those at the core of response of most emergencies. They are required to

- Assess the risk of emergencies occurring and use this to inform contingency planning
- Put in place emergency plans
- Put in place business continuity management arrangements
- Put in place arrangements to make information available to the public about civil protection matters and maintain arrangements to warn, inform, and advise the public in the event of an emergency
- Share information with other local responders to enhance coordination
- Cooperate with other local responders to enhance coordination and efficiency

In the United Kingdom, the EA is a Category 1 responder, as the lead authority for modeling, forecasting, mitigating, and warning of flooding in England and Wales, along with the Scottish Environment Protection Agency and Rivers Agency in Northern Ireland.

3.4.2   MULTI-AGENCY FLOOD PLANS

Although many emergency plans exist in the United Kingdom, Local Resilience Forums are encouraged to develop a specific multi-agency flood plan to both complement other plans and to provide more detail to other, more generic plans. Figure 3.2 shows how plans interact with each other. Multi-agency flood plans are required because of the complex nature of flooding and resulting consequences, often requiring a comprehensive and sustained response from multiple organizations.

The plans cover actions that are activated in response to floods, covering the following items:

- The flood risks affecting the plan coverage area
- Communication during a flood
- Activation of the plan and triggers for doing so, such as the receipt of a severe weather warning or a flood warning

- Actions, roles, and responsibilities—these are both generic to organizations and specific to actions that need carrying out in the community
- Vulnerable people and groups—identification of people and sites at risk, such as schools
- Critical infrastructure at risk—identification and mitigation measures
- Evacuation and sheltering of people—an outline of plans in place
- Recovery plans—documenting the actions that will take place after flooding, such as supply of waste skips for flooded properties
- Plans for training and holding exercises

3.4.3 Exercising Flood Response Scenarios

Exercises are vital for testing out new or existing plans and for building up relationships between agencies. Many exercises are held at different spatial scales and testing different flood scenarios each year throughout the United Kingdom. The first national flooding exercise, "Exercise Triton," was held in 2004. It was a control post and tabletop event designed to test on a national scale the systems and management necessary to cope with an extreme flood and the subsequent recovery. More than 60 organizations and agencies took part based at 35 locations across the United Kingdom, over a 3-day period. Many lessons were learned and embedded into response plans and procedures during the following years. The next UK national-scale exercise, "Exercise Watermark," is planned for March 2011 in response to a direct requirement from the review of the summer 2007 floods: "A national flooding exercise should take place at the earliest opportunity..." (Pitt, 2008). The aim of

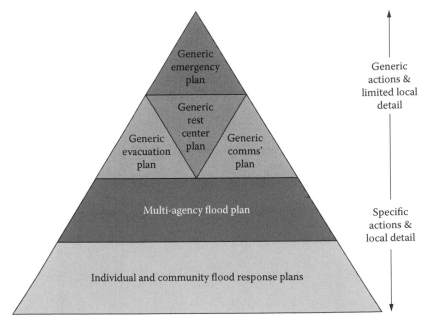

FIGURE 3.2 How multi-agency flood plans fit in with others (Cabinet Office, 2008).

the exercise will be to test arrangements across England and Wales to respond to all aspects of severe, wide-area flooding. The exercise will be developed in such a way to allow for a central management structure to test core national principles, but also allows additional scenarios to be created and run locally, which will feed into the core structure. It is also intended to encourage and include community and voluntary participation, and to publicize the event through the media in order to increase awareness of flooding and people's roles.

3.5 FUTURE CHALLENGES

The landscape of flood warning services and incident management has changed significantly over the past decade, through the introduction of new duties and technologies. The presence of a 24/7 media increases interest in flooding and its effects, bringing greater scrutiny of the services that organizations provide. Increasingly more frequent flood events throw up new problems or nuances that previously might have been overlooked, the greatest of which in the United Kingdom is the need to address the problem of surface water floods and develop warning and response services that can perform well in these situations. The impact of climate change being more frequent and intense periods of heavy rainfall means that problems of this sort will only multiply.

Significant investment in new science is currently underway in England and Wales to identify solutions for detecting, forecasting, and warning for flash floods. The most noteworthy of these improvements is the establishment of a single flood forecasting center joining together staff from the Meteorological Office and the Environment Agency. The introduction of this 24/7 operating center has been vital in providing early forecasts of periods of heightened flooding, up to 5 days ahead. There is also the development of an early alerting service for periods of extreme rainfall at a localized scale that could lead to problems from surface water flooding. The establishment of the center has resulted in a more detailed scrutiny of the linkages between meteorology and hydrology, so that areas requiring improvement or rationalization are more apparent and a seamless service ensues.

Like many other organizations, the Environment Agency currently provides a "deterministic" flood warning service, which means flood warnings are only issued when it is certain that a location will flood. Many rivers in England and Wales have a very short lead time from rainfall to flooding, making it incredibly difficult to make decisions on the best course of action in the time available. Therefore, a move from deterministic to probabilistic flood forecasting is currently being advanced through multiple research and development projects, and incremental changes will be made over the next 5 years. Potential use of probabilistic or uncertainty flood forecast information is being examined along with models of cost–loss for use in emergency response and business continuity.

The adoption of a probabilistic approach will provide stakeholders with better information for managing flood events and for making informed risk-based decisions—for example, on the need to evacuate properties, operate flood defense structures, or close rail or road links. Whether the development of such information finds its way into the content of warning messages sent to the public remains to be seen. However, the benefits should be realized by people at risk, as a clearer grasp on uncertainties in flood forecasts will lead to more timely, reliable, and accurate warnings.

REFERENCES

Bye, P., and Horner, M. (1998). *Easter 1998 Floods*. Report by the Independent Review Team to the Board of the Environment Agency (two volumes). London: Environment Agency.

Cabinet Office (2008). Developing a Multi-Agency Flood Plan (MAFP) Guidance for Local Resilience Forums and Emergency Planners. London: Cabinet Office.

EA (Environment Agency) (1999). The Flood Warning Service Strategy for England and Wales, Bristol: Environment Agency.

EA (Environment Agency) (2001). Lessons Learned: The Autumn 2000 Floods. Bristol: Environment Agency.

Fernandez-Bilbao, A., and Twigger-Ross, C. L. (2009). Improving Institutional and Social Responses to Flooding—Improving Flood Warnings Work Package 1a. Bristol: Environment Agency.

Ipsos MORI (2009). Flood Codes—Research with 'At Risk' Residents and Businesses about Flood Warning Codes; report prepared for the Environment Agency. London: Ipsos MORI

Pitt, M. (2008). *The Pitt Review—Learning Lessons from the 2007 Floods*. London: Cabinet Office.

Tunstall, S., Tapsell, S., and Fernandez-Bilbao, A. (2006). Development of Economic Appraisal Methods for Flood Management and Coastal Erosion Protection. Objective 13: The Damage Reducing Effects of Flood Warnings: Results from New Data Collection. R&D Technical Report FD2014/TR1. London: Defra.

UN/ISDR (United Nations Inter-Agency Secretariat of the International Strategy for Disaster Reduction) (2006). *Global Survey of Early Warning Systems*. Geneva, Switzerland: UN/ISDR.

4 Impacts of Flooding in the Built Environment

Derek Bell

CONTENTS

4.1 INTRODUCTION

The Metropolitan Borough of Barnsley was one of the many areas affected by the 2007 flooding. It was flooded twice in June 2007 in quick succession and, in common with many areas, the flooding affected many districts that had not experienced flooding before and were poorly prepared. This chapter describes the flooding event from the perspective of the Metropolitan Borough Council (MBC). It covers the impact on communities and council properties, the response of the borough council, and the community reaction to it. Finally, the chapter deals with the post-flood analysis and recovery actions and the future aims of the council generated from lessons learned from the flooding.

The Metropolitan Borough of Barnsley is situated in the north of England and is one of the four unitary local authorities within the county of South Yorkshire. It has a population of approximately 225,000 inhabitants spread across a number of distinct urbanized and rural communities, straddling the rivers Dearne, Don, and Dove that run through the borough (see Figure 4.1). Of these rivers, the entire 22-kilometer length of the river Dearne and the upper reaches of the Don, which crosses the west of the borough, are classified as "main river" by the Environment Agency (EA) and, as a result, the EA is responsible for their management. The River Dove with its source at Worsbrough Reservoir and its confluence with the River Dearne in the Low Valley area to the south of the Darfield village differs from the rivers Dearne and the upper Don in that the responsibility for its management, along its 8-kilometer length, falls to a variety of agencies, including the EA, Barnsley MBC, and local

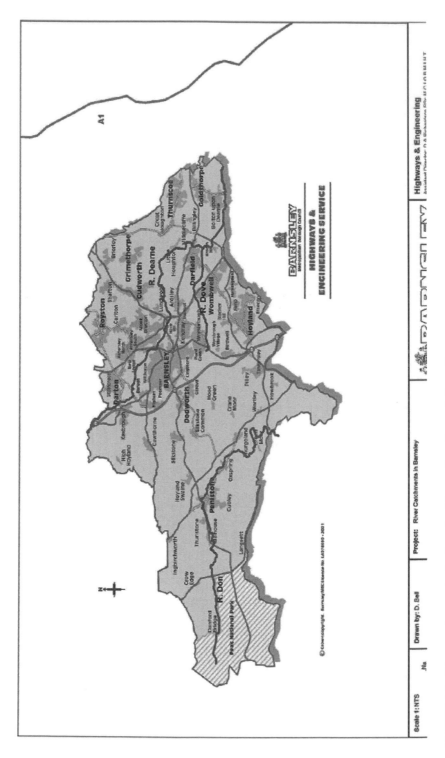

FIGURE 4.1   River catchments in the Barnsley Metropolitan Borough Council area.

landowners. This in itself brings a number of related issues that will be discussed later in this chapter.

Prior to the June 2007 floods, it was thought that the River Dearne posed the main threat of flooding within the Barnsley area. The Environment Agency's Local Flood Warning Plan identified a total of 352 properties as being at risk within the 1 in a 100-years indicative flood zones. Historically there had been a few small-scale flooding events within the borough; usually these affected the Darton village, which is known to be the first community to be affected by high water levels in the River Dearne and acts as an informal "barometer" for flooding across the borough. This area is known to have flooded during the 1950s, 1960s, 1970s, and also in the autumn of 2000. The impact of these events was relatively small scale. A handful of properties in the upper Don catchment were flooded during flooding events in the 1950s and 1960s.

In addition to this, very little was known about the flood risk posed by the River Dove, and there were no records of flooding affecting the Barnsley catchment area. Although the EA has a level gauge at its source at Worsbrough Reservoir, this was installed for operational monitoring and did not form part of the flood warning systems during the June 2007 floods. There was, in fact, no formal flood warning for any part of the Dove catchment. As a result, residents and businesses along the catchment were not forewarned of the threat of flooding in this area.

As a unitary local authority, Barnsley MBC has well-established emergency, resilience, business continuity plans, and planning arrangements (Barnsley MBC, 2007a,b), in accordance with statutory obligations, as detailed in the Civil Contingencies Act (CCA) (Cabinet Office, 2004).

The MBC participates in the South Yorkshire Local Resilience Forum (LRF), which manages the preparation and management of emergency planning arrangements for Category 1[*] and Category 2[†] organizations. These are known as the key responding agencies, across the entire LRF area, which coincides with the South Yorkshire Police Force area. The LRF engages on all levels with members of the respective organization, including chief officers, middle management, and operational and technical staff in order that appropriate strategic and policy arrangements are put in place for all types of emergency scenarios, including flooding. Once issues are identified, they are managed through distinct phases, which include planning, emergency response, and recovery; this ensures that an effective response is delivered on the ground.

The two flooding events of June 15 and June 25, 2007, tested these arrangements to the maximum. The rainfall experienced during June 2007 was approximately four times the seasonal average for the area. During a 24-hour period between June 14 and 15, a total of 118 millimeters was recorded at a local weather station, with 68 millimeters recorded in the northwest part of the borough at Cannon Hall, Cawthorne. The intense rainfall continued during the days leading up to June 25 when a further

---

[*] CCA Category 1 agencies are from the public sector and have a statutory duty to risk assess and plan for all types of emergencies within their area; they include the Environment Agency, local authorities, and police, fire, and ambulance services.
[†] CCA Category 2 agencies are from the private sector and other bodies and have a statutory duty to cooperate with the Category 1 agencies; they include the gas, water, and electricity utility companies and transport operators.

80 millimeters was recorded at Cannon Hall. This additional rain fell onto ground that was already saturated. These events constituted a civil emergency on an unprecedented scale, across the whole of the borough.

4.2 COMMUNITY IMPACT

The people of Barnsley were affected by heavy and sustained rainfall that continued for a 10-day period leading up to the first recognized flood event, starting early in the morning of Friday, June 15, 2007. All 352 properties identified within the Environment Agency Local Flood Warning Plan (EA, 2007) were inundated; a significant number of other parts were also affected.

This was by no means the only effect; floodwaters made a number of roads impassable, which resulted in traffic congestion on all arterial routes in and around the town. All major road river crossings were also affected. This caused residents the added inconvenience of having to take long detours to avoid flooded roads. These longer journeys were inevitably made worse by traffic jams that had built up during the morning rush hour. The result was severe delays for people attempting to get to work and school, and impaired movement for the emergency services.

It was evident that the Dearne, upper Don, and Dove river catchments were all seriously affected by such unprecedented rainfall. Incidents of flooding were being reported by members of the public in all parts of the borough, many of which were a considerable distance from the main rivers. These included minor watercourses that were being overwhelmed, groundwater discharge, and surface water runoff into areas that would not normally be affected by flooding.

Six distinct communities were severely affected, these being Darton, Lundwood, Darfield Bridge, and Bolton-on-Dearne on the River Dearne, and also Low Valley and Aldham Bridge areas on the River Dove. The impact ranged from just a few inches to several feet of water flowing into properties. In many cases, properties were affected by diluted, untreated sewage that had contaminated the floodwaters when wastewater treatment installations and combined sewers were overwhelmed.

In an overwhelming majority of cases, residents were poorly prepared for responding to this level of flooding, but not because they had received no warning. There had been many public messages issued throughout the event via every available communication channel. The Met Office provided severe weather warnings while the Environment Agency issued Flood Warning messages for the area within their river catchments at risk of flooding. In most cases, those affected did not heed the warnings as there had not been a history of flooding in these areas in living memory. As a result, they took very little in the way of preparatory steps to either prevent the floodwater from entering homes or mitigate the effect of water once it had entered. Matters were not helped by residents of unaffected areas who went along to affected parts on "sightseeing trips." Quite often they would use 4 × 4 vehicles to travel through deep accumulations of water on the roads, causing bow waves that broke through doors and windows of properties that would not have otherwise been flooded.

Throughout the June 15th floods, the gaze of local, regional, and national media spotlights fell onto the small number of areas that had been affected; this included Barnsley. The attention of politicians and responding agencies focused on these areas, resulting in resources being made available and distributed to them. The affected residents were frequently interviewed by the media and regularly seen on regional and national television bulletins. However, this all changed after the second flood, which occurred only 10 days later.

The effect of the subsequent flooding event that occurred on Monday, June 25, 2007, had a much more devastating effect. Not least because all of those who were inundated on June 15th were once again flooded, but additional properties, similar in numbers to the original flooding event, were also flooded. In total there were an estimated 660 residential and 72 commercial premises flooded on this date. These figures represent the numbers that were made known to the Barnsley MBC, and around 40% of this figure represents properties that were affected by surface water flooding. However, during the recovery phase of the event, it became clear that "official figures" were just the tip of the iceberg. Many homeowners and businesses did not make themselves known to the Council but dealt with the clean-up themselves, or directly through their insurance company, without needing to inform the Council. In other cases they actively chose not to contact any agency or their insurance company because they did not want to appear on an "at-risk" list, which they perceived would blight their property or business. Therefore, a truly accurate figure of those affected by the flooding will never be known.

4.3 IMPACT ON THE BARNSLEY COUNCIL

There is a complex network of Council-owned buildings, sites, premises, and assets that are controlled and managed by many Council directorates; each service area was adversely impacted by the torrential storms of June 2007. The Council's call center was barraged with requests for assistance, particularly for the distribution of sandbags. There was also an excessive volume of calls from other Council services requesting help in relation to flooding of buildings or sites under its control. The demand for such flood resilience measures grossly outstripped the capability to supply them, and the available staffing resource found it difficult to keep up with the demands being made on them.

Flooding affected many buildings under Council ownership, such as its offices, schools, depot sites, and community centers. The rainstorms during these days also caused structural damage to the Council's building portfolio, resulting in partial roof collapses, ingress of water through roofs, and surcharging of drainage systems that could not cope with the volume of water trying to pass through the system.

In addition, there were a number of shared sites, operated in conjunction with the Barnsley Primary Care Trust that operates residential care homes for the elderly. These sites had to contend with the flood damage while also trying to maintain an appropriate level of service to their clients. Throughout this episode they had to carry out an ongoing dynamic risk assessment to determine whether to continue to occupy the building or to evacuate. They would then need to deal with the issues associated with the movement of elderly people to an alternate site and manage the associated problems such as organizing the transport arrangements and finding a suitable site to accommodate those residents.

Surface water was also the cause of a large amount of infrastructure damage to the primary and local road networks, with some roads being completely washed away. Furthermore, there was substantial damage to the borough's road bridges, and in some cases they were washed away or rendered unsafe for traffic. Debris washed into the river blocked the watercourse and exacerbated the over-land flows. The combination of river channel erosion and surface water flooding caused other problems to the highway drainage systems. Silt deposited on the roads was washed into road gulleys, causing them to block, so that the water on the highway could not flow into the drainage systems. This resulted in increased levels of surface water flooding of highways, which in turn led to larger amounts of property-level flooding and additional amounts of traffic disruption due to road closures.

There were other impacts on the functioning of Council services, including many public rights of way (PROWs) being closed due to flood damage. Footpaths and footbridges were washed away or suffered sufficient damage to prevent pedestrian traffic from passing over them. Significant lengths of PROWs were damaged to such a degree by the flood that they had to be completely replaced at considerable cost to the Council.

Throughout the flood event and during the aftermath, the Council's call center experienced a 25% increase in call volumes from the public that continued for several weeks after the flood. Many of these calls were to request assistance to limit or mitigate the effects of flooding, while others were to report damage caused by the floods to buildings, bridges, roads, other infrastructure, and assets. In many cases, the call center was recording details of damages and blockages on the river networks that were not within its remit, thus causing further delay while the appropriate agency was identified and call details passed on. Although the Council and these other organizations were working to full capacity to deal with the aftermath of the flooding incident, delays in response times were experienced. This was perceived by the general public as a poor level of service and resulted in repeat calls that further exacerbated the situation.

There were a total of 48 separate localities affected by both June flooding events, comprising single properties, small groups of properties, and entire neighborhoods. Among the groups that became victims of the flooding was a community of traveling families living in seven mobile homes in the Low Valley who had their homes totally destroyed by the floods. It was necessary to procure temporary accommodation to house them while their caravan site was inundated and for the duration of the recovery process.

4.4 THE COMMUNITY PERSPECTIVE

In the majority of cases, those affected felt that the responsibility for their property flooding fell to the local council, irrespective of the cause of flooding or the management responsibility of the flooding source. As a result, their first point of contact was the local authority to either request assistance or report defects, points of damage, and blockages in the river and watercourse channels. This was due, in part, to their lack of knowledge and understanding of the roles of key agencies such as the Dearne and Dove Internal Drainage Board, the Environment Agency, the Council, and Yorkshire Water (the local private water company). Responsibility for the transfer of information was therefore concentrated within the local authority.

Under a well-established common law principle known in legal terms as a "riparian owner," the responsibility for sections of watercourse on private land falls to the owner of the land adjacent to the river or watercourse. The homeowner living on land abutting the river or watercourse is required to maintain riverbeds and riverbanks, allowing the flow of water to pass without obstruction. Local residents in the affected areas were oblivious to their riparian duties and in many cases they incorrectly assumed that these responsibilities fell to the local authority or the Environment Agency. Consequently, the riparian owners were surprised to find out that they had these responsibilities and in some cases refused to accept this fact. Landowners wanted to apportion the blame for any lack of maintenance to another party, when in fact the responsibility was theirs and that of their neighbors.

Initially, the main request for assistance was for sandbags, which in many cases would have been futile, as the source of flooding rendered them ineffective due to the depth or velocity of the floodwater. Nevertheless, many thousands of sandbags were deployed over the two flooding events and some proved effective. Where possible, this was done on a priority basis to those who had the least capability to help themselves, such as the elderly, the disabled, and caregivers. The sandbags were deployed from a single, central depot, thus suffering delays as the deliveries were caught up in the traffic jams that ensued as a result of the flooding and road closures. Those directly affected expected the Council to provide them with sufficient sandbags to provide adequate protection for their properties, and they became increasingly frustrated that their requests were not dealt with promptly. Most owners seemed unaware that the responsibility for the protection of a property falls to the owner. They wrongly perceived that the local Council has a duty to carry out this function, whereas local authorities within the United Kingdom do not have any statutory obligation to provide sandbags or any other flood defense materials. In fact, many local authorities have made the strategic decision not to provide sandbags at all, as they have a limited effect and are costly to provide and deploy.

Those who were registered with the Environment Agency Flood Warning Direct Scheme, which provides advance warnings of a potential flood, were able to take steps to save some of their belongings by moving them to the upper floors of their homes. In fact, of the 352 properties known to be at risk for flooding from rivers within Barnsley, only 180 (51%) had signed up to the scheme. This level of disregard was further evident when, despite an intensive publicity campaign through meetings and face-to-face contact immediately after the floods, this number increased only marginally to 187 (53%). Many of those who were not registered with the Flood Warning Scheme relied on word of mouth. Usually this meant obtaining flood warning information from one of their immediate neighbors who were registered with the Environment Agency scheme. The informal and unreliable nature of this arrangement rendered it of little benefit.

A further point of consternation for flood-affected communities was road closures. Many local residents incorrectly assumed that it was the responsibility of the Council to close flooded roads, when in fact it was the responsibility of the police to do so. Councils position the "Road Closed" signs and barriers following a request from the police, but it is the responsibility of the police to enforce the closure. Normally, a police officer is posted at the point of closure to prevent vehicular access to the closed section. However, due to the large number of roads needing to be closed across all the authorities within the South Yorkshire police force area and other emergency duties during the floods, it proved an impossible task and usual practice could not be maintained.

4.5 THE RESPONSE EFFORT

The Barnsley MBC is reliant on warnings from the Met Office (weather warnings) and the Environment Agency (flood warnings), which are issued in advance of severe weather and potential flooding events. Heavy rain had been forecast on the days leading up to Friday, June 15, 2007. The first Flood Watch was deployed in the upper catchment of the River Don and the River Dearne during the early hours of the morning of June 15th.

Throughout the early part of the morning, customer call volumes increased to such a level that South Yorkshire police established a local command post to deal with the emergency incident. The Barnsley MBC set up its tactical command point, known as the Emergency Control Team, to deal with the many requests for assistance from local residents and internal service areas. The key functions of the Emergency Control Team were to

- Collate and coordinate all requests for assistance as they were received
- Liaise with service managers to ensure they could be resourced
- Liaise with other key responders to coordinate the response effort
- Report to the chief executive to provide an update on the evolving situation

This led to the declaration of a Major Incident by the Council that morning, and convening the Senior Emergency Management Team, consisting of the executive directors from all directorates, headed by the chief executive. The role of the Emergency Management Team was to provide strategic leadership in response to the flooding event by

- Defining the strategic response to the incident by the Council
- Agreeing to a corporate media strategy for dealing with the impact of the incident
- Agreeing on the approach to be adopted for warning and informing the general public
- Developing a recovery phase implementation strategy for the incident

During the early stages of the event, it was critical that information from the affected areas was fed to the Emergency Control Team, to allow an appropriate response to be provided and to also gauge the scale of the event. The Emergency Control Team consisted of assistant directors from across the Council incorporating all affected service areas. This was a 24-hour process dealing with the specific aspects of managing the flooding emergency.

Throughout the following weekend, the focus was to identify the need for temporary accommodation and reduce hardship to those who had to be evacuated from their flooded homes. Many did not use the specially established reception center but chose instead to either stay in their homes on the uppers floors or to stay with neighbors or relatives who lived nearby. Although they were extremely traumatized, it was found that most residents were reluctant to leave their property after the flooding. It was unclear whether this was due to the shock of the event or concern for the safety of their homes, as the street lighting and domestic electricity supplies had failed in many communities. Liaison took place between the Barnsley MBC and South Yorkshire police to arrange additional patrols in the flood-affected areas to alleviate these fears, although at the time it was noted that no thefts were reported.

It was quickly recognized that the homeowners preferred to utilize respite centers, which provided food and hot drinks along with washing and changing facilities. Unlike other types of emergencies, the residents still had access to their homes as they were wanting to start the clean-up process themselves rather than rely on external agencies to carry out this on their behalf. However, they did expect help from the Council with the provision of cleaning materials and the disposal of flood-damaged property.

In the immediate aftermath of June 15th, much of the focus for the Council and emergency services was in the following areas:

- *Road closures:* Manage the large number of road closures across the borough, many of which had suffered significant damage
- *Dangerous structures:* Inspection of damaged buildings
- *Pumping operations:* Coordination with the fire service to pump out flooded premises
- *Restoration of utilities:* Carry out safety checks to enable reconnection of domestic gas, water, and electricity supplies
- *Highways:* Temporary repairs to make roads safe for use
- *Intelligence gathering:* Obtain other essential information regarding the function of affected communities and services

The servicing of these key priorities was an essential part of the response during the early part of the emergency phase of the response effort; this continued throughout the weekend. By midday on Monday, June 18th, the emergency phase of the flood response was formally closed down and the Emergency Control Team disbanded.

The responsibility for the ongoing management of flood-related issues was handed over to the newly established Flood Recovery Team. This consisted of appropriate managers from the affected service areas, typically assistant directors, who were able to authorize the activities necessary to bring about a return to normality. The Flood Recovery Team was split into four separate teams to manage the recovery process on a loosely geographical basis. Their purpose was to

- Coordinate clean-up operations
- Ensure that infrastructure reinstatement took place
- Assess housing needs
- Assess hardship needs

- Ensure the safety of the public
- Ensure clear communication to facilitate the Flood Recovery Team objectives
- Liaise with local councilors
- Report on the ongoing progress and issues to the Emergency Management Team

The strategic approach from the Emergency Management Team was to ensure that the recovery of the affected parts of the borough was carried out in a manner that was "tenure neutral," and to ensure that neither council tenants nor private residents were discriminated against during the clean-up operation.

There was intense activity during this time, with staff from many Council departments being redeployed from their normal duties to the flood recovery response. These staff were used to organize gas and electricity safety checks to be carried out at 200 of the 350 properties known to have been inundated, and to arrange for the removal of flood debris and damaged white goods and furniture from these homes. Water sampling was also carried out in the localities that had suffered floodwater contaminated with diluted untreated sewage when waste treatment works or combined sewerage systems were inundated.

Site visits were carried out in each area, and Council staff made door-to-door contact with the residents of flooded homes. This allowed important information to be gathered from the residents, including that in Table 4.1.

TABLE 4.1
Resident Post-Flood Information Transferred

Information	
Obtained from Residents	**Guidance Passed to Residents**
Barnsley MBC: Tenant visit contact details:	Barnsley MBC: Advice for owners whose homes have been affected by flooding
• Occupant details	
• Alternate accommodation details	Barnsley MBC: Advice for private tenants/private landlords whose homes/properties have been affected by flooding
• Insurance arrangements	
• Medical requirements	
Homeowner request for assistance form	Barnsley MBC: Frequently asked questions about flooding
	Barnsley MBC: Advice for council tenants whose homes have been affected by flooding
	Barnsley MBC: Protect your house from mold
	Barnsley MBC: Welfare rights contact numbers
	Advice from Environment Agency: General flood advice
	Environment Agency: Flood Warnings Direct (application form)
	Health Protection Agency: Guidelines on the Public Health Implications of Flooding
	Utility: Electricity (supply device)
	Utility: Gas (supply device)
	Utility: Water (supply device)

Approximately 60% of the residents known to have been affected were contacted by this method and were supplied with key contact information for the agencies responding in the aftermath of the incident. It was evident that this contact was effective because the number of calls into the call center began to fall as residents affected by the flood began to contact the appropriate agencies directly, rather than via the Council. This also reduced the frustration caused to residents by being informed that they had called the wrong organization.

All the work carried out by the Flood Recovery Team during the week after Monday, June 18th, proved to be in vain as the second flood struck during the morning of Monday, June 25th, when even heavier rainfall affected the borough. The 352 properties that were flooded a week earlier were inundated for a second time, along with a further 300 homes and 72 businesses. All the hard work and effort had to start all over again after this event.

The June 25th flood required a second major incident to be declared and the Emergency Control Team to be reconvened. This was unprecedented in terms of the Barnsley MBC, and there was no provision made for this within the Corporate Resilience Plan. The South Yorkshire Local Resilience Forum established its Strategic Coordinating Group. This is a multi-agency response forum consisting of all the relevant Category 1 and 2 bodies needed to formulate a single coherent response to the flooding event. Flooding was now affecting all local authorities and countywide agencies within the remit of the South Yorkshire Local Resilience Forum.

As with the first flood, a similar pattern emerged whereby the floodwater had passed through the borough's river catchments within a 24-hour period, and the activities in the aftermath of the second flood also followed a familiar pattern. Again, the Emergency Control Team was disbanded after 72 hours, leaving the Flood Recovery Team to manage the clean-up operation. They were able to coordinate more than 400 gas and electricity safety checks and organize Council staff to carry out the following:

- Collection of 500 tons of domestic flood-damaged debris from houses
- Visited all properties known to have been flooded
- Delivered more than 600 information packs
- Conducted safety checks at dams to reservoirs in Worsbrough and Elsecar
- Provided twice-daily press briefings to local media

During this process it was identified that 15% of all households in Barnsley did not have adequate insurance coverage for flooding events. As a result, a variety of hardship measures were put in place to assist those worst affected, which included

- Hardship payments from local charitable foundations
- Provision and distribution of 100 stoves, fridges, chairs, and tables
- Obtained government grant for part funding of replacement caravans
- Refurbished apartments to accommodate flood victims
- Organized local community "drop-in" sessions with representatives from all responding agencies
- Supported business claims to Yorkshire Forward

The Flood Recovery Team operated for one month and then was dissolved, with any outstanding residual issues being handed over to the relevant service area for ongoing management. Throughout the process of collection and disposal of flood-damaged white goods, the Council's capacity to store the vast amounts of appliances was outstripped. It was necessary for the Council to apply to the Environment Agency for derogation from its licensing arrangement to stockpile the extra "white" goods prior to their disposal.

4.6   POST-FLOODING REPAIRS AND ACTIVITIES

The identification of the causes of flooding-related issues was a priority during the emergency phase of the event, but this became even more important in the aftermath of the floods. In many cases it was obvious what the flooding mechanism was, but what had to be established was whether this still posed a flood risk in the days immediately afterward. For the affected residents, the removal of river-borne debris, blocked road gulleys, uprooted trees in the river channel, and blockages at bridge crossings were obvious issues. These needed to be targeted for urgent remedial action to prevent a reoccurrence of the flooding and therefore much of the Council resources were deployed in the removal of obstructions in watercourses and drainage systems for which it was responsible. This was necessary to alleviate flooding in areas that were still submerged after the initial floodwaters had receded. The intelligence-gathering exercise continued well into the recovery phase of the event, with much of this information coming from the residents in the affected locations and from partner agencies carrying out remedial work on their assets and reporting defects on Council-owned assets (a reciprocal arrangement).

Across the borough, many miles of road had been submerged by floodwaters and, as a result, had to be swept to remove considerable amounts of silt and debris and ensure that they were sufficiently clean to allow them to be reopened. By redeploying much of the direct work staffing resources to flood alleviation tasks, the Council was able to carry out a great deal of small-scale action; however, some of the remedial work identified could not be done quickly. This required specialized staff, equipment, and planning; there had been a number of road and footbridges destroyed and washed away, requiring significant capital investment for debris to be removed and replaced. During the recovery, these assets and other significant capital projects were identified in the overall cost of the flooding event, which was reported to the Council's Cabinet. Once the recovery effort was in full swing, the Emergency Management Team and Council leaders decided that the return to normality for the flood-hit communities was among the key priorities for the Council. The removal of flood-damaged property was sometimes hampered by the delay caused by insurance company loss adjusters insisting on viewing the property prior to its removal and disposal. This meant that in some cases the flood-damaged property was left outside homes for several weeks before it could be taken away by Council staff.

The Cabinet Report (Barnsley MBC, 2007/2008) detailed the impact of the floods on the borough and the Council's emergency response to the event, while highlighting the impact on its services and service provision. In addition, this report quantified the damage the floods caused to the Council's directorates and the estimated costs of the remedial work needed to remove the debris from watercourses on Council-owned land. It was recognized that although there were many responding agencies, the Council played a key leadership role in the response to the emergency and an even more pivotal role in the recovery operation.

The process of examining the mechanisms and causes of flooding allowed each of the agencies to determine who was responsible for the required corrective action. As in some cases, it was unclear whether the responsibility lay with the Council, Environment Agency, Dove and Dearne IDB,* Yorkshire Water (the privately owned water company), or the individual landowners from the private sector. Through this process, the Council identified 26 projects on its own land that required attention within the next 12 months to prevent further flooding in the future. Much of this work was associated with blocked watercourses, culverts, and replacing or erecting new trash screens that had been damaged during the floods. To carry out this work, the land drainage function within the Council had to design and then procure the equipment and staffing resources to complete these small-scale civil engineering projects.

Engaging with the residents of the communities affected was essential; this provided an opportunity to raise their awareness of the work that the Council was undertaking and also to respond to their needs during the recovery phase. In the immediate days after the floods, the residents were quite hostile toward all responding agencies, including the Council, as many thought the whole event could have—and should have—been avoided by preventative actions taken beforehand.

Essentially, the residents were looking to apportion the blame for the flooding event to the responding agencies. They failed to acknowledge that June 2007 was the wettest since records began in 1776, and therefore it was an unprecedented event and the impact could not have been foreseen. A series of "Flood Roadshows" were organized where all responsible agencies and key stakeholder functions within the Council and other agencies were on hand. The agency representative was able to discuss with those affected by the flooding, any issues regarding the event and any assistance they required. From this a network of Council-led Local Residents' Flood Groups was established to facilitate the ongoing need of the communities and to keep them appraised of the progress on remedial work and flood alleviation projects that were taking place or being considered. The local residents were able to contribute by identifying areas that required remedial work to address potential new flood risks and inform the development of Community Emergency Flood Plans.

Funding of the flooding emergency was borne initially by Council reserves, but there was the potential for some of this work to be retrospectively funded by the central government via a claim made by the Bellwin Scheme† (CLG, n.d.). However,

---

* An Internal Drainage Board (IDB) is an operating authority that is established in areas of special drainage need in England and Wales with permissive powers to undertake work to secure clean water drainage and water level management within drainage districts. The area of an IDB is not determined by county or metropolitan council boundaries, but by water catchment areas within a given region.

† The Bellwin Scheme provides emergency financial assistance to local authorities; it may be activated in any case where an emergency or disaster involving destruction of or danger to life or property occurs and, as a result, one or more local authorities incur expenditure on, or in connection with, the taking of immediate action to safeguard life or property, or to prevent suffering or severe inconvenience, in their area or among its inhabitants, but there is no automatic entitlement to financial assistance.

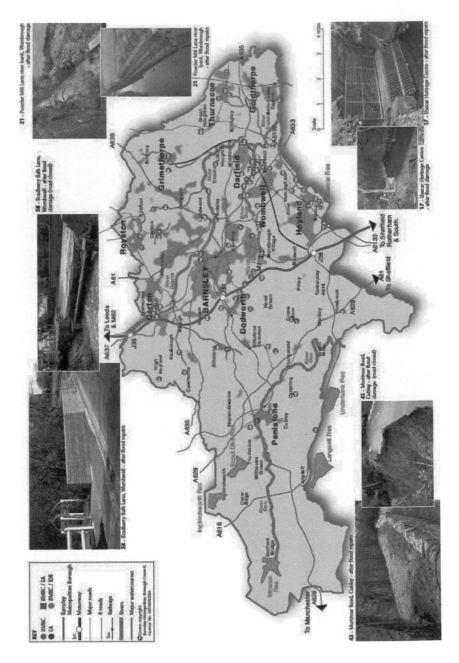

FIGURE 4.2 Damage and remedial work in Barnsley.

there was no guaranteed funding from the Bellwin Scheme or any other government department. Therefore, the Council reserves would be drawn on to finance the cost of the emergency and the subsequent recovery until any alternate funding streams could be identified and realized.

The Cabinet Report goes on to recommend the next steps to enable ongoing recovery from the flood, which include the establishment of a multi-agency Strategic Flood Forum to deal with Barnsley's flooding issues, an examination of the interim recommendations arising from the *Pitt Review* (Pitt, 2008), the development of specific flood emergency plans, an assessment of the Council's riparian ownership issues, ensuring that new Council house tenants within the floodplain are aware of the Flood Warnings Directs Scheme, and the promotion of flood prevention and awareness to all at risk of flooding.

Figure 4.2 summarizes the considerable damage and renovation work required across the district. Since the floods of June 2007, the Barnsley MBC has continued to work hard to deliver the recommendations and continues to do so.

REFERENCES

Barnsley MBC (2007a). Business Continuity Plan. Barnsley MBC.

Barnsley MBC (2007b). Corporate Resilience Plan. Barnsley MBC.

Barnsley MBC (2007/8). Cabinet Report—A Review of Flooding in Barnsley. Barnsley MBC.

Cabinet Office (2004). Civil Contingencies Act, London, Cabinet Office. Available from http://www.cabinetoffice.gov.uk/ukresilience/preparedness/ccact.aspx.

CLG (Communities and Local Government) (n.d.). The Bellwin Scheme [online]. London: CLG. Available at http://www.communities.gov.uk/localgovernment/localgovernment-finance/bellwinscheme/ [Accessed May 2010].

EA (Environment Agency) (2007). Local Flood Warning Plan (Barnsley MBC Area). Bristol: Environment Agency.

Pitt, M (2008). The Pitt Review: Lessons Learned from the 2007 Floods. London: Cabinet Office.

5 Health Impacts of Flooding

Harriet Caldin and Virginia Murray

CONTENTS

5.1 INTRODUCTION

Flooding worldwide accounted for four of the top five deadliest natural disasters in 2007 (Subbarao, Bostick, and James, 2008). The aggregate impact of flooding has a very wide scope and can affect food supplies and the economy months later, and subsequently can have a long-term impact on health (Subbarao et al., 2008). The potential for flooding to cause harm to human lives and health is great. Recent events remind us that harm occurred within England and Wales during the 2007 floods (Pitt, 2008), and more recently at Cockermouth, in northwestern England, in November 2009—and at the time of writing, already in 2010, in France and in many other areas of the world including the United States, Brazil, India, Czech Republic, and Pakistan.

Flooding is a complex natural hazard. The Office of Emergency Preparedness and Disaster Relief Coordination reported (Noji, 2005) that the short-term effects common to flooding disasters are

- Few deaths and injuries
- Small potential for an increase in communicable disease
- Common occurrence of food scarcity and major population movement

A recent detailed literature review shows that there are longer-term "hidden" health impacts that have only recently been recognized and still need to be understood further, such as the effect on mental health or the impact on vulnerable groups. Several key studies published within the past five years point to useful evidence and include the epidemiological literature review by Ahern et al. (2005), plus Few and Matthies (2006). Therefore, the data presented in this chapter are predominantly drawn from recent peer-reviewed literature, legal acts, and gray literature of import published since 2004.

The health effects have complex determinants that include the characteristic of the flood, the temporal association with the flood event, and a direct and indirect causal pathway between health and response to the flood. Additionally, the devastation of property and infrastructure is immediately apparent after a flood event but health impacts (excluding injury and death) may be less apparent, and could last longer than that of building reconstruction (Fewtrell and Kay, 2008). The effect on health facilities and the ability for health services to continue to provide care must also be considered part of the health consequences caused by flooding. Flood defense, building and planning improvements, the economic impact, community action, and recovery processes all interact with the health of the population in some way. Of note, statistics from the Environment Agency (EA, 2009) that 7% of hospitals in the United Kingdom (UK) lie on the floodplain, and in London alone there are 16 hospitals on the floodplains, heightening the necessity for infiltrating health into all dimensions of building resilience against floods.

The increase in the frequency and severity of flooding incidents has raised awareness in the health community of the current availability, quality, and extent of health guidance in the event of a flood. Reports such as the *Pitt Review* (Pitt, 2008) and the recent Department of Health report entitled "New Horizons" (HM Government Department of Health: Mental Health Division, 2010) have identified where there are gaps and set goals in order for these to be filled. These include

- The effective dissemination of Department of Health advice by local response and recovery groups (Pitt, 2008)
- The development of environmental resilience interventions against flooding (HM Government Department of Health: Mental Health Division, 2010)

Public health management challenges during a flood disaster include the safety and survival of the afflicted, provision of medical services and safe food and water, relocation of those affected if necessary, and implementation of health protection initiatives (Kshirsagar, Shinde, and Mehta, 2006). In a disaster there can be a mismatch between available resources and medical health service need; the public health response needs to reconcile this as appropriately as possible (Ahmad et al., 2008). Equally as important is that the complexity of flood documentation of adverse health effects arising from flooding is difficult. Various methods have been used but none, from the UK experience, provide a complete picture. Using three recent examples, the differences in methods are apparent; they are

- An *epidemiological study* into the health impacts after flooding in 2002 in Lewes found that having been flooded was associated with a significant increase in gastroenteritis and a four-times higher risk of psychological distress in adults (Reacher et al., 2004).
- A *real-time sociological study* from Hull, in northeastern England, in the immediate aftermath of the 2007 floods provided a clear picture of the self-reported health effects. Many participants reported that their health had been adversely affected, reporting respiratory problems, skin reactions, infections, and stress (Whittle et al., 2010).
- *Qualitative research by in-depth interviews and group discussions* among homeowners, businesses, and farmers affected by the flooding, including those flooded and those at risk of flooding, was undertaken by Growth from Knowledge National Opinion Polls (GfK NOP) after the 2007 floods for the Pitt Review. Results from this survey reported that 39% of respondents stated that the flooding had had an effect on their physical health and 67% stated that it had had an effect on their emotional health (GfK NOP, 2008).

This chapter addresses available data on the known health impacts of flooding by considering the issues around definitions of flooding for health use, the value of epidemiology, and the direct and indirect health impacts from flooding. These are described by mortality, injury, infection from chemical hazards, mental health, vulnerable groups, and health facility impacts.

5.2 DEFINITIONS OF FLOODING FOR HEALTH USE

It has become apparent that defining what constitutes a flood is difficult; furthermore, what defines a flood that activates an emergency health response is also equally complex. Flood definitions are useful for assessing the health impacts of floods and the infrastructure and financial toll they can cause, as well as providing a trigger for the activation of an emergency response. Examples of currently used definitions include

- Flood: The presence of water in areas that are usually dry, and for flood disaster—a flood that significantly disrupts or interferes with human and societal disaster (Jonkman and Kelman, 2005).

- Flood: Any case where land not normally covered by water becomes covered by water (Flood and Water Management Act, 2010). Furthermore, these new definitions of flood and coastal erosion issues describe flood risk as the relationship between the probability of occurrence with the associated consequences. These are then listed with health as the first concern, followed by social and economic welfare. This Act reflects that floods can be caused by
  - Heavy rainfall
  - A river overflowing or banks breached
  - Tidal waters
  - Groundwater
  - Anything else (including a combination of factors)

Three ways of defining a flood for health purposes have arisen, through

1. *Scientific thresholds:*
   - Depth of water: a specified level reached determines the qualification of a flood
   - Temporal and spatial: the length of time or the area that land is flooded for
2. *Population effects:*
   - Broad: medical, social, economic disruption to normal life
   - Specific: number of deaths or people affected
3. *Temporal health perspective* (WHO, 2002):
   - Immediate outcomes: during or immediately after the flooding
   - Short-term outcomes: in the days or early weeks following the flooding
   - Long-term outcomes: may appear after or last for months or years

As far as health effects are concerned, the *temporal health perspectives* are probably most helpful. However, all these definitions reflect the complexity of finding an adequate way to describe holistically the impacts of flooding from a health perspective.

5.3   EPIDEMIOLOGY OF THE HEALTH IMPACTS OF FLOODING

Assessment of the health impacts of flooding cannot generally occur through controlled prospective epidemiological studies; therefore, much of the literature on the health impacts and advice available comes from opportunistic retrospective analysis and case-study "lessons learned" (Fewtrell and Kay, 2008).

Natural disasters, of which flooding is one, represent unique health circumstances and have received increased research interest over the past few years. This particularly applies to flooding from American hurricanes such as Katrina and Ivan. However, existing epidemiological research on the impact of flooding on health is frequently limited by small sample sizes and nonrepresentative samples (Ruggiero et al., 2009).

Good baseline data are often difficult to obtain because relevant data were not collected prior to the flood incident. In addition, the attribution of health outcomes to the flooding is not always recorded in medical or health care notes, and therefore the association between health complaint and cause is not always made.

The Pitt Review has evidence from two good surveys; however, many of the impacts reported are based on "anecdotal" evidence gathered for the report, which is the case for much of the published work. While there can be no doubt of these indirect health impacts, quantifying them through good evidence-based epidemiology is difficult (GfK NOP, 2008; Hull City Council, in Pitt, 2008).

The epidemiological data on health impacts of flooding are incomplete, and further work on future events covering the pre-, during, and post-flood circumstances would be of significant value in increasing our understanding.

5.4 HEALTH IMPACTS

The term *health* is used in its broadest sense in this chapter, covering all aspects of physical and mental health.

The health effects of flooding can be divided into those associated with the immediate event and those arising post flooding event. Immediate health effects are caused by the floodwater and include drowning and injury. The health effects from a flood continue to occur after the immediate event and can still be observable for months or years afterward (WHO Regional Office for Europe, 2003). The longer-term health effects include those related to exposure to floodwater, the clean-up process, and mental health effects.

In addition, the health impacts can also be thought of as direct and indirect. Direct health effects are those caused by the actual effect of floodwater, including drowning and trauma (WHO Regional Office for Europe, 2003). Indirect health effects are the consequence of flooding and include the impacts from damage to infrastructure, and food and water supplies (WHO Euro and WHO, 2003; WHO Regional Office for Europe, 2003; Jonkman and Kelman, 2005; Kirch, Menne, and Bertollini, 2005; Fewtrell and Kay, 2008).

From Table 5.1 the direct and indirect health impacts from flooding can be summarized by mortality, injury, infection, chemical hazards, mental health, and by vulnerable groups and the impact on health services and hospitals.

5.4.1 MORTALITY

In the final decade of the twentieth century, it was stated that floods killed about 100,000 people worldwide and over 1.4 billion were affected (Jonkman and Kelman, 2005). However, there is an issue about exactly what a death from flooding is. It is suggested that a flood fatality or flood-related fatality is a fatality that would not have occurred without a specific flooding event.

A flood fatality raises questions regarding the timing of death. Indirect and direct are not useful terms; however, some deaths are immediate (drowning) and others could be delayed (deaths due to psychological effects). Therefore, to accommodate possible separation between the flood disaster and the potentially associated death, flood disaster could be categorized into three phases: pre-impact, impact, and post-impact (Jonkman and Kelman, 2005). Jonkman and Kelman (2005) postulated that

TABLE 5.1
Impact Factor of Potential Health Effects of Flooding

Direct: Those who live in flooded areas or travel through floods	Drowning and injuries, from cars driving through floods to children falling into hidden manholes and other hazards, and injuries sustained during flood recovery
	Electrocution
	Diarrheal, vector-, and rodent-borne diseases
	Respiratory, skin, and eye infections
	Chemical contamination, particularly CO poisoning from generators used for pumping and dehumidifying
	Stress and longer-term mental health issues, including the impacts of displacement
	Water shortages leading to drought during floods, due to loss of water treatment works and potentially sewage treatment plants
	Damage to and destruction of property
	Loss of access to and failure to obtain continuing health care
Indirect: Those who live near flooded areas and whose local community environment is damaged or limited as a consequence of flooding	Damage to healthcare infrastructure and loss of essential drug access
	Damage and destruction of property, including hospitals and other vital community facilities
	Damage to water and sanitation infrastructure
	Damage to crops or disruption of food supplies
	Disruption of livelihood and income
	Population displacement
	Delayed onset of mental health diagnosis due to longevity of flood recovery or fear of reoccurrence

Source: Adapted from Ahern, M., Kovats, R. S., Wilkinson, P., Few, R., and Matthies, F. (2005). Global health impacts of floods: Epidemiologic evidence. *Epidemiologic Reviews*, 27, 36–46; Few, R., and Matthies, F. (2006). *Flood Hazards and Health: Responding to Present and Future Risks*. London: Earthscan.

two thirds of deaths from flooding worldwide are from drowning, and one third are from physical trauma, heart attack, electrocution, carbon monoxide poisoning, or fire. Furthermore, 70% of deaths are males (Jonkman and Kelman, 2005).

Currently there are two global data sets that record disasters and deaths arising from them: (1) The United States Office for Foreign Disaster Assistance (OFDA) and Center for Research on the Epidemiology of Disasters in Brussels (CRED) International Disaster Database (EM-DAT) and (2) the Dartmouth Flood Observatory (DFO) at the University of Colorado. Although they have important limitations due to their data sources, they do provide useful indications of the mortality rate from flood disasters and also the differing rates between countries. Significantly, they have criteria for a flooding disaster:

- EM-DAT classifications are 10 or more people killed, 100 people or more affected, declaration of a state of emergency, or a call for international assistance.

- The DFO uses wide inclusion criteria: "large" floods are counted, defined as those inflicting significant damage to structures or agriculture and fatalities.

This, therefore, means that floods that do not classify as a disaster are not counted, whether or not people have died, so the total number of deaths globally is unknown.

It is apparent that high-income countries are more likely to collect reliable national statistics, but as yet there has been no review of these. No comprehensive mapping of data has been created but recent work points to the value of trying to show by geographical information systems the countries that have suffered most from flood-related deaths in Europe, using DFO data (Figure 5.1). These data show that the death rate associated with flood events may also provide a marker for flood severity. These appear highest in central Europe and the former Soviet Republics for 2000 to 2009. There are, however, a number of issues that arise when attempting to identify which deaths are associated with flooding. In many cases, only the immediate traumatic deaths are recorded and longer-term health events associated with the flood event are not.

In the aftermath of a flood disaster, deaths can be caused not only by the physical dimensions of the flood, but also by health and socioeconomic conditions (Ohl and Tapsell, 2000). The longer-term effects of floods on mortality were investigated in the United Kingdom following the Bristol floods of 1968 (Bennet, 1970), and it was shown that during the 12 months post-flood, there was a 50% increase in the population mortality rate in the flooded part of the city compared to no appreciable increase in the nonflooded area.

Further work is needed to understand the immediate and the longer-term mortality from flooding and to confirm the findings by Bennet (1970).

5.4.2 Injury

Injuries are likely to occur during the initial flood as moving water or deep water is dangerous. For example, water can displace vehicles, trees, and other materials (such as chemical drums) from their original location, and water can hide unseen hazards such as debris, which can cause injury.

Further injuries occur as activity shifts to recovery and clean-up (WHO Regional Office for Europe, 2003). Here, the type of injuries can include wounds caused by sharp objects, concealed hazards, and electrical hazards. Also, hazards such as chemicals used for clean-up have caused respiratory tract, skin, and eye irritations. Other surface injuries may arise from animal, reptile, or other vector contact.

Additionally, people's behavior can influence the likelihood of injury, particularly if they do not comply with evacuation orders, as was observed in New Orleans during Hurricane Katrina (in 2005) (Lave and Apt, 2006), or when the facilities and management at the evacuation center are inadequate.

Information on the causes and types of injury remains incomplete, and further work is required to prepare for and respond to these events and to document the hazards more completely.

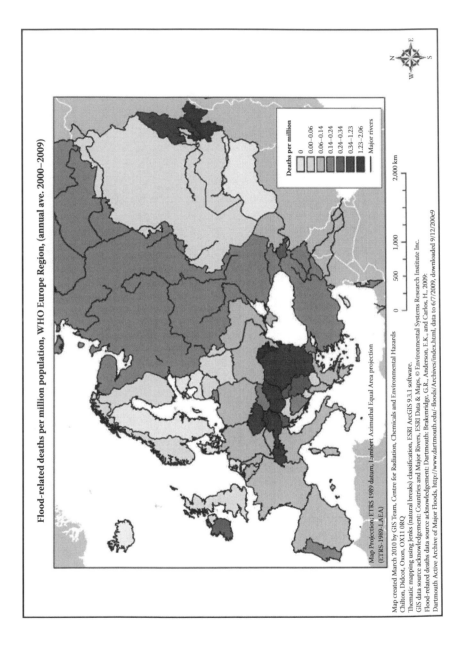

FIGURE 5.1   Flood-related deaths per million populations in Europe.

5.4.3 INFECTIONS

Flooding can cause many different health impacts by altering the balance of the natural environment and ecology, which allows vectors of disease and bacteria to flourish. Outbreaks of cholera and a higher incidence of malaria, among others, have been reported in developing countries. However, the brief summary contained herein focuses on data from Europe and North America.

Natural disasters do not usually result in outbreaks of infectious disease; however, under certain circumstances they can increase disease transmission. The risk of epidemics is proportional to population density and displacement (Noji, 2005) and to the extent to which the natural environment has been altered or disrupted.

The most frequently observed increases in communicable diseases are through fecal–oral transmission due to fecal contamination of water, and also by respiratory spread. In the longer term, vector-borne disease incidence can increase in some areas due to the disruption of control efforts (Noji, 2005).

More recently the Health Protection Agency (HPA) stated that infection from flood is rare in the United Kingdom as pathogens get diluted and provide low risk. Indeed, there was "no evidence of increased outbreaks of illness" following the 2007 floods in the United Kingdom (HPA, 2008). The National Health Service (NHS) in the United Kingdom reported that the risk to health from infectious diseases in floodwater is small, and there is no reported need for booster immunizations or antibiotics (NHS, 2008). However, vector-borne and water-borne diseases require special note.

5.4.3.1 Vector-Borne Disease

Flooding can give rise to environmental conditions that may increase the incidence of mosquito-borne infectious viruses; these have been reported in Central Europe. In the Czech Republic, antibodies were detected after the 2002 flood for *Tahyna*, *Sindbis*, and *Batai* viruses, with the only activity found for the *Tahyna* virus when there was one seroconversion among 150 residents (Hubálek et al., 2005).

After extensive flooding following Hurricane Katrina in 2005, waters contaminated with infective organisms in many areas of New Orleans resulted in exposure to *Vibrio vulnificus*: seven were infected and four died. Those at greatest risk were vulnerable with liver disease or compromised immune systems. The route of exposure was found to be via skin infection in 60% of the cases (Rhoads, 2006).

A seroepidemiological study to evaluate the risk of infection with *Leptospirae* in a population exposed to floods in northeastern Italy in 2002 found a 6.8% seroconversion rate, which suggested a low risk of infection. Walking barefoot in floodwater, contact of an injury with floodwater, rats in the household, and days (more than four) spent in cleaning activities are all risk factors for leptospirosis (Bhardwaj, Kosambiya, and Desai, 2008).

5.4.3.2 Water-Borne Disease

Setzer and Domino (2004) reported that there was a statistically significant increase in outpatient visits for *Toxoplasma gondii* and adenoviruses post Hurricane Floyd (in 1999) in severely affected areas, although the magnitude was small. No other increase was seen for other water-borne pathogens or in moderately affected areas (Setzer and Domino, 2004).

Relatively few epidemiological studies have been undertaken on populations exposed to infectious diseases after flooding, but those reported so far appear to be reassuring.

5.4.3.3   Food-Borne Disease

There is very little research about the transmission of disease through contaminated food. However, the risks are from food directly touched by the floodwater, food prepared using dirty surfaces, and food prepared with unclean hands.

5.4.4   CHEMICAL HAZARDS

In particular, two types of risks are associated with chemical hazards and flooding: unexpected carbon monoxide exposure, which has significant public health impacts, and risks from pollution in floodwater.

5.4.4.1   Carbon Monoxide

One of the greatest health risks occurs in the aftermath of the flood when generators or fuel-powered equipment are used indoors to dry buildings (Euripidou and Murray, 2004; HPA, 2008).

Carbon monoxide (CO) is produced in the exhaust of diesel generators or other fuel-driven equipment if used inside enclosed places such as domestic housing (e.g., portable grills, pressure washers, camp stoves, paraffin-fueled heaters, or other devices using gasoline, propane, or natural gas). These can build up inadequate ventilation (e.g., in garages), and there is a potential for fumes to accumulate.

An investigation using medical notes conducted by the Centers for Disease Control (CDC) (Sniffen et al., 2005) into 6 deaths and 167 patients with CO poisoning after the four hurricanes in Florida in 2004 reported that presentation peaked 3 days after landfall of the hurricane, and most symptoms occurred during the night. Portable, gasoline-powered generators were used in 96% of the incidents, five of the six deaths were male, and most of the generators were misplaced (i.e., in garages or outdoor near household windows).

Van Sickle et al. (2007) also undertook research following the 2004 Florida hurricanes using self-reported data and found that 86% of the events occurred in the person's own home, 12% in other homes, and 3% at work. Factors that affected the decisions on generator placement were 43% theft, 34% exhaust, 26% length of extension, 26% protect from elements, 11% noise, 9% protect from flooding, and 26% other—including setup by salesman. And 74% of the incidents occurred in households that had not had generators before the event.

5.4.4.2   Chemical Pollution in Floodwater

There is a risk of pollution in floodwater (EA, 2010) but chemicals in the floodwater are likely to be diluted and probably pose little acute risk (HPA UK, 2008). However, chemical release can pose a more serious threat to health if waste storage facilities or industrial plants are flooded.

Depending on the severity and extent of flooding, we might expect uncontrolled release of various chemicals into the environment, which potentially carry public health risks. However, in a typical flooding scenario, the release of chemicals is diluted with the floodwater, causing the toxicity level to decline. In unusual circumstances such as high-intensity, short-duration flooding, there may be a possibility of public exposure to elevated levels of toxic chemicals. Moreover, in the event of continual release such as from acid mine drainage (AMD), there is also potential for chronic exposure to humans that could occur through either direct contact or indirectly in the event of contamination of the food chain. As many of these chemicals (e.g., lead, arsenic, polycyclic aromatic hydrocarbons [PAHs], pesticides) are non-threshold, these are theoretical risks of health hazards from exposure.

A principal component of concern in studies reviewed thus far has been the use of proxy exposure measurements rather than direct measures. From current information it is clear that, in general, very little if any environmental sampling has been undertaken following flooding events. Effective environmental data collection and monitoring would greatly enhance future health research studies (Euripidou and Murray, 2004).

There is information to suggest that the source of the contaminated floodwater may predict the type of chemical contamination that can occur (Euripidou and Murray, 2004). Three types of risk are summarized below:

1. *Stormwater floods:* Chemical contamination due to stormwater flooding could vary quite significantly, depending on the land use and associated infrastructure. In the case of runoff from roads, (interstate) highways, and bridges, the nature of chemical release could reflect typical pollution caused by traffic on the road and can include a range of chemicals such as heavy metals, petroleum hydrocarbons, and PAHs. In a rural catchment, the nature of runoff is more expected to be eroded soil materials containing fertilizers, herbicides, and pesticides. The intensity of this pollution is very much dependent on land geology and farmland management practices. In urban and semi-urban land uses, the nature of contamination might be very diverse and can include various pharmaceutical residues; domestic, industrial, and commercial chemicals; and road runoff.

2. *Overloaded sewers (backflow) in combined sewer system:* May cause sewer overflows containing a variety of residential, industrial, and stormwater waste, including chemicals within sewers as a residue of consented industrial discharges (national and local regulatory bodies such as the Environment Agency and Local Authority in the United Kingdom will regulate discharges and are a source of information) and chemical contamination within residential sewers, canals, or residential rivers that have collected (as a sludge) as a result of runoff that may be remobilized.

3. *Acid mine drainage (AMD) and public health risk from flooding:* In the United Kingdom there is a long legacy of mining, and risks exist of AMD from abandoned mines—especially coal mines. For example, Devon and Cornwall alone have approximately 1,700 abandoned mine workings, affecting 212 kilometers of rivers, and metal mining for centuries has impacted

rivers; the impacts remain significant in some rivers many decades after the mines have closed (Whitbread-Abrutat, 2004). Under normal operating conditions, abandoned or closed mines were constantly drained with large pumps; however, post-closure they are susceptible and may flood. Where mine water is exposed to fresh air at the face, sulfides may be allowed to oxidize, leading to the formation of sulfuric acid with pH values ranging between 2 and 3 being common. Heavy metals may dissolve under such acidic conditions and thus potentially become more mobile and available. Typically, the minerals and metals found in mines include aluminum, arsenic, cadmium, cobalt, copper, iron, lead, and manganese.

Hazardous landfill sites and wastewater lagoons may also potentially contaminate floodwater. For example, Itai-itai disease in Japan, where cadmium-contaminated river water from mining (1910 to 1945) was used to flood rice fields, resulted in long-term harm to those eating contaminated rice (Pearson, 1999). Another example of melting snow and ice in Bashkiria resulted in a toxic waste lagoon having its contents washed into the river at Ufa just above the intake for the water supply for 600,000 people in 1990 (Tejima et al., 1996).

Data on chemical contamination are incomplete and probably do not provide enough information to confidently reassure; further work is required, particularly to alert clinical colleagues about the risk of carbon monoxide and with environmental experts to agree on sampling protocols if required.

5.4.5 MENTAL HEALTH

Several methodological difficulties should be considered when evaluating epidemiological mental health studies following flooding (Galea, 2009). Taking Galea's research forward, it is helpful to take into account the following methodological limitations:

- The assessment of studies on mental health impacts from flooding requires a clear understanding of the scale of the event investigated. To develop this, it is difficult to separate the effect of a disaster from the effect of the subsequent flooding, as was seen with Hurricane Katrina that was followed by widespread flooding.
- Many different tools are used to assess mental illness, most notably for posttraumatic stress disorder (PTSD). Such tools are not always comparable or of adequate validation. Furthermore, the period of PTSD assessment in relation to the event is often unclear. The diagnostic tool may not be able to explicitly link PTSD to a disaster and may require clarification.
- The measurement of association of mental health prevalence and incidence with disasters is difficult because it is difficult to ensure that research carried out to assess mental health impacts post-flooding uses a population sample that has not had previous mental illness prior to the event.
- The range of correlates and statistical models used to document associations is very wide and difficult to generalize from in an effort to provide a single summary of broad application to public health.

Despite these methodological issues, the evidence that floods adversely affect mental health and well-being is well established. The size of the effect, however, appears to vary across studies, and the nature of the effects is subject to differing cultural interpretations and definitions. The following studies provide excellent examples of some of the mental issues identified:

- A case-control study of both directly and indirectly flood-affected households and nonaffected controls found that up to 75% of the population affected by a flood will experience mental health impacts, and that higher age correlated with more severe health impacts (Green et al., 1985).
- A cohort study undertaken in southeast England following severe river flooding in 2000 found that adults had a four times (95% CI 2.6, 6.4; $p <$ 0.0005) higher risk of psychological distress, and suggested that this could explain some of the excess physical illness reported by flooded children and adults some 9 months after the flood (Reacher et al., 2004).
- Tunstall et al. (2006) also found that two-thirds of the flood victims in thirty locations in England and Wales were found to have scores on the General Health Questionnaire-12 scale indicative of mental health problems (i.e., scores of 4+) after flooding.
- Displacement from the home and the post-flood disruption associated with this has been reported as the most significant stressor from flooding (Tunstall et al., 2007).
- The quality and speed of response from insurers and from contractors involved in reconstruction are factors that can exacerbate mental distress (Tapsell, 2009).
- A randomly generated survey of 1,510 people who survived the Hurricane Katrina floods in New Orleans found significant racial and gender differences in psychological impacts, including sleeplessness, anxiety, depression, and worries about the future (Adeola, 2009).

From these data, further research on mental health is required to identify the most effective tools and measures to investigate this complex and worrying consequence of flooding events.

5.5 VULNERABILITY AND SPECIFIC GROUPS FOCUSING ON CHRONIC DISEASE SUFFERERS

This section considers vulnerability in relation to flooding and additionally identifies high-dependency chronic diseases where patients may suffer disproportionately in a flood due to their reliance on constant medical treatment and clinical care. There are known factors and population determinants that increase the risk of adverse health impacts of flooding—for example, the elderly, children, and the socially and economically disadvantaged (Table 5.2).

TABLE 5.2
Vulnerable Group Indicators

Indication of Vulnerability	Demographic Group at Risk
Limited physical capacity	Elderly, preexisting chronic conditions, reliance on home care
Limited mobility	Elderly, preexisting chronic conditions, reliance on home care
Reliance on important medication	Elderly, preexisting chronic conditions, reliance on home care, substance misusers
Reliance on regular home care	Elderly, preexisting chronic conditions
Reliance on regular care at health facility	Elderly, preexisting chronic conditions, substance misusers
Weak social networks	Elderly, preexisting chronic conditions, reliance on home care, homeless, substance misusers, ethnic minorities, rural inhabitants
Poor flood awareness	All vulnerable health groups, those living in high-risk flood areas, those of low income, ethnic minorities
Lack of resources for resilience and response	All vulnerable health groups, those living in high-risk flood areas, those of low income, ethnic minorities
Little access to public warnings and guidance	Elderly and ethnic minorities
High-risk built environment	Those living in high-risk flood areas, areas of deprivation

5.5.1 VULNERABILITY

Recent results from a longitudinal study that followed flood victims for 18 months after a major flooding episode in Hull, United Kingdom, in 2007 (Whittle et al., 2010) found that these are indicators of vulnerability but do not necessarily lead to a direct causal pathway between a flood event and those most at risk from suffering from the event. The findings suggest that it is a complex interaction between these preexisting conditions and the specific circumstances of the flood and in a person's life that determine how, when, and if they become vulnerable.

When a number of the defined vulnerable groups intersect—for example, in deprived communities—the problems caused by flooding can intensify due to links with poor flood awareness; a lack of resources to protect, repair, and insure property; weak social networks; poorer health baseline; and a lack of mobility or physical capacity. The negative impact of this situation can be exacerbated if the area flooded is poorly maintained. Indeed, it may lead to a built environment at high risk of flood damage. Whittle et al. (2010) stated that these factors do not necessarily determine vulnerability to experiencing a flood hazard itself, but may influence vulnerability to the *impacts* of flood hazards.

The Marmot Review (2010) stated that those living in the least favorable environmental conditions, including flood risk, in the United Kingdom are also those of greatest deprivation, an indication of where the greatest health impacts will be felt. Those in society who are already disadvantaged are likely to experience even more severe consequences following a flood (WHO Regional Office for Europe, 2003).

Ethnic minorities are at risk during a flood due to possible communication barriers, which could result in delayed response or misinformation. Disaster preparedness needs to fully integrate factors related to race, culture, and language into risk communication, and public health action and policy at every level (Andrulis, Siddiqui, and Gantner, 2007).

5.5.2 Specific Groups Focusing on Chronic Disease Sufferers

People who have high-dependency chronic diseases may incur greater suffering because of their reliance on heavy medical and nursing care—for example, those with sickle cell anemia, diabetes, renal failure, cystic fibrosis, directly observed therapy for tuberculosis, HIV/AIDS cancer, and mental illness. The following examples of these impacts are summarized below:

- *Diabetics:* A statistically significant increase in presentations of diabetic foot to a surgical ward in Grenada General Hospital was reported post-Katrina (Sjoberg and Yearwood, 2007). The availability of health care and lack of electricity makes it difficult to store insulin properly, thus heightening the problem.
- *Renal dialysis patients:* Kutner et al. (2009) reported that Hurricane Katrina did not cause statistically significant excess death for dialysis patients; however, the event may have led to increased incidents of hospitalization.
- *Patients with cystic fibrosis:* A disaster can disrupt the routine, displace people from their homes, interrupt the electricity needed to run equipment (i.e., nebulizers, oxygen tanks, and refrigerators), and disrupt the supply of clean water needed to disinfect equipment (Flume et al., 2005).
- *The disabled:* If displacement occurs or evacuation is needed, the disabled may suffer because of their need for wheeled mobility, specialized transfer techniques, and specific medical supply needs, especially for spinal cord–injured people (e.g., catheters) (Bloodworth et al., 2007). The treatment of pain and the handling of controlled substances may be difficult to address and plan for.
- *Mental health of chronic disease sufferers:* The mental health of those suffering from a chronic disease can be affected by evacuation and displacement. Hyre et al. (2008) reported that 45% of patients receiving hemodialysis in one of the nine clinics in New Orleans suffered from depression one year after Hurricane Katrina. A longer displacement time was associated with lower mental health status and a higher prevalence of depressive symptoms.

More research is needed to understand the effects that flood resilience and recovery processes have on vulnerability and the way they may change the dynamics and boundaries of vulnerability.

5.6   HEALTH FACILITIES AND SERVICES

There has been little research into the impact of flooding on health facilities. This is partly because epidemiological research would be difficult to achieve; however, it is a very important part of understanding the health consequences that flooding can have. Literature in this area usually focuses on a specific population group, such as diabetics, and how continuity of care is achieved, or the impact on capacity and demand for specific clinics or services. There are case studies and reflective "lesson learned" reports describing how departments coped in the wake of a flood, but these are mostly hurricane generated from the United States. The United Nations International Strategy for Disaster Reduction (UN/ISDR) initiative "Safer Hospitals" and the World Health Organization (WHO) "Hospitals Safe in Emergencies" are both campaigns to protect hospitals and have produced tools to assess the structural and functional safety of hospitals in emergencies, such as the "Hospital Safety Index" developed by Pan American Health Organization (PAHO) and WHO.

Specific points of concern for health—including capacity, infrastructure, and water availability—are identified below:

- *Capacity*: There is little evidence in Europe that health facilities have come under extreme medical pressure in terms of increased patient caseload as a direct consequence of a flooding event. However, given the demographic changes and the need for more housing and development, more health infrastructure is being developed in at-risk areas, which could have serious consequences. If the flooding event is widespread and has caused infrastructure disruption, health facilities may find that patients whose care is normally controlled in the community are admitted to the hospital. Furthermore, if transport infrastructure is disrupted, patient transfer into and out of the hospital can be affected, as can emergency services.
- *Facilities/infrastructure within health facilities*: Power in a healthcare facility is very vulnerable to failure during a flood event. This is exacerbated by the usual practice of having emergency generators in the basement and many medical physics departments are downstairs. This leads to many expensive and vital pieces of equipment failing or being ruined in a flood, thus disrupting the healthcare facility's ability to function properly—both in terms of diagnostic patient care and the electricity supply to the entire hospital. In 2007, Tewkesbury Hospital (United Kingdom) was under great threat, and the decision was taken early to evacuate a hospital site and transfer approximately 20 patients to other hospital facilities (Whiteley, 2008). Both internal and external communication systems may be affected, either inhibiting the internal functioning of the hospital or the external communication between facilities. Patient records may be at risk when the power supply goes down (Dalovisio, 2006; Nasman, Zetterberg-Randonn, and Brandstram, 2007).

- *Availability of water:* Despite an excess of water, flooding can lead to a shortage of clean, drinkable water. This not only affects health systems and patients, but also all members of the public, and can cause outbreaks of *Legionella* (Whiteley, 2008). Accessing people who need bottled water can be difficult because the transport systems have flooded. During the 2007 floods in Gloucestershire (United Kingdom), a local main water pumping station was affected, causing mains water to become unavailable, and all hospital sites were affected as the floodwater contaminated the county-wide pipes. The maintenance of adequate fresh water supplies, essential for ensuring the safety of patients and staff at the hospital, was put at great risk (Whiteley, 2008). If fresh water supply becomes unavailable, these patients may have to be transferred elsewhere to maintain critical treatment (Whiteley, 2008).

5.7 PUBLIC HEALTH AND EMERGENCIES

This chapter focused on the health impacts from flooding. Indeed, the WHO and the United Nations advocate an "all-hazards" approach to emergency response (see Table 5.3).

Large-scale natural disasters often require a multisectoral response, the planning of which can be extremely difficult as these sectors can run in parallel, work together, or be a mixed model that evolves as the response to the disaster unravels. Indeed, health still needs to be central to, and considered part of, all developments for resilience improvements against flooding. Within the United Kingdom, health protection and public health response professionals act as the interface between emergency response, clinical healthcare, and the overall management of a flooding or any type of incident response (Cabinet Office, 2010). It is they who, with infectious disease, toxicological, environmental, extreme event, emergency planners and responders, and other experts, will contribute to the many decisions, including sheltering and evacuation and the setting up of an incident control team if required.

TABLE 5.3

The "All Hazards" Approach to Emergency Response

All-hazard is a concept acknowledging that, while hazards vary in source (natural, technological, societal), they often challenge health systems in similar ways. Thus, risk reduction, emergency preparedness, response actions, and community recovery activities are usually implemented along the same model, regardless of the cause. Experience shows that a substantial part of essential response actions is generic (health information management in crises, emergency operations center, coordination, logistics, public communication, etc.), irrespective of the hazard, and that prioritizing these generic response measures generates synergies to address the hazard-specific aspects better (WHO, 2008).

The **whole-health** approach promotes that the emergency preparedness planning process, the overall coordination procedures, surge, and operational platforms are led and coordinated by an emergency coordination body at the central and local level, which involves all relevant disciplines of the health sector and deals with all potential health risks (WHO, 2009a,b).

Depending on the scale and impact of the incident, it is they who will attend to the operational, tactical, or strategic incident arrangements:

- Operational command (also referred to as "bronze" command): On-scene incident command
- Tactical command (also referred to as "silver"): Usually off-scene command to control multiple incident commands, complex incidents with a large affected area, overall logistics, and so on.
- Strategic command (also referred to as "gold"): Off-scene command post with public or government officials in charge of strategic decisions (WHO, 2010)

Within the United Kingdom, additional resources have been identified as providing value in the emergency phase of the response to flooding, and these are the Science and Technical Cell (Cabinet Office) or, perhaps, even support from the Cabinet Office Briefing Rooms (2010).

5.8 CONCLUSIONS

What is imperative is that health is embedded through the entire flooding response by all sectors: Ultimately, all actions of any sector will have an impact on health. All natural disasters are unique and may impact each country differently, due to varied economic, social, and health backgrounds. Some similarities exist, however, among the health effects, so that good cross-sectional planning and preparedness at all levels, with effective routine training, can enable the effective management of health and emergency relief in any given disaster (Acharya, Upadhya, and Kortmann, 2006).

Research needs have been identified and these include

- The epidemiological data on health impacts of flooding is incomplete, and further work on future events covering the pre-, during, and post-flood would be of significant value in increasing our understanding.
- Further work is needed to understand the immediate and the longer-term mortality from flooding, and to confirm the findings by Bennet (1970).
- Information on the causes and types of injury remains incomplete, and further work is required to prepare for and respond to these events and to document the hazards more completely.
- Relatively few epidemiological studies have been undertaken on populations exposed to infectious diseases after flooding, but those reported so far appear to be reassuring.
- Data on chemical contamination are incomplete and probably do not provide enough information to confidently reassure; further work is required, particularly to alert clinical colleagues about the risk of carbon monoxide and with environmental experts to agree on sampling protocols if required.
- More research is needed to understand the effects that flood resilience and recovery processes have on vulnerability and the way they may change the dynamics and boundaries of vulnerability.

- From these data, further research on mental health is required to identify the most effective tools and measures to investigate this complex and worrying consequence of flooding events.
- From this research it is apparent that the difficulties that arise from water shortages during flooding need to be understood further, and consensus on quantity, quality, and delivery of water needs to be developed.
- Emergency planning for health facilities and services for flooding is thin, and not necessarily incorporated into national emergency plans. Health needs to be part of all emergency planning and response.

REFERENCES

Acharya, L., Upadhya, K. D., and Kortmann, F. (2006). Mental health and psychosocial support aspects in disaster preparedness: Nepal. *International Review of Psychiatry*, 18, 587–592.

Adeola, F. (2009). Mental health and psychological distress sequelae of Katrina. *Human Ecology Review*, 16, 195–210.

Ahern, M., Kovats, R. S., Wilkinson, P., Few, R., and Matthies, F. (2005). Global health impacts of floods: Epidemiologic evidence. *Epidemiologic Reviews*, 27, 36–46.

Ahmad, R., Mohamad, Z., Noh, A. Y. M., Mohamad, N., Saharudin, M., Che Hamza, S., Mohammed, N. A. N., Baharudin, K. A., and Kamauzaman, T. H. T. (2008). Health major incident: The experiences of mobile medical team during major flood. *Malaysian Journal of Medical Sciences*, 15, 47–51.

Andrulis, D. P., Siddiqui, N. J., and Gantner, J. L. (2007). Preparing racially and ethnically diverse communities for public health emergencies. *Health Affairs*, 26(5), 1269–1279.

Bennet, G. (1970). Bristol Floods 1968. Controlled survey of effects on health of local community disaster. *British Medical Journal*, 3, 454–458.

Bhardwaj, P., Kosambiya, J., and Desai, V. (2008). A case control study to explore the risk factors for acquisition of leptospirosis in Surat city, after flood. *Indian Journal of Medical Sciences*, 62(11), 431–438.

Bloodworth, D. M., Kevorkian, C. G., Rumbaut, E., and Chiou-Tan, F. Y. (2007). Impairment and disability in the Astrodome after Hurricane Katrina: Lessons learned about the needs of the disabled after large population movements. *American Journal of Physical Medicine and Rehabilitation*, 86(9), 770–775.

Cabinet Office, UK. UK Resilience Provision of Scientific and Technical Advice to Strategic Co-ordinating Groups during a Major Incident. Available at http://www.ukresilience.gov.uk/news/stac_guidance.aspx. [Accessed 2010]

Dalovisio, J. R. (2006). Hurricane Katrina: Lessons learned in disaster planning for hospitals, medical schools, and communities. *Current Infectious Disease Reports*, 8(3), 171–173.

EA (Environment Agency) (2009). Flooding in England: A National Assessment of Flood Risk. Bristol: Environment Agency.

EA (Environment Agency) (2010). Cleaning up after a flood. Available at http://www.environment-agency.gov.uk/homeandleisure/floods/54871.aspx [Accessed February 3, 2010].

Few, R., and Matthies, F. (2006). *Flood Hazards and Health: Responding to Present and Future Risks*. London: Earthscan.

Fewtrell, L. and Kay, D. (2008). An attempt to quantify the health impacts of flooding in the UK using an urban case study. *Public Health*, 122, 446–451.

Flume, P., Gray, S., Bowman, C. M., Kerrigan, C., Lester, M., and Virella-Lowell, I. (2005). Emergency preparedness for the chronically ill. *American Journal of Nursing*, 105(3), 68–72.

GfK NOP (Growth from Knowledge National Opinion Polls) (2008). Qualitative Research Undertaken during October 2007 in Areas Affected by the 2007 Summer Floods. Growth from Knowledge National Opinion Polls (GfK NOP) Social Research, 451492. COI Cabinet Office Flooding.

Green, C., Emery, P., Penning-Rowsell, E., and Parker, D. (1985). The Health Effects of Flooding: Survey at Uphill, Avon. Flood Hazard Research Centre. Enfield: Middlesex Polytechnic.

HM Government Department of Health: Mental Health Division. New Horizons. Confident Communities, Brighter Futures. A Framework for Developing Well-Being. 25-3-2010.

HPA UK (Health Protection Agency) (2008). Health Advice—General Information Following Floods. Available at <http://www.hpa.org.uk/web/HPAwebFile/HPAweb _C/1194947339369>. [Accessed July 28, 2008].

Hubálek, Z., Zeman, P., Halouzka, J., Juřicová, Z., Šťovíčková, E., Bálková, H., Šikutová, S., and Rudolf, I. (2005). Mosquitoborne viruses, Czech Republic, 2002. *Emerging Infectious Diseases*, 11(1), 116–118.

Hyre, A. D., Cohen, A. J., Kutner, N., Alper, A. B., Dreisbach, A. W., Kimmel, P. L., and Muntner, P. (2008). Psychosocial status of hemodialysis patients one year after Hurricane Katrina. *American Journal of the Medical Sciences*, 336(2), 94–98.

Jonkman, S. N., and Kelman, I. (2005). An analysis of the causes and circumstances of flood disaster deaths. *Disasters*, 29(1), 75–97.

Kirch, W., Menne, B., and Bertollini, R. (2005). *Extreme weather events and public health responses,* Springer-Verlag, Berlin.

Kshirsagar, N. A., Shinde, R. R., and Mehta, S. (2006). Floods in Mumbai: Impact of public health service by hospital staff and medical students. *Journal of Postgraduate Medicine*, 52(4), 312–314.

Kutner, N. G., Muntner, P., Huang, Y., Zhang, R., Cohen, A. J., Anderson, A. H., and Eggers, P. W. (2009). Effect of Hurricane Katrina on the mortality of dialysis patients. *Kidney International*, 76(7), 760–766.

Lave, L., and Apt, J. (2006). Planning for natural disasters in a stochastic world. *Journal of Risk Uncertainty*, 33, 117–130.

Nasman, U., Zetterberg-Randonn, B., and Brandstram, H. (2007). KAMEDO Report No. 88: Floods in the Czech Republic and Southeastern Germany, 2002. *Prehospital and Disaster Medicine: The Official Journal of the National Association of EMS Physicians and the World Association for Emergency and Disaster Medicine in Association with the Acute Care Foundation*, 22(1), 90–92.

NHS (National Health Service, UK). http://www.nhs.uk/Livewell/Summerhealth/Pages/ Floodsafety.aspx [online]. Available at <http://www.nhs.uk/Livewell/Summerhealth/ Pages/Floodsafety.aspx>. [Accessed June 27, 2008]

Noji, E. K. (2005). Public health issues in disasters. *Critical Care Medicine*, 33(1), S29–S33.

Ohl, C., and Tapsell, S. (2000). Flooding and human health. *British Medical Journal*, 321, 1167–1168.

Pearson, A. (1999). Itai Itai Byo—Japan, 1940s. *Chemical Incident Report 13*. Chemical Incident Response Service, pp. 15–16.

Pitt, M. (2008).The Pitt Review: Learning Lessons from the 2007 Floods. Pitt Review.

Reacher, M., Mckenzie, K., Lane, C., Nicholas, T., Kedge, I., Iverson, A., Hepple, P., Walter, T., Laxton, C., and. Simpson, J. (2004). Health impacts of flooding in Lewes: A comparison of reported gastrointestinal and other illness and mental health in flooded and non-flooded households. *Communicable Disease and Public Health*, 7, 56–63.

Rhoads, J. (2006). Post-Hurricane Katrina challenge: *Vibrio vulnificus. Journal of the American Academy of Nurse Practitioners*, 18(7), 318–324.

Ruggiero, K. J., Amstadter, A. B., Acierno, R., Kilpatrick, D. G., Resnick, H. S., Tracy, M., and Galea, S. (2009). Social and psychological resources associated with health status in a representative sample of adults affected by the 2004 Florida hurricanes. *Psychiatry*, 72(2), 195–210.

Setzer, C., and Domino, M. E. (2004). Medicaid outpatient utilization for waterborne pathogenic illness following Hurricane Floyd. *Public Health Reports*, 119(5), 472–478.

Sjoberg, L., and Yearwood, R. (2007). Impact of a category-3 hurricane on the need for surgical hospital care. Prehospital and Disaster Medicine. *The Official Journal of the National Association of EMS Physicians and the World Association for Emergency and Disaster Medicine in Association with the Acute Care Foundation*, 22(3), 194–198.

Sniffen, J. C., Cooper, T. W., Johnson, D., Blackmore, C., Patel, P., Harduar-Morano, L., Sanderson, R., Ourso, A., Granger, K., Schulte, J., Ferdinands, J. M., Moolenaar, R. L., Dunn, K., Damon, S., Van Sickle, D., and Chertow, D. (2005). Carbon monoxide poisoning from hurricane-associated use of portable generators—Florida, 2004. *Journal of the American Medical Association*, 294(12), 1482–1483.

Subbarao, I., Bostick, N. A., and James, J. J. (2008). Applying yesterday's lessons to today's crisis: Improving the utilization of recovery services following catastrophic flooding. *Disaster Medicine and Public Health Preparedness*, 2(3), 132–133.

Tapsell, S. (2009). Developing a Conceptual Model of Flood Impacts Upon Human Health. Report T10-09-02 of Floodsite Integrated Project, Enfield: Flood Hazard Research Centre.

Tejima, H., Nakagawa, I., Shinoda, T., and Maeda, I. (1996). PCDDs/PCDFs reduction by good combustion technology and fabric filter with/without activated carbon injection. *Chemosphere*, 32, 169–175.

The Marmot Review (2010). Fair Society, Health Lives. Strategic Review of Health Inequalities in England post-2010.

Van Sickle, D., Chertow, D. S., Schulte, J. M., Ferdinands, J. M., Patel, P. S., Johnson, D. R., Harduar-Morano, L., Blackmore, C., Ourso, A. C., Cruse, K. M., Dunn, K. H., and Moolenaar, R. L. (2007). Carbon monoxide poisoning in Florida during the 2004 hurricane season. *American Journal of Preventive Medicine*, 32(4), 340–346.

Whiteley, D. (2008). 2007 NHS Flood Response Report.

Whittle, R., Medd, W., Deeming, H., Kashefi, E., Mort, M., Twigger Ross, C., Walker, G., and Watson, N. (2010). After the rain — Learning the lessons from flood recovery in Hull. Final project report for 'Flood, Vulnerability and Urban Resilience: A real-time study of local recovery following the floods of June 2007 in Hull'. Lancaster, UK: Lancaster University.

WHO Euro and WHO (2003), Extreme Weather Events: Health Effects and Public Health Measures.

WHO (World Health Organization) (2002). Flooding: Health Effects and Preventative Measures.

WHO (World Health Organization) (2008). Global Assessment of National Health Sector Emergency Preparedness and Response.

WHO (World Health Organization) (2009a). Strengthening WHO's Institutional Capacity for Humanitarian Health Action. A Five-Year Programme 2009–2013.

WHO (World Health Organization) (2009b). *Manual for the public health management of chemical incidents*. ISBN 978 92 4 159814 9. Available at <http://www.hpa.org.uk/web/HPAwebFile/HPAweb_C/1243467938380>. [Accessed 2010].

WHO Regional Office for Europe (2003), Extreme Weather Events: Health Effects and Public Health Measures. Copenhagen, Rome, September 29, 2003, Fact Sheet EURO/04/03.

6 The UK Sewer Network
Perceptions of Its Condition and Role in Flood Risk

Lee French, Victor Samwinga,
and David G. Proverbs

CONTENTS

6.1 INTRODUCTION

Flooding continues to be one of the top 10 "most important" disasters worldwide in terms of the numbers of people killed, the numbers of people affected, and the economic damages incurred. Of the five natural disaster categories annually monitored by the Centre for Research on the Epidemiology of Disasters (CRED), the hydrological disasters category, which includes floods, remained the most common in 2009. A total of 180 hydrological disasters were reported, accounting for more than 53% of the total natural disaster occurrences in 2009. With more than 57.3 million people affected, the hydrological disaster events of 2009 were largely caused by floods (82.8%), with the remainder (17.2%) being classified as "wet mass movements" (Vos et al., 2010).

CRED also highlights that the number of people affected by hydrological disasters increased by 27.4% in 2009 compared to 2008. In addition, CRED's statistics show that although the economic damages from hydrological disasters in 2009 were lower than in previous years, they still accounted for 19.1% of all economic damages from natural disasters worldwide (Vos et al., 2010). Flooding, which is the main component of hydrological disasters, is clearly a very significant problem worldwide.

Within the United Kingdom, flooding has received increasingly greater attention than ever before due to a number of significant flooding events recently, the resultant damage, and associated costs. Flooding associated with the UK sewer network has arguably been ignored as a flood risk factor for a long time (Gray, 2007). However, sewer flooding has received increased attention in recent years, probably due to its potential to be more distressing and disruptive than any other source of flooding (Defra, 2004).

There are concerns over the aging UK sewer network's ability to cope with the risk of flooding due to a combination of factors, including the predicted effects of climate change and increased housing development (ABI, 2004). This chapter therefore examines the current state of the UK sewer network in order to highlight its role in flood risk, the main reasons for any inadequacies in the sewer network, and the most realistic remedies to the problems identified. Apart from local authorities, the Environment Agency (EA), internal drainage boards, and water and sewerage companies are arguably the main partners as far as flooding associated with the sewer network is concerned. This chapter presents findings of exploratory in-depth interviews with representatives of water and sewerage companies in the United Kingdom on the theme of the sewer network.

6.2   THE UK SEWER NETWORK

The UK sewer network is an essential part of the national infrastructure, comprising 189,000 miles (302,000 kilometers) of underground piping (NAO, 2004). The sewers collect and remove waste as well as surface water during rainfall via separate or combined systems (Defra, 2004).

The Water Services Regulation Authority (Ofwat), which is an independent economic regulator of the water and sewerage sectors in England and Wales, has been critical of the number of parties involved as well as the length of time that the sewerage system has evolved on the less than perfect knowledge and records that exist (Ofwat, 2002). Balmforth et al. (2006) also identified the vast array of organizations responsible for what is essentially one network as the biggest problem facing the UK drainage network. The responsibilities are split between the 10 main sewerage undertakers, the local authorities, and private companies (Balmforth et al., 2006). More recently, the Pitt Review also highlighted this challenge, arguing that the splitting of responsibilities makes it difficult to manage the network in the way it should be—as one entity (Pitt, 2008).

Approximately 7% of the total UK network was built prior to 1885, and the majority built before World War II. Most of the systems generally work well although in extreme events there is clear evidence of undercapacity in some cases, which has resulted in flooding (Pitt, 2008). The following sections look at flooding from the sewers, the causes, and the potential remedies.

6.3 FLOODING FROM SEWERS

Flooding from sewers occurs when waste and surface water escape or cannot enter a drain or sewer system, and either remains on the surface or enters buildings (Ettrich, Schmitt, and Thomas, 2004). This can be caused by a network's insufficient capacity to cope with the level of flow, particularly during heavy rainfall or because of blockages or collapses in the system (NAO, 2004).

The role of sewers in flood risk has arguably received limited attention until recently. The unusually wet weather that resulted in extensive sewer flooding in late 2000 and early 2001 has been credited with raising the profile of the problem of sewer flooding (NAO, 2004). More recently, the 2006 flooding in Hull and Yorkshire (United Kingdom) was also largely blamed on drainage problems, prompting some to conclude that sewers have been ignored for too long as a flood risk factor (Gray, 2007).

When flooding occurs in built-up areas, it can cause significant damage to the infrastructure as well as untold misery for business owners and household occupants. Sewer flooding has been identified as more distressing and disruptive than any other type of flooding. As a result, it is not surprising that it has gained more visibility and has raised increasing levels of concern (Defra, 2004).

Despite its recent rise to prominence, sewer flooding is still relatively misunderstood (Defra, 2004). Because flooding from sewers cannot be accurately quantified and is in most cases directly associated with rainfall, there is still some skepticism over whether there is any real problem with the condition and capacity of the United Kingdom's sewer network. Both Ofwat (2002) and Ettrich, Schmitt, and Thomas (2004) argued that there is insufficient evidence to say conclusively that there is a problem with the United Kingdom's sewer network.

Although it is nearly impossible to predict the future with any real certainty, the Association of British Insurers (ABI) fears that flooding from sewers could become even worse as "the impacts of climate change and increased housing development add further strain to the capacity of our ageing sewer systems" (ABI, 2004). It has been predicted that by 2080, the number of homes at high risk of flooding will quadruple and that the current annual cost of flooding and flood management at £2.2 billion could rise to £27 billion in the same time period (Ockenden, 2004).

As a result, there is overwhelming pressure on UK sewer network stakeholders to act to avoid underperformance of the network and thereby contribute to reducing the flood risk associated with the sewer network.

6.4 CAUSES OF FLOODING FROM SEWERS

Flooding from sewers can be a contentious issue, especially because in most cases it is directly associated with rainfall. However, there are numerous factors that have been associated with sewer flooding, the majority of which can be placed within one of the following categories:

1. Condition/age (NAO, 2004; Ockenden, 2004; Balmforth et al., 2006; EA, 2007; Gray, 2007)
2. Blockages (NAO, 2004; Gray, 2007; Cornelius, 2008; UW, 2008)
3. Overloading (insufficient capacity) (Ofwat, 2002; ABI, 2004; Defra, 2004; NAO, 2004; Ockenden, 2004; Balmforth, et al., 2006; Pitt, 2008)
4. Climate change (Ofwat, 2002; ABI, 2004; ICF, 2004)
5. Lack of concern (Gray, 2007)
6. Increased development (Howe and White 2001; Lancaster, Marshall, and Preene, 2004; Defra, 2008; Pitt, 2008)
7. Illegal connections and the right to connect (NAO, 2004)
8. Underinvestment (Gray, 2007; Cornelius, 2008)
9. Exceptional storms and intense rainfall (Ofwat, 2002; Ettrich, Schmitt, and Thomas, 2004; ICF, 2004)

Some insist that the problem of sewer flooding is mainly due to the condition and age of the networks (Balmforth et al., 2006; Gray, 2007). However, some credit has been attributed to Victorian over-engineering as the main reason why sewer flooding is not even more widespread (Ockenden, 2004) despite the network's age.

Although they may have different causes, both blockages and overloading often prevent surface water from being able to enter the drainage systems. Blockages have been primarily blamed on materials that are not meant for discharge being placed down the sewers (NAO, 2004). Cornelius (2008) even argues that had it not been for the placement of inappropriate material down the sewers, many of the Victorian sewers would be perfectly adequate to cope with modern-day demands. Unlike blockages, overloading is primarily caused by a combination of insufficient capacity (Defra, 2004), increased development (Howe and White, 2001), and more intense rainfall events possibly due to climate change (Defra, 2008). While it would seem logical to upgrade sewers to cope with current demand, many disagree (NAO, 2004; Balmforth et al., 2006; Pitt, 2008), arguing that it is completely impracticable to do so. It is argued that other, more practical remedies and solutions can be effected at the management and investment levels, and these are discussed in the next section.

6.5   REMEDIES FOR THE PROBLEM OF SEWER FLOODING

Gray (2007) points out that not enough concern has been shown toward sewer flooding, and ignorance over recent years has made the problem almost impossible to counter-act. This may be partly due to a perception of a slight bias toward river flooding and yet sewer flooding should probably receive more attention (Gray, 2007) due to its potential health effects over and above the impacts shared with other forms of flooding.

Climate change and increasing exceptional storms are more difficult to quantify and control, and tackling the sewer capacity only to deal with the challenge is deemed unsustainable and excessive (Ettrich, Schmitt, and Thomas, 2004; ICF, 2007).

Other factors such as lack of concern, underinvestment, and dealing with the new connections to the system are all management-related issues, and critics argue that there is great scope for improvement in the management of each of these aspects (NAO, 2004; Gray, 2007; Cornelius, 2008).

A number of practical remedies to the problems highlighted in the previous section with respect to sewer flooding have been suggested in literature. These include

1. Better understanding of the existing network (NAO, 2004; Pitt, 2008)
2. More investment to upgrade old sewers (Ockenden, 2004; EA, 2007)
3. Increasing awareness and understanding of associated issues (Defra, 2004)
4. Better planning and management of new development (Ofwat, 2002; NAO, 2004)
5. Management of connections and urban creep (Ofwat, 2002; Pitt, 2008)
6. Sustainable urban drainage systems (SUDS) (Balmforth et al., 2006; Defra, 2008; Pitt, 2008)
7. Having one governing water body (Cornelius, 2008; Pitt, 2008)

The monitoring and maintenance of the sewers is extremely difficult and understanding is relatively low, partly due to 99% of the network being located underground (NAO, 2004). As a result, there is a suggestion that surface flows should be mapped and quantified with sufficient detail to enable local and strategic flooding problems to be easily dealt with (Gill, 2008; Pitt, 2008). The NAO (2004) also stresses the importance for companies to survey selected sewers through industrywide or local initiatives and then place reliance on this information in managing the network. This way, water and sewerage companies would develop a clearer understanding of the rate of deterioration of their existing sewerage network assets, which is essential.

Although additional investment to upgrade the sewer network would seem to be the logical mitigation strategy, it is unlikely that simply throwing money at the issue will solve all the problems. The Environment Agency does highlight the need for more investment in the existing system; however, constantly increasing the system's capacity to cope cannot continue indefinitely (EA, 2007).

Better planning and management of new development and urban creep is another mitigation strategy, especially for dealing with future sewer flooding problems before they become apparent. However, the NAO (2004) does not see any improvement in this area as long as the responsibilities continue to cut across so many organizations. The application of sustainable urban drainage systems (SUDS) also comes within the management of the planning process and government legislation; and if the negative impact of urban drainage is to be minimized, it will need to mimic natural drainage processes more closely in the future than it does at present (Balmforth et al., 2006).

The NAO (2004) and Pitt (2008) both agree that numerous authorities having responsibilities in very similar areas is a big problem. Pitt (2008) recommended that a single organization should have an overarching responsibility for all types of flooding, as the lack of transparency in ownership and the complexity involved is the main cause of the shift in attention to different causes and types of flooding. One regulatory body would be able to look at the flooding picture as a whole and apportion investment and concentration on the most needy or problem areas.

Following the recommendations by the Pitt Review (Pitt, 2008), Defra (2010) produced a comprehensive document with detailed technical guidance on the preparation of surface water management plans (SWMPs). This guidance is primarily intended for use by local authorities with input from external stakeholders such as the Environment Agency, Ofwat, sewerage undertakers, and the public. It will empower local authorities with a framework for coordinating and leading local flood risk management activities. The new SWMP guidance provides an overarching framework for the various stakeholders to collaborate and develop a shared understanding of the most suitable solutions to problems associated with surface water flooding. Such an approach would remedy what is often perceived as disjointed efforts by a myriad of stakeholders dealing with aspects of surface water management.

6.6   INTERVIEW FINDINGS

This section presents a summary of interviews conducted with six senior water authority employees. The questions were informed by the literature review, and the following interview findings are structured around three main themes—namely, sewer flooding in the United Kingdom, the factors affecting the network's performance, and the remedies.

6.6.1   SEWER FLOODING IN THE UNITED KINGDOM

Sewer flooding is quite obviously a significant problem, and the majority of the interviewees (refer to Table 6.1) see it as more of a problem than flooding from any other source. Even one interviewee, who thought it was less of a problem, admitted that sewer flooding was a big problem although river flooding was more so as it has the ability to cause sewer flooding.

One interviewee stated that sewer flooding was the largest flooding problem, claiming that many people may not agree simply because river flooding is more "glamorous," which worsens their perception. Further reasoning given as to why sewer flooding was more of a problem was due to the physical damage caused, the issues with environmental health, and because it is a lot less predictable than fluvial flooding.

TABLE 6.1
Sewer Flooding and Factors Affecting Performance of Sewers

Variable	Values	No.
Significance of flooding problem	More of a problem	4
	Less of a problem	1
	Equal problem	1
Network's coping with present demands	Very well	5
	Not very well	0
	Satisfactorily	1
Main factors affecting network's performance	Increased development	4
	Restricted outfalls	2
	Inadequate sections of drainage	2
	Increased rainfall	4
	Lack of investment	1

These results mirror most literature sources and, in particular, Defra's (2004) statements that sewer flooding is more disruptive and distressing than any other type.

The majority of interviewees, somewhat surprisingly, thought that generally the UK sewer network was coping very well with modern-day demands. As shown in Table 6.1, only one interviewee thought the network only coped satisfactorily, adding that the network was already suffering from the effect of increased severe weather patterns.

The apparent positive verdict on the United Kingdom's sewer network was clearly blended with a degree of hesitation, owing to concerns that small sections of the network performed poorly. In addition, predictions of future increases in extreme weather events were also a source of concern over the network's ability to continue its performance. The ABI (2004) and ICF (2007) seem to share similar concerns, stressing that the main focus should be on future planning and looking at long-term measures.

When asked for their views regarding the various stakeholders' efforts and efficiency in addressing sewer flooding, the interviewees' verdict was almost unanimous. Five out of six interviewees viewed the efforts and performance of stakeholders in addressing sewer flooding as unsatisfactory or poor, with one interviewee stating that the stakeholders are performing satisfactorily.

The main cited reasons why the stakeholders performed poorly were poor effort, inefficiency, and unavailable finances. Pitt's (2008) recommendations made in the Foresight Report were cited on three separate occasions, and each interviewee thought that those recommendations should be followed rigorously by the stakeholders although they doubted this would ever happen despite it having the potential of raising the profile of sewer flooding.

6.6.2 FACTORS AFFECTING THE UK SEWER NETWORK'S PERFORMANCE

In keeping with the interviewees' responses in the previous subsection, future performance of the network seems to be the primary worry. Table 6.2 shows that the two most cited factors affecting the performance of the sewer were increased development and increased rainfall. Howe and White (2001); Lancaster, Marshall, and Preene (2004); and the ABI (2004) all support the view that increased development or urban creep is the main factor affecting the UK sewer network's performance. Without any mitigating measures, the future impact of the two factors, increased development and increased rainfall, will only continue to increase.

TABLE 6.2
Main Factors Affecting Sewer Network Performance

Main Factors Affecting Network's Performance	Occurrence
Increased development	4
Increased rainfall	4
Restricted outfalls	2
Inadequate sections of drainage	2
Lack of investment	1

It was rather surprising that only one of the six interviewees discussed the issue of lack of investment, considering that Gray (2007) and Cornelius (2008) both strongly suggest that the UK sewerage systems suffer from chronic underinvestment. However, the interviewees did acknowledge in a previous question that some sections of the network were in disrepair and required attention.

6.3.3 REMEDIES TO THE UK SEWER NETWORK'S PROBLEMS

Based on findings from a literature review, a number of potential remedies to the sewer flooding problem were put forward to interviewees to rank according to importance (1 being most important and 5 being the least important). Table 6.3 shows that "Better understanding of the existing network," "Better planning and management," and "More investment to upgrade old sewers" were all jointly ranked the most important remedies by the six interviewees. According to one interviewee, modeling of the existing network would be essential both for present-day and future management:

> Fundamental to preventing sewage flooding is knowing how the existing system performs. A fully modelled network would highlight areas of concern either at present or for proposed developments. (Project Manager, Water and Sewerage Company)

Such an emphasis on the need for quantitative data regarding the network, with surface flows mapped and quantified to such detail as to enable local and strategic tackling of flooding problems, has support in the literature (Gill, 2008; Pitt, 2008). Sir Michael Pitt (2008) and the NAO (2004) also advocated the need for better understanding of the infrastructure in tackling the challenge of flooding.

"Increasing awareness and understanding" was, by far, ranked as the least important measure by the interviewees. This may be due to interviewees focusing more on pro-active measures to tackle sewer flooding rather than raising awareness of how flooding occurs and its impacts on drainage assets.

Having ranked the importance of the seven potential remedies, interviewees were also asked to rank the potential remedies in terms of how realistic they were to implement (1 being the most realistic and 5 being the least realistic). The most

TABLE 6.3
Potential Remedies Ranked According to Importance

Potential Remedy	Ranking Frequency							Weighted Score	Rank
	1	2	3	4	5	6	7		
Better understanding of existing network	3	0	0	0	1	1	1	21	1
Better planning and management	2	0	1	0	2	1	0	21	2
More investment to upgrade old sewers	0	2	1	2	0	1	0	21	3
Having one governing body (no split)	0	2	1	1	1	0	1	23	4
Sustainable urban drainage systems (SUDS)	1	0	1	2	1	1	0	23	5
Management of connections and urban creep	0	2	1	1	0	0	2	25	6
Increasing awareness and understanding	0	0	1	0	1	2	2	34	7

TABLE 6.4
Potential Remedies Ranked According to Importance Based on Their Realistic Implementation

Potential Remedy	Ranking Frequency							Weighted Score	Rank
	1	2	3	4	5	6	7		
Better planning and management	4	0	1	0	1	0	0	12	1
Increasing awareness and understanding	1	3	1	1	0	0	0	14	2
Sustainable urban drainage systems (SUDS)	1	1	2	1	0	1	0	19	3
Better understanding of existing network	0	1	0	2	2	1	0	26	4
More investment to upgrade old sewers	0	0	2	1	2	0	1	27	5
Management of connections and urban creep	0	1	0	1	0	4	0	30	6
Having one governing body (no split)	0	0	0	0	1	0	5	40	7

striking result in Table 6.4 was the ranking of "Better planning and management" in first position. This response is well aligned with the earlier assertion by interviewees that increased development and increased rainfall are the two major factors affecting sewer flooding. "Better planning and management" was seen as a realistic measure to minimize future risks associated with flooding from sewers.

In addition, "Increasing awareness and understanding," despite being perceived as the least important measure, was ranked highly in terms of realistic measures. This may partly be due to the option being low cost and relatively easy to implement.

"Increasing awareness and understanding" was ranked as the second most realistic measure to take in managing the problem of sewer flooding. This is consistent with the emphasis by Defra that

> Flood risk, especially in built-up areas, can be managed most effectively if there is an understanding of the way floods arise and have an impact on the various drainage systems. (Defra, 2004, p. 68)

At the other end of the scale, "Having one governing body" and "Management of connections and urban creep" were the least-favored realistic options for dealing with the problem of sewer flooding. Based on these findings, it would appear that Pitt's (2008) vision of the future—of having one regulatory body in charge of the management of the network—may be a distant hope. Although having one governing body was seen as relatively very important, interviewees pointed out that it is unlikely to ever happen. The reasons cited include the lack of support in government for such a measure and the numerous implications such a change in structure would have.

Although "More investment to upgrade old sewers" was raised as one of the top-three most important solutions, interviewees recognized the limitations and hence saw it as an unrealistic expectation. Instead of upgrading the network to take increased flows, which may not be sustainable, interviewees felt that it is more realistic for investment to be targeted at improving specific poorly performing parts of the network:

Investment in the network will be required over a prolonged period to resolve flooding at locations which are currently at risk. (Project Manager, Water and Sewerage Company)

SUDS were perceived as a relatively realistic measure. This may be partly due to the continuing increasing use of SUDs in England although much more needs to be done if it is to be a significant factor in the reduction of flood risk.

6.7   SUMMARY

The United Kingdom continues to suffer the consequences of severe flooding events with their attendant significant costs that are directly associated with such events. Although sewer flooding is only one source of flooding, its effects can potentially be more distressing and disruptive than any other source of flooding.

The visibility of the United Kingdom's aging sewer network as a risk factor in flooding has increased in recent years and questions have been raised, especially with respect to its future performance. This is due to concerns over the potential impact of increased development combined with predictions of extreme weather events in the future.

While there is a degree of confidence in the network's current performance, there is a level of concern over its ability to cope with increased capacity demands due to factors such as the predicted impact of climate change and increased infrastructure development. The chapter stressed that water and sewerage companies see the problem of sewer flooding as being primarily due to some sections of the network underperforming. However, it is essential to have a better understanding of the existing network to identify these areas and thereby target investment where it is most needed.

In addition, the option of increasing the sewer network to cope with increased future demands was not seen as practical, feasible, or even sustainable. Instead, management-related remedies such as better planning and management and increased awareness and understanding were arguably among the most realistic remedies. Therefore, most of the attention should focus on better management of the network and its associated problems. Defra's guidelines on SWMP offer some hope in alleviating the major factors associated with sewer flooding discussed in this chapter.

REFERENCES

ABI (Association of British Insurers) (2004). *The case for increased investment in sewer infrastructure.* Association of British Insurers (ABI), July 2004. Available at <http://www.abi.org.uk/_Policy_Issues/15272.pdf> [Accessed February 10, 2009].

Balmforth, D., Digman, C., Kellagher, R., and Butler, D. (2006). *Designing for Exceedence in Urban Drainage—Good Practice.* London: Construction Industry Research and Information Association.

Cornelius, D. (2008). David Cornelius talks to Stephen Battersby about the UKs ageing sewers. *Environmental Health Practitioner*, 116, 14–15.

Defra (Department for Environment, Food and Rural Affairs) (2004). *Making space for water.* London: Department for Environment, Food and Rural Affairs.

Defra (Department for Environment, Food and Rural Affairs) (2008). *Future water: The Government's water strategy for England.* London: Department for Environment, Food and Rural Affairs. February 2008 (Online). Available at <http://www.defra.gov.uk/environment/quality/water/strategy/pdf/future-water.pdf> [Accessed December 10, 2008].

Defra (Department for Environment, Food and Rural Affairs) (2010) *Surface Water Management Plan Technical Guidance.* London: Defra.

EA (Environment Agency) (2007). Hidden Infrastructure. The Pressures on Environmental Infrastructure. (Online). Available at <http://publications.environment-agency.gov.uk/pdf/GEHO0307BMCD-E-E.pdf> [Accessed December 10, 2008].

Environment, Food and Rural Affairs Committee. (2004). *Climate Change, Water Security and Flooding.* London: The Stationary Office.

Ettrich, N., Schmitt, G., and Thomas, M. (2004). Analysis and modelling of flooding in urban drainage systems, *Journal of Hydrology*, 299, 300–311.

Gill, E. (2008). Making Space for Water: Urban Flood Risk & Integrated Drainage, A Defra IUD pilot summary report (HA2), June 2008. Available at <http://www.defra.gov.uk/environment/flooding/documents/manage/surfacewater/urbandrainagereport.pdf> [Accessed December 10, 2008].

Gray, R. (2007). Flooding blamed on failure to upgrade ancient sewers. *The Sunday Telegraph*, July 8, 2007, p. 23.

Howe, J., and White, I. (2001). Flooding: Are we ignoring the real problem and solution? *Policy Review Section*, pp. 368–370.

ICF International and RPA. (2007). The Potential Costs of Climate Change Adaptation for the Water Industry. London: ICF International.

Lancaster, J., Marshall, C., and Preene, M. (2004). *Development and flood risk,* London: CIRIA.

NAO (National Audit Office). (2004). Out of sight—Not out of mind: Ofwat and the public sewer network in England and Wales. London: The Stationary Office.

Ockenden, K. (2004). Flood risk homes to quadruple by 2080. *Utility Week*, 21(15), 22.

Ofwat (Water Services Regulation Authority) (2002). *Flooding from sewers: a way forward. Response to consultation.* September 2002. Available at <http://www.ofwat.gov.uk/regulating/res_stk_floodsewers.pdf> [Accessed December 10, 2008].

Pitt, M. (2008). The Pitt review: Learning lessons from the 2007 floods, London: COI.

UW (2008). Money Down the Drain, *Utility Week*. 26(21), 43.

Vos, F., Rodriguez, J., Below, R., and Guha-Sapir, D. (2010). Annual Disaster Statistical Review 2009: The Numbers and Trends. Brussels: CRED.

Section II

Recovery, Repair, and Reconstruction

7 Flood Insurance in the United Kingdom
The Association of British Insurers' View

Swenja Surminski

CONTENTS

7.1 INTRODUCTION

In light of rising flood risks, insurers are continuing to call for better risk management practices to keep homes and businesses insured. The Association of British Insurers (ABI), which represents over 90% of the insurance market in the United Kingdom and 20% across the European Union (EU), wants to ensure that flood risk is managed effectively and that as many people as possible can continue to obtain competitively priced insurance to protect themselves from the financial cost of flooding. Under a 2008 flood agreement with the UK government and the devolved administrations, insurers committed to continuing to provide flood insurance to the vast majority of customers until 2013. In return, the government gave an overarching commitment to ensure that flood risk is appropriately tackled. Over the next 3 years, the insurance industry will continue to work with the government to put in place long-term solutions that will enable flood insurance to be as widely available as possible through effective long-term management of flood risk, backed by adequate funding.

The availability of flood insurance varies between countries, reflecting the nature of weather risks affecting the market and the nature of any public/private partnerships in operation. The United Kingdom is one of very few countries that has a fully functioning private insurance market for flood insurance. In the United Kingdom, flood risk is usually covered as a standard part of business and household insurance. The UK government is not the insurer of last resort, unlike in many other countries.

Without insurance, property owners must assume financial risks themselves, mortgages are difficult to obtain, and properties are virtually unsellable. Recent experience has demonstrated that the United Kingdom is poorly prepared to deal with flooding, with a range of very severe and costly events causing tragic loss of life, devastating property damage, and disruption to families, communities, and the entire economy.

The floods of summer 2007 were the most severe weather-related event that the United Kingdom had experienced in decades. Insurers paid out more than £3 billion, dealing with 180,000 claims for damaged homes, businesses, and vehicles. But the full economic and social costs were many times higher, with thousands of people forced to leave their homes, schools closed, and companies unable to operate. Insurers dealt with claims quickly and efficiently and, overall, customer feedback has been positive. The Financial Ombudsmen Service has reported a very low level of complaints. The Pitt Review on the 2007 floods (Pitt, 2008) concluded that the wide majority of insurance customers were satisfied with the claims service received after the floods, but highlighted the need to improve the claims handling process for future surge events:

> The scale of the summer 2007 floods were a challenge for insurers and loss adjusters, and many rose to that challenge. However, a small but significant number of households did not experience the good service received by many. Issues arose in the immediate aftermath, with conflicting information on clear-up and evidence levels for claims. Most insurance companies were in touch relatively quickly but there were delays for some in terms of contact and face-to-face visits from loss adjusters (some of over a month), which then led to delays in the onset of work. Those that were dissatisfied with their insurers raised concerns around information availability (difficulty in getting any information and length of time to get it), length of time to repair properties and issues around money.

The insurance industry has taken on board the need to learn the lessons. The ABI has produced guidance on what to expect from an insurer and on the claims handling process in response to the Pitt Review's recommendation No. 32:

> The insurance industry should develop and implement industry guidance for flooding events, covering reasonable expectations of the performance of insurers and reasonable actions of customers. (Pitt, 2008)

The leaflet entitled "Responding to Major Floods: What to Expect from Your Home Insurer" explains how an insurance company will respond in an emergency situation where many thousands of homes have been affected by flooding. The guide sets out the support that customers can expect from their insurer in the days, weeks, and months after a major flood (ABI, 2009d).

The concern about rising flood risk has intensified with the demand for new homes, a realization that urban areas are becoming more intensively developed, and the predicted impacts from climate change. British insurers are most concerned about the predicted increase in coastal, fluvial, and pluvial flooding. Recent studies conducted by the ABI and other stakeholders and the new UK Climate Impact Projections 09 all show that climate change is likely to increase the risk of flooding across the United Kingdom (ABI, 2006; ABI, 2009a,c).

7.2   THE STATEMENT OF PRINCIPLES

The insurance industry has argued that more risk management and investment in resilience are necessary in order to maintain widely available flood insurance at an affordable price. To achieve this, insurers and the UK government signed a voluntary agreement in 2002, known as the Statement of Principles (SoP) (ABI, 2002). Under this agreement, ABI members commit to provide flood insurance for as many customers as possible in return for the adequate management of flood risk by the government, including sufficient investment in new and existing defenses and other risk reduction measures. The SoP commits insurers to continuing to offer flood insurance to existing customers where the flood risk is adequately managed.

Following the floods of summer 2007, and in light of the projected increase in flood risk, the UK government and the insurance industry determined that a revised SoP was needed (ABI, 2008b). The new agreement sets out a long-term approach to the provision of flood insurance. It reflects the need to put in place measures to ensure that flood coverage can continue to be widely available and competitively priced in the future. Under the renewed SoP, both government and the industry commit to certain actions. These include

- Improving understanding of flood risk
- Government putting in place a long-term investment strategy that will set out strategic flood prevention aims and assess future policy options and funding needs
- Ensuring that the planning system prevents inappropriate development in flood-risk areas
- Raising awareness in areas where flood risks are significant, encouraging property owners to take sensible precautions, and providing more information about how to obtain flood insurance
- Promoting access to home insurance for low-income households
- Insurers making flood insurance for homes and small businesses available under household and commercial insurance, where the flood risk is no worse than a one in 75 years (1.3%) annual risk
- Insurers offering flood cover to existing domestic and small business customers at significant flood risk, with the provision that there are plans to reduce that risk to an acceptable level within 5 years

Similar agreements have been reached with the devolved Administrations in Scotland, Northern Ireland, and Wales.

Although the SoP commits insurers to continuing to offer flood coverage under certain scenarios, it has serious distorting effects on the market. It undermines incentives for homeowners to take steps to improve the flood resistance and resilience of their properties, and it hinders the development of specialist flood insurance for seriously at-risk properties. Therefore, the SoP will end in 2013. The implementation of the agreed measures over the next 3 years should ensure that flood insurance continues to be widely available, without the need for the SoP. Until then, we will be monitoring closely progress toward implementing the measures we have agreed to under the renewed SoP.

7.2.1   A Long-Term Strategy for Flood Risk Management

The key achievement of the agreement is the commitment from the UK government and the devolved administration that it will put in place a long-term investment strategy to reduce flood risk. This marks a move away from the current short-term approach characterized by allocating spending to a problem without knowing the full extent of the risk to a full assessment of the scale of the flood risk in the medium term, taking into account the impact of climate change, with an investment program to match. Under the agreement, the Environment Agency has published the impact, outcomes, and funding implications of a range of policy options for the next 25 years. The Long Term Investment Strategy produced for England shows that significant investment would be required to reduce the levels of flood risks. A doubling of expenditure over the next 25 years would only lead to the maintenance of the status quo in terms of numbers of properties at significant risk—in other words, we need to double the amount we spend on flood defenses over the next 25 years to stand still (EA, 2009). For Wales, the Environment Agency concluded that current spending levels would need to rise threefold over the next 25 years to maintain the level of flood risk in Wales broadly as it is today. To reduce the present-day risk level (138,000 properties at significant and moderate risk reduces from 138,000 to 107,000), spending would need to increase fourfold over the next 25 years (EA, 2010).

Publication of these options will allow the government to publish its long-term strategy for managing river and coastal flood risk for the next 25 years, based on the Environment Agency's range of options. A public debate about a vision for the future of flood risk and flood risk investment needs to take place now. The long-term investment options presented by the Environment Agency do provide the evidence base for this debate. Despite the substantial fiscal challenges facing the new liberal democrat–conservative government in the United Kingdom, the ABI will be pressing for a suitably ambitious long-term government plan backed up by sufficient funding.

7.2.2   Improved Legislative Framework for Flood Risk Management

The floods in summer 2007 demonstrated the need for proper coordination of flood risk management. The ABI conducted a major research project looking at the need to improve surface water management in order to reduce flood damage in urban areas (ABI, 2009e). This research draws evidence from recent projects that have aimed to manage surface water in urban areas, supplemented by two projects (in

Peterborough and Bristol) where surface water management plans were developed through stakeholder groups including the key institutions involved. The research has highlighted a range of institutional and process issues that need to be addressed in order to better manage the increasing risk of urban surface water flooding in the United Kingdom. The new Floods and Water Management Act 2010 has given the Environment Agency overall responsibility for overseeing new powers and responsibilities for local authorities to draw up and implement plans to assess and reduce surface water flooding. Now we need to see the government commit to providing local authorities with adequate powers and funding to deliver this important new role. We would also like to see legally binding targets for reducing the number of properties and businesses at risk of flooding, and a requirement for the Environment Agency to publicly report against these. And we recommend an independent audit of local authorities' performance against flood risk management plans and making this information publicly available.

7.2.3 FLOOD RESILIENT AND RESISTANCE MEASURES

The agreement also reinforces the partnership between the industry and the government to encourage homeowners and businesses to take flood resilient and resistance measures and to promote financial inclusion. The ABI continues working closely with government to promote better the benefits of insurance among low-income households. Experience from the 2007 floods shows that property owners tend to be reluctant to complete resilient and resistant repairs, as they fear this will lead to a devaluation of their property or that the features of such modifications will be unsightly. Homeowners also have to pay any additional costs. To inform our work on how to promote property-level flood protection and flood resilience, the ABI conducted a research project into the cost of resilient repair, resulting in a guidance document published in cooperation with the Environment Agency, the National Flood Forum, and the Chartered Institute of Loss Adjusters (ABI, 2010). This guide is intended to encourage flooded customers to discuss their options with their insurer and loss adjuster. This work is being followed up by another ABI research study that looks at how insurers are promoting resilient repair and how take-up among customers could be increased.

7.3 NEW PROPERTY DEVELOPMENT

The SoP does not apply to new property development. Here, both government and industry feel that the planning system should ensure that new development only goes ahead where flood risk is at an acceptable level.

The most efficient way to avoid creating new flood risk is to stop building in high flood–risk areas. If development is needed in high flood–risk areas, higher minimum standards should be applied to any buildings constructed. These should include design features to protect against flooding and minimize damage should a flood occur. Applying the routine national standards is not sufficient in high-risk areas. In 2006, the ABI welcomed the introduction of the new Policy Planning Statement 25 (PPS25) that was intended to strengthen and clarify policy on developments and flood risk. While it appears to be working in most areas, further work is needed to spread best practice among all local planning authorities. This highlights the importance of developers and planners complying with the ABI's recently published guidance on insurance for new developments, published in January 2009 (ABI, 2009b).

With the guide, we want to help developers, architects, engineers, planning authorities, and prospective buyers to ensure that they design, plan, build, and buy new developments that rise to the challenges presented by climate change; and buildings that are sustainable, energy efficient and of low impact, attractive to occupy, and come with the information necessary to access affordable insurance.

The key messages of the guide for property developments can be summarized as follows:

- Climate change means that buildings will be increasingly vulnerable to severe weather.
- This will have an impact on the cost and availability of insurance unless steps are taken to reduce the risk.
- Buildings must be located and designed to ensure that they are able to withstand increased flood risk.
- Insurers will only be able to insure buildings—vital to ensure that they are sellable—if this risk is managed to acceptable levels.
- We recommend that developers
  - Follow National Planning Policy Statements
  - Provide buyers with information on flood resilience measures
  - Develop publicly available standards or kitemarks that certify enhanced resilience to climate change impacts
- Before buying a property in a new development, prospective owners should check the flood risk and obtain information on measures taken to reduce it.

7.4   IMPROVING INSURERS' RESPONSE TO FLOOD EVENTS

The insurance industry recognizes that we cannot just make more demands on others. We must do all we can to help tackle rising flood risks. The ABI has been leading the work on learning from past events and developing innovative solutions to reduce future risks.

As soon as the scale of the 2007 summer floods became apparent, insurers implemented their emergency plans designed to cope with major events. Thousands of additional staff were brought in from across the United Kingdom and overseas to handle calls from affected customers and visit their homes and businesses. The insurance industry recognized that customers wanted to be given as much information as possible quickly, for promises to be kept, and to be treated as individuals. A survey commissioned by the ABI soon after the event showed that customers rated the response of the industry above that of local authorities, the Environment Agency, and the UK government (ABI, 2008c). However, while the insurance industry mobilized quickly to respond to the size of the challenge and made every effort to meet customers' expectations and requirements, in many cases the nature of the repairs needed has taken time. Repairing a typical house with severe flood damage can take 12 to 18 months.

Insurers recognize that we have lessons to learn and we want to respond to the needs of our customers even more effectively should disaster strike again. The ABI has therefore taken steps that reflect feedback received from insurance customers:

1. The ABI has published new guidance for customers affected by severe flooding (ABI, 2008a) called "Responding to Major Floods: What to Expect from Your Home Insurer." This sets out clearly and concisely the key steps and timelines involved in repairing homes after a major flood event to ensure that this information is readily available and will be widely distributed should another major flood occur.
2. The ABI wants to ensure that its customers have the right insurance coverage. So it has published a new leaflet entitled "Is Your Home Underinsured?" (ABI, n.d.) to provide better guidance for customers on assessing their home insurance needs. Research by the Zurich insurance company in 2005 revealed that one in five households was at risk of being underinsured because they were unsure of the value of their home's contents, while a recent AXA survey concluded that 90% of small businesses were underinsured for buildings coverage (Zurich, 2005; AXA, 2007).

Home insurance is divided into two separate policies: one for buildings and one for contents. While the buildings policy covers the structure of a house, the content applies to just about everything you would take with you if you moved from your house. For both policies it is important to make sure that the sum insured is adequate. For a buildings policy, the sum insured is the total cost of rebuilding your house; for the content part, the sum insured should be in line with how much it would cost to replace every item insured with one of similar quality, new, at today's prices. The guidance document gives homeowners advice on how to calculate a correct sum insured.

7.5 THE NEED TO IMPROVE THE UNDERSTANDING OF CURRENT AND FUTURE RISKS

We need to create an open culture of sharing information between all those involved in increasing risk awareness. Comprehensive, clear, and accessible flood risk information is a prerequisite for effective management of flood risks. We all need to improve our understanding of flood risk through assessing both the probability and consequences of flooding from all sources, including surface water.

The ABI continues to work with the Environment Agency to ensure that more detailed flood risk information is made publicly available and updated annually. The provision of free flood risk surveys to owners of high-risk properties would also increase understanding and acknowledgment of risks. It is vital that improved information on flood risk is available to insurers. This will allow them to continue to provide insurance to as many customers as possible on terms that reflect the risk of flooding as accurately as possible.

But insurers also need to share their knowledge and expertise. We are committed to providing industry data that will help improve the understanding and modeling of risks. In November 2009, the ABI published the results of a major research project conducted by the catastrophe modeling firm AIR and the Met Office, examining the financial implications of climate change using climate models and insurance catastrophe risk models (ABI, 2009a). The research team has used groundbreaking modeling techniques, combining the latest climate models with insurance risk models for the first time.

In assessing the impact of climate change scenarios on insurance pricing, the research team applied a typical cost-pricing algorithm to average annual loss models. The research has examined the implications of 2°C, 4°C, and 6°C changes in global mean temperature on inland flooding in Great Britain, windstorms in the United Kingdom, and typhoons in China. For each of these, the research assesses the impact of our changing climate by modeling

- How these likely temperature changes will affect weather hazards and the resulting insured loss;
- The flow-through impact on insurance prices; and
- The impact on insurance capital requirements.

Increases in global mean temperature will impact most directly on rain in Great Britain and the associated inland flooding. The risk of extreme flood events similarly increases under each of the three temperature increases. As a result of these changes in rainfall, average annual insured losses and the impact of a severe event are likely to increase. For example, the inevitable 2°C temperature change will increase average annual insured loss in Great Britain from inland flooding by 8% (i.e., £47 million) to £600 million. All areas of Great Britain show increased losses from any of the possible global temperature increases. These increased losses, and other factors such as growth in the number and value of properties and increasing population density in at-risk areas, put upward pressure on insurance pricing. These factors have been considered alongside conservative growth projections to show, for example (assuming 2.5% annual growth in GDP), that insured losses from an extreme 1 in 100-years flood could rise by 18% (ABI, 2009a).

7.6   CONCLUSION

While risks and exposures are increasing, the industry is committed to continue providing flood insurance coverage to as many people as possible. This will depend on the implementation of the agreed steps under the SoP, supported by a suitably ambitious long-term funding commitment from the UK government and devolved administrations. Spending reviews completed every three years are not sufficient. Tackling flooding will require a strong partnership between central and local government, the private and public sectors, and communities and individuals to assess and reduce flood risk. As demonstrated in the past, the insurance industry is willing and able to play its role within this partnership. Its aim is that should a severe flood strike the United Kingdom again in the future, the insurance industry's work will ensure that loss of life, damage to property, and the impact on communities are as minimal as possible. That is what our customers need and deserve.

REFERENCES

ABI (Association of British Insurers) (no date). Is Your Home Underinsured? London: Association of British Insurers.

ABI (Association of British Insurers) (2002). ABI Statement of Principles on the Provision of Flooding Insurance. London: Association of British Insurers.

ABI (Association of British Insurers) (2006). Coastal Flood Risk — Thinking for Tomorrow, Acting Today. *Summary Report.* London: Association of British Insurers.

ABI (Association of British Insurers) (2008a). Responding to Major Floods: What to Expect from Your Home Insurer. London: Association of British Insurers.

ABI (Association of British Insurers) (2008b). Revised Statement of Principles on the Provision of Flood Insurance. London, Association of British Insurers.

ABI (Association of British Insurers) (2008c). The Summer Floods 2007: One Year On and Beyond, 2008. London: Association of British Insurers.

ABI (Association of British Insurers) (2009a). Assessing the Risks of Climate Change: Financial Implications. London: Association of British Insurers.

ABI (Association of British Insurers) (2009b). Climate Adaptation — Guidance On Insurance Issues for New Developments. London: Association of British Insurers.

ABI (Association of British Insurers) (2009c). The Financial Risks of Climate Change. ABI Research Paper 19. London: Association of British Insurers.

ABI (Association of British Insurers) (2009d). Responding to Major Floods. London: Association of British Insurers. Available at <http://www.abi.org.uk/Publications/ ABI_Publications_Responding_to_major_floods_What_to_expect_from_your_home_ insurer_92a.aspx>.

ABI (Association of British Insurers) (2009e). Urban Surface Water Management Planning — Implementation Issues. ABI Research Report 13. London: Association of British Insurers.

ABI (Association of British Insurers) (2010). A Guide to Resistant and Resilient Repair after a Flood. London: Association of British Insurers. Available at <http://www.abi.org.uk/ Publications/ABI_Publications_A_guide_to_resistant_and_resilient_repair_after_a_ flood_670.aspx.>.

AXA (2007). Preparing for Climate Change, A Practical Guide for Small Business. London: AXA Insurance. Available at <http://www.axa.co.uk/assets/documents/axa.co.uk/busi- ness/insurance/advice-guidance/axa-preparing-for-climate-change.pdf>.

EA (Environment Agency) (2009). Investing for the Future. Flood and Coastal Risk Management in England — A Long Term Investment Strategy. Bristol: Environment Agency.

EA (Environment Agency) (2010). Future Flooding in Wales: Flood Defences — Possible Long-Term Investment Scenarios. Bristol: Environment Agency. Available at <http:// www.environment-agency.gov.uk/research/library/publications/116654.aspx>.

Pitt, M. (2008). The Pitt Review: Learning Lessons from the 2007 Floods. London, Cabinet Office. Available at <http://archive.cabinetoffice.gov.uk/pittreview/thepittreview.html>.

Zurich (2005). Underinsurance. *Moneyweekly* March 2, 2005. Available from <www.uk.biz. yahoo.com/moneyweekly/underinsurance.html 2 March 2005>.

8 A Practical Guide to Drying a Water-Damaged Dwelling

Bill Lakin and David G. Proverbs

CONTENTS

8.1 INTRODUCTION

The drying of flood-damaged property is an essential and integral part of the flood recovery process (Proverbs and Soetanto, 2004) as appropriate and adequate drying is key to reinstating the property in a timely and effective manner (Soetanto and Proverbs, 2003). There exists a considerable body of information and guidance on the drying of flood-affected property; see, for example, the BRE (Building Research Establishment) (1974), CIRIA (Construction Industry Research and Information) (2005), BSI (British Standards Institution) (2005), Flood Repairs Forum (2006), and EA (Environment Agency) (2007). Tagg et al. (2010) also provide a comprehensive review of the available guidance and advice on drying flood-damaged buildings, incorporating references to detailed technical guidance, research findings, and international reports.

Despite this body of information, it seems that many of those involved in the repair and recovery process are failing to make use of such guidance (Rhodes and

Proverbs, 2008) and seemingly rely on their experience (or indeed ignorance) of what is a highly complicated and skilled process. Certainly, evidence from the UK government's review of the 2007 flooding (Pitt, 2008) indicated that improper drying was a major contributor to the delay in reinstating flooded properties following the summer 2007 floods.

The purpose of this chapter is to provide a practical and user-friendly guide on the drying of water-damaged property, incorporating specific advice on flood-affected properties, but also giving due consideration to other types of causes, known within the damage management sector as "escapes of water," that might emerge through burst pipes or leakages and spillages of water. In doing so, the author draws on his extensive experience gathered over 35 years of undertaking domestic and commercial restoration and his role as a former president of the National Carpet Cleaners Association (NCCA), the National Institute of Disaster Restoration, and the Restoration Industry Association (RIA, formerly the Association of Specialists in Cleaning and Restoration).

8.2 AN INTRODUCTION TO WATER DAMAGE AND DRYING

When a building has been exposed to a volume of water from a flood, pipe burst, or overflow, the property will require a thorough survey to determine the extent of the water damage. If the survey undertaken is not thorough and does not cover all areas of the property—apparently affected or not—there is a real possibility that the drying process will be incomplete. This can lead to rot, mold, and degradation of the building fabric. The building occupants can also experience health issues. A well-designed and executed plan of action must be created that will allow the drying process to be both efficient and adaptable as the work continues. Most water losses can be a very rapid affair with water levels receding as quickly as they appeared.

It is very important to ensure that before commencing any drying-out processes, all health and safety issues have been addressed. Safety is paramount, and it is important to always ensure that occupants, technicians, and visitors to the property are not in any possible danger. Warning notices should be posted, and all work should be undertaken in strict accordance with relevant health and safety regulations and legislation. All electrical power should be isolated and made safe. It is wise to check for the odor of leaking liquid propane or natural gas and turn off these services. A fully trained and qualified person must be employed to make safe and deal with all utility services.

A drying standard is uniform and uses the normal equilibrium conditions in a building as its benchmark. Using the information that is gathered on-site, targeted moisture levels of the affected areas are determined based on established principles and criteria. An RIA drying standard (RIA, 2010) clearly illustrates that a proper response and drying regimen will and can prevent primary water damage to a building from becoming anything more. If, however, a drying plan or standard is not adhered to, problems can occur and move quickly from an initial building dry to structural damage. If left unchecked, the next stage would be water vapor movement to unaffected areas, followed by some form of microbial damage that could also cause indoor air quality issues.

Although drying out a building can sometimes be a very time-consuming exercise, the following important facts will greatly aid a successful process.

8.3 TYPE OF WATER

Contaminated water presents health risks to dwelling occupants and water loss technicians alike. Those involved in the aftermath of flooding should always be aware that floodwaters may be contaminated with sewage or animal waste and therefore present a health hazard. Similarly, if allowed to stand, clean water from burst pipes can also become contaminated by contact with surfaces. The Institute of Inspection, Cleaning and Restoration Certification (IICRC, 2010) states in its standard for professional water damage restoration that it classifies water into three categories. The first is deemed to be clean from an internal potable source; the second has a biological or chemical contamination; and the third is grossly unsanitary, containing pathogens, sewage, etc. The guide recommends that during clean-up, technicians must protect their eyes, mouths, and hands, and use disinfectants to wash hands before eating. It is advisable not to enter a property where there may be issues regarding personal safety or before a risk assessment has been undertaken and recorded by a professionally trained person. Only then can it be considered safe to proceed.

8.4 AMOUNT OF WATER AND DIRECTION OF FLOW

The quantity of water and indeed its direction of flow can greatly influence the approach to drying a building. The time taken to return the building to its pre-incident condition is also affected by the permeability of structural materials, the dwell time, and the complete removal of standing water.

Vertical water flow will usually cause more damage to a building. It spreads out floor by floor as it travels down throughout the structure, eventually settling at the lowest level (gravity being the driver). This process can produce pockets of reservoired or pooled water that, if not located and dried thoroughly, can result in long-term damage to the fabric of the structure.

By its nature, horizontal damage spreads throughout a property, affecting all areas on that level. The damage is usually confined to flooring and the structure at the depth of water. However, moisture will wick upward in porous building materials. Moisture readings and the location in which they are taken should be recorded on the drying plan.

8.5 SURVEY AND DRYING PLAN

Once the extent of water (burst or ingress) has been determined, a drying plan can be drawn up and implemented. Moisture readings taken in an unaffected area provide an excellent guide as to the expected completion readings. The survey will identify areas that require immediate action: *water extraction* and *containment*. Containment will prevent moisture vapor from moving into unaffected areas and causing secondary damage. The plan will greatly aid efficient drying by allowing the water loss technician to monitor the progress of the drying process. By creating and maintaining the correct environment using industry-recognized testing equipment, together with a psychrometric chart, the technician will ensure that a balanced, controlled drying plan is successful.

The plan will also include a *triage approach* to ensure that secondary loss as a result of water and water vapor migration does not occur. The whole of the building must be surveyed, keeping the following points in mind:

1. *What areas are affected and require immediate attention?* This would include extraction and removal in areas of pooled or reservoired water in voids, ducts, ceilings, etc.
2. *What is affected and requires secondary attention?* This would include structural materials affected by contact with water but not visibly wet. These materials and areas will need to be dried using the correct dehumidifiers, air movers, etc., as part of the drying plan.
3. *What is unaffected and requires containment and protection?* This would include unaffected areas that do not "read" wet, or need further investigation, on a moisture meter but will require protection from water vapor migration.

The principles upon which a drying standard is based are now considered. An ideal initial approach is to allow natural ventilation and evaporation to "flush" the wet air from the building. This is achieved by correct placement of air movers to draw out damp air, thus replacing it with drier air from outside. It is much better for the structure if climatic conditions allow and enable the process to be kick-started to aid efficient drying. The specific humidity differential of the introduced air needs to be 5% to 10% lower between outside and inside, with a temperature of 20°C as an ideal. The initial flushing process can be repeated as the outside air conditions permit.

Great care must be taken when using high-heat forced air or air conditioning systems. Rapid drying out of historic buildings using hot-air power drying systems can cause irreparable harm to significant features of the building. Knowledgeable use of a psychrometric chart, thermometer, and hygrometer will greatly aid this process and illustrate the effectiveness of the drying plan for future reference. It is vital that on completion of the drying process, the structural materials are checked to ensure they will not enable or support mold or mildew growth.

All building materials will continue to complete their return to equilibrium moisture content (EMC) with normal ambient conditions in the building. This will happen without incurring further damage. That is, the mechanical drying process does not have to continue until absolutely every square inch of the affected materials is returned to the EMC. They will complete their return to EMC within a reasonable time under the normal conditions that will exist in the building after the drying equipment has been removed.

The IICRC water damage standard (IICRC, 2010) places water damage in buildings into one of the following four classes:

Class 1: Has a slow evaporation rate, which usually only affects part of a room or area. This includes buildings that have low-permeance structural materials (e.g., plywood, particle board, structural wood, concrete). The nature of the materials is such that moisture is absorbed, releasing moisture slowly.

Class 2: Has a fast evaporation rate and affects larger areas than Class 1 losses. It denotes wicking water up to half a meter in walls.

Class 3: Has the highest evaporation rate, with vertical water flow from above. Ceilings, plasterboard, lathe and plaster, walls (dry lined or plaster), insulation, and subfloor in the entire area are saturated. These will dry quickly due to the high permeance factor of the materials.

Class 4: The IICRC considers this class to be "specialty drying situations": These consist of wet materials with very low permeance (hardwood, plaster, brick, concrete, stone). Areas that have pooled and reservoired water can greatly affect the drying process in this class. Warm floor construction can be placed in this class.

8.6 BUILDING AND CONTENTS DRYING CONSIDERATIONS

When working in a water-damaged home, great care must be taken to ensure that the contents within are not damaged by the actions of water damage technicians in any way. Householders should be consulted in this regard to help identify any issues or concerns they may have. Contents should be lifted and moved with great care, and the stability and fragility of all items should be checked prior to being moved. Taking photographs and producing a list of all items before commencement are recommended. The house contents may also require treatment due to the effects of water damage. This may include, but is not limited to, germicidal treatment, deodorization, removal, drying, and storage of the householder's personal property. Some personal property may be beyond economic restoration (BER). Either the property may be simply uncleanable or the cost of restoration would exceed its current value. In BER situations, items should be set aside, listed, and held to await authorization for off-site storage or disposal.

Unaffected items may need to be moved to a safe dry place or to aid drying and restoration. It is important to consider and use the information from the initial survey in deciding the appropriate course of action.

8.7 MONITORING THE DRYING PROCESS

When setting up a balanced, controlled drying system, it is important to be proactive in the monitoring process. Effective monitoring provides information needed to successfully and efficiently return the property to its pre-incident condition. The initial survey not only shows the state of the building on the day of the incident, but also provides target readings to aim for, to enable completion of the task. As the job proceeds, regular monitoring visits will provide the information needed to target the drying effort, adjust the containment areas, and place equipment accordingly.

It is important to ensure that all drying equipment is working well. An understanding of the drying dynamics, together with a quick calculation, will provide the information needed. To check that dehumidifiers are operating efficiently, it is useful to refer to the manufacturer's instructions. These will state the figures for water removal for each particular dehumidifier over a 24-hour period at a given temperature.

Once these dehumidifier statistics have been ascertained, it is simple to calculate the working efficiency by carrying out a simple calculation using a psychrometric chart in conjunction with the Protimeter® Moisture Management System (MMS), or hygrometer and thermometer. To determine the efficiency of the equipment and to check that a controlled, balanced drying system has been established, the following procedure is recommended:

Step 1:
1. Measure the inlet air temperature and relative humidity going into the dehumidifier using the Protimeter MMS.
2. Plot the readings on the psychrometric chart.
3. Record the moisture content at this point (grams per kilogram, g/kg; specific humidity).
4. Measure the outlet processed air temperature and relative humidity using the Protimeter MMS.
5. Plot the readings on the psychrometric chart.
6. Record the moisture content at this point (g/kg; specific humidity).

Step 2:
7. Subtract the g/kg leaving the dehumidifier from the g/kg entering the dehumidifier.

Step 3:
8. Multiply the result of Step 2 by a constant of 1.2 kg/m³; this is a given constant for the value of air density.

Step 4:
9. Multiply the result from Step 3 by the hourly air flow rate of the dehumidifier being used (for example, for the Dri-Eaz 1200, the air flow rate is 386 m³/hr)
10. Multiply this result by 24, as there are 24 hours in a day.

As an example:
Dehumidifier model: Dri-Eaz 1200 (386 m³/hr)

$$(10 - 6.4) \times 1.2 \times 386 \times 24 = 40.02 \text{ kg/day}$$

One liter of water weighs approximately 0.999 kg. Therefore, from these calculations, the dehumidifier used in this example, when working at 100% efficiency, would remove 40.02 L of water per day.

If the calculated result is appreciably lower than 40 L, it would suggest that either the dehumidifier is not working correctly or the balanced drying system has collapsed. Both of these issues will require attention.

Many points must be considered:

• Is the drying equipment set up properly?
• Is the area contained and closed properly?
• Is the water loss technician qualified to adjust, review the drying process, and utilize new techniques?

- Are the dehumidifiers in good working order, and are they maintained well?
- Are the monitoring checks being taken and recorded properly?
- When is the building "dry"?

To calculate the number of dehumidifiers that will be initially required to be deployed, the first step is to decide how many air changes per hour will be required to use the drying equipment efficiently. Table 8.1 provides some guidance in this respect.

The next step is to calculate the volume of the area that needs to be dried:

$$\text{Length} \times \text{Width} \times \text{Height} = \text{Volume (m}^3)$$

Then calculate the volume of air that needs to be changed per hour; that is,

$$\text{Air Volume} \times \text{Air Changes}$$

Finally, decide on the type of dehumidifier to initially install.

As a working example:

Room dimensions are: $12 \times 8 \times 2.5 \text{ m} = 240 \text{ m}^3$

Dehumidifier: Airflow: Dri-Eaz ($386 \text{ m}^3/\text{hr}$)

Materials affected are porous; therefore, three air changes per hour are required.

Therefore, an air volume of 720 m^3 must be changed each hour. In this example, two Dri-Eaz 1200 dehumidifiers would be adequate to commence the drying process.

Moisture readings must continue to be taken and recorded on the drying plan. As progress with the drying develops, containment and adjustment of equipment will be guided by the information on the drying plan.

Finally, it must be decided when the property is dry following the incident. Listed below are standard industry phrases to answer this sometimes vexing question:

1. When the materials in the building are returned to their pre-incident condition
2. When the materials in the building are at equilibrium
3. When the materials in the building will not support or sustain mold growth
4. When the materials in the building are returned to their normal state

These statements may seem a little unclear or vague and are further explained in the following.

TABLE 8.1
Recommended Air Changes per Hour

Air Changes: 3 per Hour	Air Changes: 2 per Hour	Air Changes: 1 per Hour
Materials like carpet and underlay, along with surface moisture, heavy wall moisture, and heavy penetration into flooring.	Porous materials like carpet, along with some structural saturation in walls and flooring materials.	Low-porosity materials like wood and concrete with minimum saturation; wet for less than 24 hours.

If the building is always damp in some areas, usually due to a preexisting situation such as a rotten window frame or a broken downpipe, returning the property to its pre-incident condition will not result in a dry building unless repairs are made that will prevent further ingress of water. Simply having the building materials at equilibrium does not produce a dry building because it may be constantly wet. If the building materials do not support or sustain mold or mildew growth, it is a good indicator that the property is dry. It is expected that the normal state of the building materials, once dry, would maintain a healthy living environment within the property.

The appropriate use of diagnostic equipment can help identify where and in what quantity there is water within the building. A floor and wall plan drawn on graph paper recording all initial readings will be a great help in recognizing improvement as time progresses. Taking readings of affected and unaffected areas will provide a good start and finish point. Knowing the exposure (dwell) time to water, the different materials affected and their permeability factors will be a good guide for the drying task ahead.

Removing pooled and reservoired water is critical to minimizing drying times and reducing consequential loss. Dehumidifiers remove water vapor from the air—not liquid water—so the more removed with extraction and wet pick-up vacuums, the better. Leaving any standing water delays the process while waiting for the water to evaporate.

Targeting the drying effort will increase drying efficiency. It is an invaluable method of assisting drying in the affected area. Increasing the temperature of a given body of air decreases the relative humidity, thereby increasing the moisture-carrying capacity of the air. Directing the dry, warm air from a dehumidifier into a "contained" tented area greatly enhances the process. Drying is more efficient and speedier in this way as materials that are wet always migrate to dry and reach a state of equilibrium (either by absorption or vapor movement). This method is ideal when faced with a wall that is wet at the bottom above a baseboard (skirting board). A plastic tent can be taped along the wall above the moisture line and again to the floor to create an angle. The skirting may be removed, thus allowing the hot dry air to pass over the wet wall, collecting the wet air, and delivering it back toward the dehumidifier for reprocessing.

If containment is set up between affected and unaffected areas, consideration should be given to practical factors such as customer access.

8.8   TYPES OF DEHUMIDIFIERS

There are two types of dehumidifiers commonly used in the United Kingdom: (1) condensing/refrigerant dehumidifiers and (2) desiccant/adsorption dehumidifiers.

8.8.1   Condensing/Refrigerant Dehumidifiers

A simple explanation of the principle of dehumidification is as follows: a machine that has the capability to extract moisture from the air as it is drawn through it by changing water vapor to liquid.

In flood or water loss situations, the fabric of the building, walls, floors, ceilings, etc. become saturated or moisture laden. This in turn causes the air to become moisture laden as evaporation takes place. The dehumidifier takes the moist air and draws it across refrigerated coils, causing it to condense. The dehumidifier lowers the air temperature below the dew-point temperature, enabling the water vapor to become a liquid. The moisture runs off into a receptacle or is pumped into a sink or drain. The air that has been "dried" or "processed" is then returned to the room to repeat the process. The condensing dehumidifiers are considered the workhorse of the water restoration industry and are affordable, rugged, and having low power consumption. A proper maintenance program will ensure that the dehumidifier remains at peak efficiency. By cleaning the coils regularly and ensuring that dirty and clogged filters are replaced, the machine will provide many years of good use.

8.8.2 Desiccant/Adsorption Dehumidifiers

This dehumidifier operates on the adsorption condenser basis. It takes moisture from the air as it is blown through a honeycomb desiccant material wheel. The principle is the same as with silica gel or crystals; it takes and can hold moisture on the surface until it is removed. The adsorption dehumidifier works well at very low temperatures and low relative humidity and permeability factors.

The dehumidifier's regeneration process works continuously. The wheel or rotor in the dehumidifier is a honeycomb and is coated with highly active silica gel, which has an extremely long lifespan. As with all equipment, a good maintenance plan will prolong successful operation.

The processed air is sucked into the desiccant dehumidifier via a filter using a radial fan, and thereafter it passes through the silica gel–covered honeycomb rotor. The processed air is dehumidified in the rotor. The rotor turns completely approximately 12 times per hour and is driven by a drive belt. It revolves through wet air slowly but continuously. The moisture taken up by the rotor is removed in the regeneration sector where a flow of heated air passes back through the rotor and then out of the desiccant dehumidifier. A much drier and lower relative humidity air is then returned to the area to recommence the process.

Drying time is directly linked to air movement, humidity control, and temperature control. It is dangerous to assume that dry air is equivalent to dry materials. For example, a wall in one corner of the room can be wet, yet the room relative humidity reading may be acceptable. The relative humidity of the air in a room relates to the humidity in that room, not to the moisture content of the walls, floor, or ceiling.

Moisture content relates to an amount of water in a particular material, measured by weight expressed as a percentage. Conversely, materials that read dry do not indicate a low relative humidity in the surrounding environment. Materials that are hygroscopic absorb water vapor from the surrounding air in an attempt to reach equilibrium. As relative humidity increases, the drier hygroscopic material absorbs more moisture.

The hygroscopic nature of structural materials can cause serious consequential loss issues if not addressed at the primary stage of drying. Normal relative humidity levels in the United Kingdom range between 30% and 50% indoors. Therefore, internal building and decorative materials are designed to work well in this range without delaminating, swelling, or warping.

It is very important that steps are taken to understand the need for a correctly balanced drying system using dehumidification, air movement, and correct temperature. The balanced, controlled drying system must remove moisture at the rate that the fabric of the building gives it up and the equipment installed can cope. This will prevent secondary damage from occurring to unaffected areas.

A balanced, controlled drying system means just that. The drying process must be monitored and adjusted regularly to ensure complete control of the drying process.

8.9 FINAL INSPECTION AND COMPLETION

Once the temperature, humidity, and moisture content are deemed acceptable and safe according to industry standards, the water damage restoration equipment can be removed and the water damage restoration process will be complete.

Some homeowners, property managers, and building maintenance companies use their own personnel to perform water damage restoration to save on the growing costs involved. It is prudent, however, to utilize the services of well-trained and qualified professional water loss technicians. They can perform these services well and fully understand the many defining criteria and complex methods to use for assessing water damage and establishing restoration procedures. Because of the unique circumstances of every water damage restoration project, it is impractical to have a one-procedure-fits-all rule. Many extenuating circumstances occur; therefore, it will be practical to adjust actions to work within the spirit of the Standard and Reference Guide for Professional Water Damage Restoration.

8.10 SUMMARY

The drying of a water-affected property is a complex process that is best undertaken by a competent and well-trained technical person such as a qualified water loss technician. This chapter set out the recommended procedures for approaching the drying of a building. In summary, the following outline plan provides a succinct guide to returning a property to its pre-incident condition.

1. Stop the flow of water. No drying process can be successfully completed until the flow or ingress of water is stopped.
2. Assess and address health and safety issues. Post warning and information notices.
3. Survey the entire building, both inside and out; take pictures of preexisting damage to the property. Take moisture readings and record on a drying plan. Take control dry readings. Check outside relative humidity and temperature readings to facilitate flushing the building.
4. Access and extract pooled and reservoired water.
5. Install correct drying equipment and erect containment to create a balanced, controlled drying system.
6. Use the target drying method to efficiently achieve aims.
7. Visit the property regularly to record moisture readings and adjust until completion.

Each stage of this process involves considerable care and attention, requiring a detailed knowledge and understanding of the many technical issues involved.

REFERENCES

BSI (British Standards Institution) (2005). PAS 64 Professional Water Damage Mitigation and Initial Restoration of Domestic Dwellings. London, UK: British Standards Institution.

CIRIA (Construction Industry Research and Information) (2005). Standards for the Repair of Buildings Following Flooding. By Garvin, S., Reid, J., and Scott, M. Construction Industry Research and Information (CIRIA).

EA (Environment Agency) (2007). After a Flood: Practical Advice on Recovering from a Flood. Bristol: Environment Agency.

Flood Repair Forum (2006). *Repairing Flooded Buildings; An Insurance Industry Guide to Investigation and Repair.* Bracknell, UK: BRE Press.

IICRC (2010). Institute of Inspection, Cleaning and Certification. Available at <http://www.iicrc.co.uk/> [Accessed October 15, 2010].

Pitt, M. (2008). Learning Lessons from the 2007 Floods. Full Report. London: Cabinet Office.

Proverbs, D., and Soetanto, R. (2004). *Flood Damaged Property: A Guide to Repair.* Oxford: Blackwell Publishing.

Rhodes, D. A., and Proverbs, D. (2008). An investigation of the current state of preparedness of the flood damage management sector in the UK: What lessons have been learnt? COBRA 2008, The Construction and Building Research Conference of the Royal Institution of Chartered Surveyors (RICS), Dublin Institute of Technology, September 4–5, 2008.

RIA (Restoration Industry Association) (2010). Restoration Industry Association (RIA), The Effects of Water on a Dwelling over a Period of Time, Minutes, Hours, Days. Restoration Industry Association Guide Drying Standard. Available at <http://www.restoration-industry.org/content/restorative-drying> [Accessed October 15, 2010].

Soetanto, R., and Proverbs, D. (2003). Methods of drying flooded domestic properties: the perceptions of UK building surveyors. *Cobra 2003, Proceedings of the RICS Foundation Construction and Building Research Conference,* Royal Institution of Chartered Surveyors, University of Wolverhampton, September 1 & 2, 2003

Tagg, A., Escarameia, M., von Christierson, B., Lamond, J., Proverbs, D., and Kidd, B. (2010). CLG Building Regulations Research Programme—Lot 13 Water Management, Signposting of current guidance. Available at <http://www.ciria.org/service/knowledgebase/AM/ContentManagerNet/ContentDisplay.aspx?Section=knowledgebase&ContentID=17599> [Accessed October 14, 2010].

9 The Art of Reinstatement

Roger Woodhead

CONTENTS

9.1   INTRODUCTION

This chapter discusses the real-life operational issues facing those with the biggest task of all—physically dealing with the emotional and physical turmoil and complexity of managing the flood victim while reinstating the property.

Service quality is provided by the capabilities and attitudes of the workmen on site, particularly when looking after domestic policyholders. There are workers who are highly competent but arrogant and those who have a wonderful attitude toward customers but are technically incompetent. It is important for flood damage management and reinstatement contractors that their employees are both competent and considerate of the customers' needs. The essential ingredient, however, is *project management*. Astonishingly, at the time of writing, it is only just becoming evident to most insurer procurement managers that they have been leaving this most important aspect of the recovery process to chance.

Following one surge event after another, the insurance industry has been criticized for its inability to support and manage its customers through the trauma of the aftermath. Historically, insurers were happy to discharge their liability by paying a sum of money in respect of damage caused by an insured event. Procurement managers did not exist; policyholders submitted repair estimates from local contractors and appointed those contractors to carry out the work on their behalf. Individual insurance companies, brokers, and loss adjusters had relatively small market shares, all had offices local to their customers, and there were no national supply chain contracts or agreed-to rates. All parties in the underwriting and claims environment competed locally on service and capability, and had strong local business relationships.

During the 1980s, there was a clamor for national merger and takeover, with many insurers amalgamating and increasing market share. With greater use of technology and competition increasingly being based on price rather than service, most local offices closed; local knowledge and close working relationships were lost. Insurers convinced themselves that customer service and reduced leakage were driven by remote working and the provision of a tender-based supply chain, managed by service-level agreements. The latest procurement methods are based on so-called e-auctions (electronic auctions) and risk-and-reward contracts, where supplier risks substantially outweigh their potential rewards. E-auctions, forcing suppliers to under-bid each other for large tranches of work, are intended to drive efficiencies on the assumption that all bidding parties will not bid below a level at which they are able to maintain service and profitability, leaving the insurer with a near-perfect supplier solution and maximum control over its spend and an ability to increase its own market share against its competitors. In theory, this means more work for its supply chain. Unfortunately, these insurers need to realize that, in a first-price auction, many suppliers will bid below what they truly believe is a realistic margin in a desperate attempt to win the business. They then seek all possible avenues to maximize savings by providing the best service they can for the least cost, while the insurer then attempts to enforce compliance to the contract by risk and reward. The inevitable result is an arm's-length relationship, lack of trust, poor service to the customer, and early termination of the supplier's contract due to either financial failure of the supplier, unsatisfactory performance, or both. In this environment, the likelihood of most insurer supply chains to deliver consistent satisfactory service in a business-as-usual environment is unlikely, let alone in a flood surge event. This problem may well be reduced by any insurers preferring a second-price auction system, where the contract is awarded to the second-lowest price. Game theory suggests that in second-price auctions, the bidder's dominant strategy is to bid a true value for the contract. Although the contract is awarded at a higher value than a first-price auction, a fairer balance of cost and service is achievable.

Regrettably, the majority of property insurance policies are heavily marketed on price rather than service. Few customers are savvy enough to study the policy detail and, because of the demise of the standard fire policy some years ago, customers only realize what they purchased when they make a claim. Cheap premiums are matched by restrictive coverage and a poorly resourced supply chain. Why is there such surprise, then, when every flood surge event is followed by extensive criticism of the insurance industry's response?

Interestingly, insurers and loss adjusters are reasonably adept at dealing with the initial response and take great pride in visiting every flood victim within a few days of the flood incident. One insurer, in particular, uses a promotional trailer with a large sign to set up an emergency control center close to the scene of every major flood event; it stays prominently on site during the media frenzy, only to leave again as soon as the last of the television cameras has departed. It is a pity that the crisis is not over quite so quickly for the policyholders. Loss adjusters go back to their home bases many miles away from the flood zone to handle the claims and suppliers at arms' length. Ultimately, the insurers leave their reputations and customers in the hands of their supply chain, often the subcontractor of a contractor to a building repair network of a loss adjuster. Many of these companies work for several networks, and it is common for the contractor on the ground to have no idea which insurer he is actually working for. So much for the millions of pounds spent on branding and brand loyalty.

There have been a few exceptional insurers that truly believe in providing service, having true strategic alliance relationships with their key suppliers and being prepared to fund the cost of superior customer service. Few of these are general market insurers, although one large insurer was praised for its commitment to providing such service following the Carlisle floods of January 2005 (Communities Reunited, 2006).

Clearly, high and ultra-high net worth insurers provide such a service at a premium cost, as one would expect. It is rare, however, for such insurers to have large volumes of customers, and they will have relatively few claims from any flood surge event, as their customers' properties are rarely highly concentrated. Furthermore, high net worth clients tend to have excellent contacts and their own preferred contractors. As such, there is little call for significant supply chain provision in this sector.

9.2   THE CUSTOMER JOURNEY

In simple terms, the physical process of recovery, post-incident, falls into three distinct phases. How these are handled has an enormous influence on the cost, time, and quality of the work and on the psychological recovery of the occupants.

1. *Mitigation: Initial Response Phase.* This relates to the period prior to, during, and immediately after the flood, before the recovery teams arrive.
2. *Recovery and Restoration: Damage Management Phase.* The period in which the damage management experts, appointed by insurers and housing associations, strip, clean, dry, and sanitize the properties.
3. *Reinstatement: Reconstruction Phase.* Once the building fabric is considered sufficiently clean, dry, and sanitized, the job of repairing the damaged building can begin.

9.2.1   Mitigation: Initial Response Phase

In proportion, most flood incidents arise from sudden and intense rainfall (flash flooding), giving property owners and occupants little, if any, opportunity to take anything other than minimal action to protect their property and mitigate damage pre-incident. Even where predicted fluvial flooding occurs following storms in upper catchments and flood warnings are given many hours in advance, there remain relatively few occupants who take significant protective action, except those who tend to be in uninsured primary flood locations. This is partly because there is never any certainty that other properties will flood. We have yet to see any insurance company attempt to preempt a flood incident and assist its customers with protective measures. Indeed, the lack of clear predictability and the logistics of such a plan would not be feasible in the practical sense. In any event, emergency services would not welcome insurance company representatives interrupting the primary function of preserving life by placing further people at risk in the attempt to safeguard property. It is inevitable, therefore, that any pre- and mid-incident mitigation measures are the sole preserve of the property owners or occupants. Surprisingly, some will stand by and watch their possessions be ruined, while others will take whatever steps they can (as required by the terms of the insurance coverage) to prevent or minimize the damage.

Accepting that no contact is made between insurers and their customers prior to the incident, insurers will normally wait for their customers to notify a claim before taking action. There are rare occasions when an insurer will make proactive contact with customers in a known flood zone, but a reactive response is the norm. Hence, in the vast majority of cases, floodwaters will already have receded and customers will have begun to take their own action before any involvement from their insurers. Ironically, this is the most crucial time, particularly where the flood depth is low and duration short. The faster any standing water, including water held within fitted carpeting, is removed, the less will be absorbed into the building fabric, thereby significantly reducing the subsequent structural drying, degree of strip-out, and reinstatement time and cost. Many liters of water are soaked up by carpets and soft furnishings.

Regrettably, there is sometimes an erroneous message, promoted by some junior insurance company staff or householders, that their customers should do nothing until the insurer's representative has visited, for fear of prejudicing a claim. There is normally a policy requirement to take steps to mitigate the damage.

The formation of the BDMA (British Damage Management Association), Flood RepairNet, and the National Flood Forum (NFF) and Scottish Flood Forum has created an environment in recent years wherein proper advice is available. NFF representatives are regularly interviewed on national television and radio stations during a flood emergency, advising householders to take mitigating measures by removing saturated items from buildings without delay and to take photographs and samples for evidence.

During this period, those businesses and individuals who do the most to help themselves find the subsequent recovery period much easier. Quite simply, the best form of help is self-help.

Nevertheless, safety is the first priority. Anyone entering a building that is potentially dangerous must ensure that they have assessed the risk and taken suitable precautions. It is rare for property to be structurally dangerous after a flood. In most cases, floodwater is relatively shallow and flow rates low. Contamination is overplayed, as most contaminants are thoroughly diluted by the floodwater, but occupants need to listen for local authority advice. In most cases, plenty of ventilation while working in the property and the wearing of boots, gloves, overalls, and occasionally a suitable facemask are more than sufficient. It is important to wash hands regularly and keep hands away from the face.

9.2.2 Recovery and Restoration: Damage Management Phase

Initial contact between customer and insurer is normally made by telephone, a process often referred to in the insurance industry as FNOL (First Notification of Loss). It is these call handlers who are charged with the task of identifying the nature and extent of damage caused by the flood event. For very minor cases, damage can be minimal and they may simply agree to allow a customer to obtain repair estimates or send an approved supplier to investigate and repair.

In the case of more serious damage, some insurers will send an approved repair and restoration (R&R) contractor as an immediate response to a call from a customer and then appoint a loss adjuster simultaneously. This was very common in the 1990s, but insurers became concerned that this was providing the R&R contractors with too much responsibility to maximize their involvement, leading to over-scoping and unnecessary costs (leakage).

It is becoming far more common for insurers to appoint their loss adjusters to visit and assess the situation first, rather than simply throwing their R&R supplier straight at the customer. The benefit is perceived to be better cost control, against the risk of secondary damage caused by delay. Different R&R companies have widely varying standards and advice for customers, and this is controlled to some extent by the "adjuster first" approach.

Most R&R contractors are now members of the BDMA, formed in 1999. In its early years, the BDMA was too heavily influenced by a few large R&R companies that were attempting to use it as a trade organization to gain commercial advantage. With much hard work on the part of its executive, it has now become a respected members' body, with the sole purpose of representing the interest of individual members, promoting standards, providing education, and working with other insurance and construction industry bodies to develop best practices. The BDMA Mission Statement (BDMA, 2010):

> To represent the interests of practitioners working in the damage management industry, to facilitate education, training, technical support, advice on standards and representation of members' interests in the public, industry and commercial domains.

Ironically, within its extensive glossary of terms, the BDMA does not define the term *damage management*. Generally, however, BDMA members' primary function is to undertake the initial recovery and restoration work, but not the subsequent building reinstatement. In summary, if left unattended, the extent of damage caused by the flood event will cause rapid and extensive deterioration of the property. Most building contractors are not trained or equipped to deal with these conditions sympathetically. Most building contractors can cope with the situation, but this is achieved by stripping a damaged building back to its bare structure, drying it in an unscientific manner (by installing "some" dehumidifiers), and then replastering and replacing the finishes once the building is dry.

The role of the professional R&R contractor is to minimize the damage, remove and restore affected contents, rapidly gain control of the specific humidity within the property, remove only those finishes that are irreparable or causing barriers to evaporation, dry the building with a thorough understanding of psychrometrics, sanitize, and hand the property over to the reinstatement contractor to complete the repairs.

A number of industry experts, in conjunction with representatives of the BDMA, worked with the British Standards Institution (BSI) in 2004 to develop a recognized standard for damage management, which was subsequently published in 2005 as BSi PAS 64 (BSI, 2005). The full title of the document is "PAS 64 Professional Water Damage Mitigation and Initial Restoration of Domestic Dwellings—Code of Practice":

> Scope
> This Publicly Available Specification (PAS) gives recommendations for the resto-ration of water-damaged buildings and contents. It is intended for use by professional water damage restoration organizations, but may also be of interest to the insurance community, others involved in the assessment and restoration of water-damaged domestic dwellings and their contents, and those persons affected by water damage.
> It is applicable to all common forms of water damage, including river flooding, surcharging of drainage systems, sewage systems, storm water run-off, inundation of seawater and bursting/overflowing/leaking of internal domestic plumbing and heating installations.It is also applicable to water damage which is the secondary result of other incidents, such as motor vehicle impact, storm or extinguishment water following a fire.
> It is not applicable to building reinstatement and redecoration methods.

Like all standards, PAS64 deals with the standards to which practitioners should carry out their work, not the specific techniques employed. It covers site health and safety, the manner in which information should be obtained and disseminated, the need to set clear goals, and the requirements to manage the process professionally. PAS 64 is being re-written in 2011, with enhanced detail and a greater emphasis on use by insurers and their customers. The scope will change to accommodate this welcome enhancement.

9.2.3   REINSTATEMENT: RECONSTRUCTION PHASE

What insurance companies have failed to realize is that, in their headlong desire to move from a traditional claim payment approach to reinstatement with their own sup-ply chain, the loss adjuster's traditional role is no longer valid. While forcing downward pressure on fees, insurers have completely lost sight of the fact that the restoration and reinstatement of severely damaged properties with an independent selection of sup-pliers must be project managed. They do not provide loss adjusters with anything like the fee necessary to carry out such work, even if adjusters were trained or experienced in construction project management, which they are not. Indeed, the loss adjuster role should be to engage such fellow professionals on an insurer's behalf.

Most loss adjusters now employ in-house surveyors and their own contractor network. These departments often use the title "surveying services" or something similar. For the most part, they provide a free or low-cost service to the insurer (at the point of sale). They produce relatively basic schedules of work for their contrac-tors and take claw-back payments of between 5% and 15%. For the higher figure, they purport to their principals that they project manage the claims. In reality, the project management is minimal, as contractors are largely left to manage their work themselves. This situation is not entirely surprising; adjusters need to utilize the claw-back payments to manage their networks and subsidize their highly competitive loss adjusting fees. A proper surveying and project management provision, created to closely manage all supply chain partners on a claim, requires visits to the site at least every 10 days and provides constant support to the customer, and costs at least 20% of the contract value.

With regard to the reinstatement phase of flood recovery, and despite a variety of expert steering groups producing some excellent textbook material, there remains no real consistency in the approach taken by those involved in the restoration of flood damage. The situation is improving, however, and publications such as *Flood Damaged Property—A Guide to Repair* by Proverbs and Soetanto (2004) and *Repairing Flooded Buildings: An Insurance Industry Guide to Investigation and Repair* by the Flood Repairs Forum (2006) have brought about far more debate. In the author's view, the main problems with setting absolute guidelines are manifold:

- Flood damage is a function of the flood and building characteristics—source, depth, velocity, contaminants, duration, component materials, age of construction, general maintenance, standard of workmanship, etc. The potential combinations of factors are almost infinite.
- Customer attitude—the extent to which one property owner will accept a "repair" rather than reinstatement solution is extremely variable and difficult to manage, particularly when influenced by the next item.
- Extraneous influences—comparison with works being undertaken at neighboring properties.
- Attitude toward customers—contractors with more interest in their pockets than their customers.
- Lack of joined-up thinking—too many R&R contractors with different attitudes and vested interests in their own competitive advantage.
- Lack of training of R&R contractors.
- Lack of building knowledge of R&R contractors.
- Preference for accelerated R&R periods by all parties—strip more, dry faster.
- Mistrust by insurers.
- Variability of structural and finishing elements.
- Complete misunderstanding of flood damage by "builders"—rip it all out and make it easier to guarantee the repair—and increase margins.

Time and time again in flood surge events, a considerate and professional approach to flood damage reinstatement is destroyed by excessive work at adjacent properties. Customers not having their homes ripped to pieces somehow believe that they are being short-changed, when the opposite is often the case. In stark contrast, there are also those who would like us to believe that just about everything can be dried, cleaned, and redecorated. In reality, there is a fair and reasonable solution somewhere between the two extremes, and each case must be considered on its individual merits.

Put simply, the individual parties (customer, loss adjuster, R&R contractor, and building contractor) have inconsistent views and no proper coordination. Even chartered building surveyors are inconsistent in their approach, although they are, in my view, the best people equipped to pull the parties together to project manage the recovery and reinstatement process. The answer requires the use of highly trained building surveyor project managers who specialize in the field of insurance-related damage reinstatement. They need to understand

- Building construction and pathology
- Diagnosis and remedying of dampness
- The effect of water on all common building finishing materials and their substrates
- The science of psychrometrics, both during the work and within the building after re-occupation
- Managing insurer and customer expectations
- A working knowledge of property insurance principles and practice
- Preparation of detailed schedules of work
- Programming and management of key critical path dates

The essential requirement to the professional and consistent reinstatement is quality project management!

9.3   PROJECT MANAGEMENT

Project management is the essential ingredient for any complex flood reinstatement project, particularly if funded by an insurance claim in a surge event. This is because

- Expert opinion needs to be seen as consistent.
- Insurers' interpretation and application of their policy coverage need to be seen as consistent and fair.
- Loss adjusters need to have a consistent approach.
- Emergency repair and restoration contractors need to work to consistent guidelines and work alongside one another, rather than openly criticizing and trying to outdo their rivals working in adjacent properties.
- The Financial Services Authority (FSA) requires insurers to actively demonstrate the FSA requirements to treat customers fairly (TCF).
- Insurers need to ensure that any contractor network they engage to provide a reinstatement service to their customers is controlled, and that those contractors are highly familiar with both water damage reinstatement and the insurer's requirements. This is currently a major cause of serious service failure among insurers.
- Insurers need to ensure that such contractor networks are discouraged from overextending their capacity and devise proper surge event planning for their customers.

With any major flood reinstatement project, different outcomes are highly likely and inconsistent approaches to the reinstatement process have been the repeated source of concern. Each individual project requires careful assessment, planning, and organization at the individual trade level to deliver a satisfactory result.

Successful project management of flood damage reinstatement requires several key processes, as follows.

9.3.1 PROJECT SPECIFICATION

The project specification should be an accurate description of what the project aims to achieve. In simple terms: to return the property to its pre-incident condition, subject to insurance reinstatement principles of "new for old" and statutory compliance.

The project manager must consult with the customer, insurer (or more likely their loss adjuster), and the contractor in formulation of the specification with relevant authorities. The project specification is essential and should involve at least one draft before it is agreed to. It creates measurable accountability at any time to assess how the project is going, or its success upon completion. A properly formulated and agreed-to project specification protects the project manager from being held accountable for issues that are outside the original scope of the project or beyond his or her control.

9.3.2 PROJECT PLANNING

To professionally manage the reinstatement process and the expectations of both the customer and the client insurer, the project manager must plan and communicate the various stages of the project. Complex projects will have both critical path items and those that may run in parallel. Some elements require lead-in times for the manufacture and supply of materials (kitchen installations being a prime example).

9.3.3 PROJECT TEAM

The project manager's role will only be successful if he or she is supported by committed team members. Whether directly employed, freelance, contractors, suppliers, consultants, or other partners, their capabilities and attitudes toward the customer and the project are crucial to the quality of the project, and the ease with which the project manager is able to manage it.

This is one of the most common failings of the insurance industry. Regrettably, almost all insurance company personnel, particularly procurement and supply chain managers, fail to recognize that their contractual promise and brand reputation rest extensively with the tradesmen delivering the reinstatement. This is not delivered by a good loss adjuster, but rather by excellent construction management and commitment of the project team. Most service failures are caused by the insurance industry's desire to either disregard or seriously undervalue one of these essential ingredients.

Appointing the project team early maximizes their ownership and buy-in to the project. Ideally, regular and repeated use of the same team on multiple projects (a factor easily achievable by a quality insurer) pays dividends time and again. This means

- A small, dedicated, technically competent large-loss team at the insurance company
- A small number of designated (named) individual loss adjusters
- Designated project managers working with dedicated trade contractors under their direct control

To plan and manage large complex projects with various parallel and dependent activities requires a critical path analysis Gantt chart. Critical path analysis shows the order in which tasks must be performed, as well as their relative importance.

9.3.4 PROJECT FINANCIAL PLANNING AND REPORTING

In the modern, closely audited business of insurance supply chain management, accurate control of expenditure activity and the recording of client approval are essential. The project management team must be able to demonstrate approval of the schedule prior to any work being undertaken, and that the cost management can be broken down into individual elements, if required. A system for allocating incoming invoices to the correct activities and showing when the costs hit the project account is required for some insurers, and all written and oral approvals must be recorded for audit purposes. Project managers must establish clear payment terms with both the client insurer and all suppliers—and stick to them. Projects develop problems when team members become dissatisfied, and non- or late payment is a primary cause of dissatisfaction.

All construction projects are too variable to avoid the need for contract variations, and this is even more evident when dealing with serious water damage, as the full effect of the incident and condition of inaccessible areas can only be estimated at the time of initial survey. A worst-case scenario approach can minimize this but most insurers prefer that reinstatement schedules be optimistic. However, particularly with this approach, they need to allow the project manager to set aside some budget for contingencies—he or she will certainly need it.

9.3.5 PROJECT CONTINGENCY PLANNING

In the area of serious flood damage restoration, the project team can always expect the unexpected with respect to both the property and the customer. Allowing for the possibility that some activities may not go as expected, some contingency planning will be needed. Contingency planning is vital when outcomes cannot be guaranteed. Contingency planning is about preparing fallback actions and making sure that appropriate allowance is made for time, activity, and resource variances. It can be difficult to plan for, but experienced flood damage surveyors develop a skill for anticipating the longer-term effects of the flood incident. Contingencies allowed for by such experts are sometimes rejected by untrusting insurers, only to be re-introduced at a later stage, after other interdependent work has been carried out, causing both time and cost overruns. This situation is not helped by overly optimistic emergency damage restoration franchisees convincing loss adjusters that buildings have been successfully dried with minimal deterioration, and then remaining in ignorant bliss when finishes delaminate and fail several weeks or months later.

It is worth noting, however, that there are many more occasions, particularly in flood surge events, where properties are far more extensively stripped out at the start than is necessary. I refer to the "stripping-out" debate later in this chapter.

9.3.6 ENABLEMENT

The project program should clearly identify those responsible for each activity. Both Excel and Bespoke programming software packages allow individual trade and supplier Gantt chart task bars to be color or pattern coded. When delegated tasks fail, it is typically because they have not been explained clearly, agreed to with other parties, or supported and checked while in progress. It is important to issue the full plan to all of the team, but not all tasks unless the recipients are capable of their own forward planning.

Major flood reinstatement often takes several months, and surge events require many adjacent reinstatement projects run side by side. If we refer to these as subprojects, it is evident that, while essentially independent, each subproject must form part of the team's master project plan.

Long-term complex projects must be planned in detail and constantly reviewed. The speed of structural drying and extent of damage between one project and another are affected by many factors, and the project program changes constantly, particularly affected by the drying phase, which can only be predicted rather than guaranteed.

While it is useful to appraise the entire project team with a high-level overview of the total project and where each subproject is expected to fall within the whole scheme, further detail at the initial enablement phase is too extensive for the entire team to absorb. The project manager should enable each subproject in turn, providing the team with more specific detail as each subproject proceeds. As long as the team continues to have the overall master project program constantly updated, they can picture where they are within this and quickly start to correctly anticipate the next subproject content prior to release.

9.3.7 PROJECT PERFORMANCE AND COMMUNICATION

It is more essential to keep the customer (insured) constantly updated than any other party, with key updates to the loss adjuster and insurer. Progress against the program should be reviewed at least weekly and slippage managed accordingly.

Policyholders have a wide-ranging level of willingness to be involved in the reinstatement process. The project manager needs to recognize this very early and manage the customer accordingly. Some are happy to have minimal input, even making it difficult to obtain approval for certain key decisions, such as kitchen unit finishes, tile choices, or decoration colors, while others (and thankfully few) actively interfere and disrupt the project, extensively undermining both the basis of the insurance contract liability and the project manager's ability to carry out his or her work effectively.

With a major flood claim, formal site meetings are preferable at certain stages in order to check the progress of activities against the plan. These should, at a minimum, take place immediately prior to commencement of the reinstatement phase (to set the project expectations) and at practical completion (to agree to the extent of any minor finishes, snagging work, and procedures for reoccupation). Ideally, a third key meeting should take place immediately after completion of trade first fixes, before any replastering or other finishing trades begin.

Whatever the formal meeting or review strategy, the project manager must keep talking to people and must make him- or herself available to all.

9.3.8 COMPLETION

For a project manager, the end of a project fully met, on time and on budget, is a significant achievement, whatever the project size and complexity. The mix of skills required is such that good project managers can manage anything.

It is often stated that an optimist will see that glass as half full, while a pessimist will consider the same glass as half empty. A project manager will see neither of these. To him or her, the glass is quite obviously twice as big as it needs to be.

9.4 SUMMARY

All too often, insurance procurement managers and their suppliers talk with passion about customer service and project management, but relatively few really deliver. All too often, the pressures of cost and volume overcome the intention, particularly in the £0 to £10,000 claim value category, which accounts for more than 90% of claims. For the most part, they can—and do—accommodate this level of service in order to maintain control of claims costs, as individual exposure to excessive compensation and negative media coverage is relatively low.

This is not something they can afford to do with larger and complex claims, where the requirement for insurers to deliver satisfactory service is more important. Larger claims require careful handling and constant monitoring. This is as much an art as it is a science. In summary, the key ingredients are

- Capabilities and attitudes of all involved in the investigation, mitigation, and repair
- Strategic alliance partnerships between insurers and key suppliers
- Empathy and appropriate support to match the individual requirements of each customer—one size does not fit all
- Early commencement of mitigation measures and reinstatement
- Correct understanding of damage caused by major perils and appropriate specification of work
- Professional project management of the reinstatement work

REFERENCES

BDMA (British Damage Management Association) (2010). Official website. Available at <http:\\www.bdma.org.uk/> [Accessed August 13, 2010].

BSI (British Standards Institution) (2005). PAS 64 Professional Water Damage Mitigation and Initial Restoration of Domestic Dwellings — Code of Practice. London: British Standards Institution.

Communities Re-united (2006). Press release, Carlisle. January 2006.

Flood Repairs Forum (2006). *Repairing Flooded Buildings: An Insurance Industry Guide to Investigation and Repair*, Watford, UK: BRE Press.

Proverbs, D.G., and Soetanto, R. (2004). *Flood Damaged Property: A Guide to Repair,* Oxford, UK: Blackwell Publishing.

10 The Development of Standards in Flood Damage Repairs
Lessons to Be Learned from the United Kingdom Example

Tony Boobier

CONTENTS

10.1 INTRODUCTION: ARE STANDARDS AND QUALITY THE SAME?

Standardization provides the "bedrock" for all flood repair activity. Without it, it is impossible to ensure consistency in repair, communication, professional advice, and management of the repair process. Consistency is essential not only to ensure an effective repair process, but also to optimize repair costs and end-user satisfaction.

Standards specify the minimum acceptable level of repair for constructed objects such as buildings affected by flood, and help avoid the situation where neighboring properties with similar damage are repaired in different ways. At their widest interpretation, standards also provide guidance and inform stakeholders as to minimum acceptable levels in all aspects of the flooding event. This extends not only to physical minimum requirements (e.g., drying or repair standards), but also to nonphysical standards (e.g., performance standards for communication, customer response times, and the like).

Generally, standards fall into several key categories:

- Physical repair standards
- Building repairs, which embrace building codes and regulations
- Drying standards
- Health and safety standards
- Service standards, including communication

In setting out the scope of this topic, it is necessary to ask whether the expression "a good standard of work" solely relates to compliance with Building Regulations (or other statutory instruments), to the way the work should be done (i.e., construction requirements), to the materials used, to the expectation of the end user, or to all of these? For example, is it possible to achieve good overall "quality" repairs without having reference to the quality of materials, which may be generally "fit for purpose" but of inadequate specification relative to the overall original standard of the original building. Furthermore, what is the link, if any, between the cost of repair and the quality of repair?

The practical application of this seemingly theoretical question is that of the appropriateness of basic standards to the original quality of the damaged property. The owner of a high-specification property is unlikely to tolerate low-specification materials being used in the repair process, even if there is a shortage (e.g., due to a high demand because of the flood). On the other hand, the use of high-specification materials on the repair of a property of basic standards would seem excessive, but may be tolerable under certain circumstances—for example, to allow homeowners to return quickly to their properties and avoid alternative accommodations. Quality therefore appears to require a degree of appropriateness.

In searching for a way to identify good quality that underpins the concept of standardization, an immediate difficulty begins to arise as one starts to consider the variety of construction types. Such an issue is difficult enough in the UK market alone, notwithstanding the many different types of properties across different geographies, all usually influenced by local building materials and construction practices.

In the United Kingdom, it is recognized that there are at least four levels of quality of property types, which links in to the issue of the standard of repair (accepting that standardization has some de facto alignment to quality). The four levels of quality are (Richardson, 1996):

1. *Economy:* Built from standard plans with the intention to sell at low cost, they meet minimum building codes, with materials and workmanship following the same basic pattern. Special-use rooms such as studies are unusual in this category.
2. *Standard:* Construction from standard plans, with rather simple designs. Workmanship and materials are slightly average or slightly above average, meeting minimum codes and exceeding them in some places. These houses often have areas such as a dining room, and also ornamentation or trim. Most post-war UK homes fall into this category.
3. *Custom:* Constructed from special plans or modified standard plans with customer alterations, these homes generally exceed building codes, use good-quality materials, and are attractively decorated. They often have special-use rooms (e.g., dining room, family room).
4. *Luxury:* Usually architecturally designed homes with custom features that exceed building codes, and feature excellent quality materials and workmanship. They often have a number of special-use rooms (e.g., media rooms, exercise rooms) and special or elaborate decoration and trim.

It can be reasonably argued that the standard of repair to a luxury property is likely to be different from the standard of repair to an economy property, although both types of repair will be required to comply with the legal standards and local codes.

10.1.1 THE BENEFITS OF STANDARDIZATION: "CONSTRUCTING EXCELLENCE" AND STANDARDIZATION

In the mid-1990s there was widespread recognition of the need for the construction industry to improve the service it provided to its clients while also ensuring future viability for the wide range of organizations that operated in the industry. Standardization formed part of that wider agenda, and there are obvious crossovers with the topic of improving the standard of flood-damage repair that are, inter alia, clearly connected to the UK construction industry.

"Constructing Excellence" was the title given to the UK organization charged with driving the change agenda in construction (Constructing Excellence, 2010). Existing to improve industry performance in an effort to produce a better built environment, it comprised a cross-sector, cross-supply chain, member-led organization operating for the good of the industry and its stakeholders.

In response to Sir Michael Latham's 1994 report entitled "Constructing the Team" and Sir John Egan's 1998 report entitled "Rethinking Construction" (Egan, 1998), a number of cross-industry bodies were formed to drive change—standardization aimed at improving quality. Their definition of quality appeared to have been mainly linked to three key metrics:

1. *End-user satisfaction*, that is, level of complaints and thus degree of rework (duplication of effort)

2. *Effective operation*, by unifying or "standardizing" the process as far as practically possible, in effect creating virtual economies of scale
3. *Improved safety standards,* recognizing at that stage the poor accident record within the construction industry

The overall ambition of Constructing Excellence was to improve not only the quality, but also the reputation of the construction industry. Although now relatively low key, Constructing Excellence recognized that standardization is a key enabler in reducing cost, improving service, and increasing the effectiveness of the supply chain. As a result, standardization has emerged as an important component in the development of best practices, a number of which underpin the flood-damage repair process in the United Kingdom today.

10.2   BRIEF REVIEW OF PHYSICAL REPAIR STANDARDS, INCLUDING PROFESSIONAL SERVICES

In this context, physical repair standards generally fall into three distinct sections: building repairs, professional services, and drying standards.

10.2.1   BUILDING REPAIRS

All building repairs are normally required to comply with codes of practice. The predominant purpose of a building code is to protect public health, safety, and general welfare as far as the repairs relate to the construction and subsequent occupancy of the buildings or structures. To that extent, there is a clear overlap between compliance with building codes in the physical reconstruction following flooding to avoid health issues, and the need to carry out the work themselves in a safe way. Building codes become the law of a particular jurisdiction when formally enacted by the appropriate authority.

Building codes are generally intended to be applied by architects and engineers, although this is not the case in the United Kingdom, where building control surveyors act to verify or regulate compliance both in the public and private sectors (known as approved inspectors). Additionally, codes are also used for various purposes by safety inspectors, environmental scientists, real estate developers, contractors and subcontractors, manufacturers of building products and materials, insurance companies, facility managers, tenants, and others.

There are often additional codes or sections of the same building code that have more specific requirements that apply to dwellings and special construction objects such as canopies, signs, pedestrian walkways, parking lots, and radio and television antennas.

10.2.2   BUILDING REGULATIONS IN THE UNITED KINGDOM

More specifically, Building Regulations are statutory instruments that seek to ensure that the policies set out in the relevant legislation are actually carried out. Building

Regulations approval is required for most building work in the United Kingdom. Building Regulations that apply across England and Wales are set out in the Building Act 1984 (Office of Public Sector Information, 1984) while those that apply across Scotland are set out in the Building (Scotland) Act 2003 (Office of Public Sector Information, 2003).

In the case of flood-damaged properties, there is no threshold (either in terms of cost or scope) below which Building Regulations do not apply. The Building Act 1984 is a UK statute and is the enabling act under which the Building Regulations, which extend to England and Wales, have been made. Sections 1(1) and 1(1A) read as follows:

Power to make building regulations.

　　1.—(1) The Secretary of State, under the power given in the Building Act 1984, may for any purposes of:

　　　　(a)　Securing the health, safety, welfare and convenience of persons in or about buildings and of others who may be affected by buildings or matters connected with buildings;

　　　　(b)　Furthering the conservation of fuel and power;

　　　　(c)　Preventing waste, undue consumption, misuse or contamination of water;

　　　　(d)　Furthering the protection or enhancement of the environment;

　　　　(e)　Facilitating sustainable development; or

　　　　(f)　Furthering the prevention or detection of crime;

—Make (building) regulations with respect to the matters mentioned in subsection (1A) below.

　　　　(1A) Those matters are—

　　　　(a)　The design and construction of buildings;

　　　　(b)　The demolition of buildings;

　　　　(c)　Services, fittings and equipment provided in or in connection with buildings.

In Scotland, the position is slightly different in that, while the spirit of the UK Building Regulations is deemed applicable, the Building Standards Division (BSD) has been reintegrated into the Scottish government and has become part of a new directorate for the Built Environment that includes Planning and Architectural Policy. The BSD is responsible for writing the Scottish building regulations. Further detailed information can be found at http://www.sbsa.gov.uk.

The Building Research Establishment (BRE, 2010) is another body that aims to improve the standard of construction in the United Kingdom. The BRE Group (BRE and BRE Global) has a history stretching back more than 90 years, which has witnessed the bringing together of a number of separate research, testing, and approvals organizations during that time. The BRE Trust is a charitable company whose objectives are, through research and education, to advance knowledge, innovation, and communication in all matters concerning the built environment for public benefit.

In the area of flooding, a good guide provided by BRE can be found at http://products.ihs.com/BRE-SEO/ap242.htm.

10.2.3 PROFESSIONAL STANDARDS

The unique issues that attach to the repair of a flood-damaged property often demand some degree of expertise in the resolution of the problem and, as a result, the term *expert* is often used. There is some correlation between expertise and professional standards, although the former is not entirely dependent on the latter.

Construction methods and materials used in domestic properties vary considerably, and many older properties need the knowledge of a qualified building surveyor or architect to specify and oversee repair strategies. The expression *qualified* usually refers to a professional who has acquired minimum professional standards such as those awarded by chartered organizations like the Royal Institute of Chartered Surveyors (RICS) and Chartered Institute of Loss Adjusters (CILA). Qualifications are usually evidenced by designated letters, typically RICS, ACILA. However, not all members of such organizations are either qualified or experienced in dealing with property repairs following flood damage.

In Europe, the presence of similar professional institutes is less developed. Professionals experienced in flood-damage repairs can be identified by the designated letters *Eur Ing* (European engineer). Nevertheless, the relative absence of professional bodies (compared to the United Kingdom, for example) should not be taken as an indication of lesser professionalism or of any compromising of standards, and in many cases there are stringent training requirements imposed on company employees.

While major professional institutions tend not to offer any flood-specific training module, nevertheless there is an increasing trend toward specific training at so-called flood schools. These are independent organizations that aim to increase the understanding of the effects of water damage on property, usually domestic building. Although they provide only limited professional qualifications for the management and repair of flood-damaged properties, they are nevertheless a good indicator of improved competency in this relatively complex area.

10.2.3.1 British Damage Management Association

In the UK, one of the leading organizations in this field is the British Damage Management Association (BDMA), a certifying body committed to setting industry standards, which describes itself as the "only route for practitioners wishing to gain professional accreditation in the recovery and restoration industry." Accredited Membership is by examination and is open to all who work in this field. In addition to the core damage management categories, there are specific accreditation categories for insurance and loss adjusting personnel, as well as claims practitioners. Associate membership is available to anyone with an interest in damage management (Table 10.1).

The BDMA has set up a number of licensed training centers, offering consistency of training in core damage management principles and processes and, alongside its examination regime, the BDMA requires members to meet a Continuous Professional Development (CPD) target each year.

In addition to a comprehensive training and reference manual for practitioners, the BDMA offers a series of flood support leaflets designed to assist victims of flooding. These are primarily distributed by local authorities, emergency planning officers,

TABLE 10.1
BDMA Membership Levels and Minimum Requirements

BDMA Membership Levels	Basic Criteria (detailed criteria and further information available from the BDMA website www.bdma.org.uk)
Associate	Damage management industry awareness.
Technician (*BDMA Tech*)	Minimum 12 months' experience in fire, flood and water damage restoration as a practitioner. Will have the ability to undertake basic risk assessments and carry out relevant practical tasks.
Senior Technician (*BDMA SenTech*)	Minimum 3 years' experience, including 12 months at BDMA Technician level. Practitioner with well-developed knowledge and competence, and the ability to scope work/undertake risk assessments.
Consultant Restorer (*BDMA ConRest*)	Minimum 10 years' experience. Degree level in at least one related discipline such as health and safety, engineering, surveying or a science-based subject, with full awareness of all practical, technical and legal aspects of fire and flood restoration and proven project management skills.
Insurance Technician (*BDMA InsTech*)	Individuals involved in property claims activities on behalf of an insurance company or loss adjusting company, for whom detailed knowledge of professional damage management practices will enhance their ability to achieve appropriate outcomes.
Claims Practitioner (*BDMA ClaimsPrct*)	Individuals who are not directly employed by an insurance or adjusting company for whom interaction with professional damage management practitioners and the insurance sector is a feature of their day-to-day activity. Awareness of insurance property claims processes and damage management industry awareness required.
Specialist Restorer (*BDMA SpecRest-Specialism*)	Experienced practitioner who is familiar with general aspects of damage management, with a specific specialty, such as furniture, document or fine art restoration, etc. Minimum 5 years' experience.
Specialist Affiliate (*BDMA SpecAffil*)	Experience and independently verified qualification as a conservator or specialist restorer, providing specialist services that would not normally be carried out by core damage management practitioners; 5 years' experience
Fellow (*BDMA Fellow*)	Recognition of outstanding contribution to the industry. Will have reached the highest levels of BDMA accreditation and be seen as an established role model by his or her peers. *Honorary Fellow*—A discretionary award for service to the BDMA that carries no implication of technical accreditation or expertise.

Environment Agency officers, insurers, loss adjusters, emergency services, damage management contractors and flood victim support agencies. The leaflets are free to order or download from the BDMA website at www.bdma.org.uk

A number of major UK property insurers and loss adjustment companies have put significant numbers of their staff through the BDMA Insurance Technician Training and Accreditation program and some insurers are now requesting the attendance of accredited BDMA technicians on relevant property claims jobs. Arguably, therefore, the BDMA is gradually creating a degree of standardization at least for drying processes across the UK.

10.2.3.2 National Flood School

The National Flood School (http://www.nationalfloodschool.co.uk), established more than two decades ago, professes to be "the only 'Institute of Inspection, Cleaning and Restoration Certification' approved school with a purpose built flood house for training and research" (National Flood School, n.d.). Believed to be the only purpose-built floodable house in Europe, it comprises 8 rooms and 60 common household materials, and can be flushed with 1,500 gallons of water to provide realistic insight into the difficulties homeowners face when their property is flooded. The National Flood School not only supports and trains restorers, but also provides information and support for many other associated industries, including insurers and loss adjusters. The National Flood School is accredited by the Chartered Institute of Insurance (CII) as part of the CII's program for CPD.

As an organization, National Flood School services are delivered through three specific divisions:

1. *Training Division:* Develops and delivers a diverse program of both technical and nontechnical training courses aimed at restorers and insurance professionals.
2. *Consultancy Division:* Provides a range of consultancy-led services aimed at making real savings by driving standards and delivering best practices.
3. *Research & Development Division:* Investigates and analyzes all aspects of the industry to identify best practices as well as develop new and innovative methods of restoration.

The school was founded in 1988 to drive the standard of restoration professionals and protect the public from the anguish and distress caused by unqualified restorers. Describing themselves as an "independent watchdog," they make the point that poorly trained restorers can give the restoration industry as a whole a bad name, and that a bad experience can directly impact the insurer, causing policyholders to move their valuable business elsewhere.

To mitigate this, the National Flood School's aims are to pass on knowledge, to encourage and promote excellence, to set and drive standards, to investigate and analyze, and to increase awareness in the field of flood repair.

10.2.4 DRYING STANDARDS

In many cases of flood-damaged buildings, the most appropriate method of drying needs to be decided by an "expert" with full knowledge and understanding of the facts and conditions. Drying is covered in more depth in Chapter 8, and detailed information is available from a Construction Industry Research and Information (CIRIA) publication (Garvin, Reid, and Scott, 2005) and in the recent CLG (Communities and Local Government) publication (Tagg et al., 2010).

One challenge is that of "What is the minimum drying standard?"; that is, "When can a building be considered dry?" The underlying principle is that the moisture levels in the repaired building should be reduced to the levels that existed before the flooding. Standardization of drying and repair processes in the United Kingdom are mostly applicable to domestic houses built after 1930. Properties built before that time ideally require the advice of a qualified building surveyor.

A property might be considered dry when

1. The condition of the internal construction materials is at or better than that "normal; considered acceptable" or compares favorably with areas not affected by flood.
2. The moisture on and in the building materials will not support the growth of mold and mildew.
3. The level of trapped or bound water within the building envelope, construction materials, or contents will not migrate or transfer to areas or surfaces that may promote mildew growth, cause failure or damage to areas previously repaired or restored, or damage to previously unaffected areas.

10.2.5 HEALTH AND SAFETY STANDARDS

Part of the overall mix of standards in flood damage repair is the topic of health and safety, an area that is complex. Organizations involved in the repair process have a duty of care to the occupants of a property, and also to their employees. As a result, health and safety standards must be taken into account.

Under the Health and Safety at Work Act 1974 (Office of Public Sector Information, 1974), the employer has a duty to ensure health and safety, in particular to

- Provide safe systems of work
- Provide training, instruction, supervision, and information to ensure health and safety
- Provide arrangements for use, handling, transport, and storage of articles and substances
- Ensure the health and safety of others affected by the work

Under the Management of Heath and Safety at Work Regulations 1999 (National Archives, 1999), it is the employer's duty to

- Carry out risk assessments;
- Identify plan implement control and monitor preventative measures;
- Provide information to employees;
- Encourage coordination and cooperation between employees where the workplace is shared;
- And it is the employee's duty to work in accordance with training and information provided for health and safety, and to notify the employer of serious or imminent danger, or health and safety shortcomings.

Under the Control of Substances Hazardous to Health Regulations 2002 (COSHH) (Health and Safety Executive, 2002), it is the employer's duty to

- Assess the risk of exposure to hazardous substances
- Avoid the exposure (or, if not possible, to control the levels of exposure) to hazardous substances

Under the Personal Protective Equipment at Work Regulations 1992 (National Archives, 1992), it is the employer's duty to

- Assess the risk of exposure to hazardous substances
- Avoid exposure (and, if not possible, to control the levels of exposure) to hazardous substances by providing and maintaining personal protective devices (PPE)
- Provide protection to employees against exposure to risks that cannot be controlled by alternative means which are more or equally effective as PPE
- Provide information and training to employees for using PPE

Risk assessments also must specifically identify issues relating to confined spaces under the Confined Space Regulation 1997 (National Archives, 1997) and how appropriate measures should be implemented.

Overall, it becomes clear, therefore, that any repair work must be undertaken in a way that is safe both for the homeowner and the repairer. Strict rules apply and stiff penalties exist for those organizations that run afoul of these standards.

10.3   SERVICE STANDARDS, COMMUNICATION, AND PERFORMANCE MANAGEMENT

As part of the overall topic of standards of service, communication is a critical part of the repair process, managing expectation throughout the life of the event, regardless of the uncertainties that inevitably arise in damages of this nature. It is a critical part of the interaction that occurs where insurers, landlords, or public authorities are involved.

It is recognized that there are traditionally six stages in the repair process:

1. Discovery of the damage and notification to the insurer and/or property owner if tenant involved
2. Communication of the next steps
3. Assessment of the scope of the damage
4. Ongoing management of the drying and repair process
5. Completion of the work
6. Completion of the claim (in the case of an insurance event)

Time-based standards can be applied to each stage. For many commercial organizations, these standards and the speed of response can provide commercial differentiation. Typical time-based standards might be as shown in Table 10.2.

The increasing trend toward performance-managed business metrics also starts to provide a platform for continuous improvement of the flood process. Competitive and best-in-class organizations are using performance dashboards and predictive analytics to

- Control costs through managing key drivers such as inflation
- Improve efficiency through effective workflow management

TABLE 10.2

Typical Time-Based Service Standards for Flood Damage Repair

Initial Communication	Initial Assessment of Scope	Management of Work— Number of Visits by Supervisor	Completion of Work	Completion of Claim
Within 24 hours of receiving notification of the claim or identifying damage	Within 5 working days of incident, subject to access being possible	Weekly or as required on site	Within 3 months of drying completed	Within 1 month of completion of work

- Improve customer satisfaction through quicker claims closures
- Provide a closed loop between underwriting and claims management
- Predict and litigate fraudulent claims earlier
- Improve regulatory compliance

In the insurance sector, typical key performance metrics that can be captured include

- Notification by customer/initial information gathering:
  - How many calls?
  - How many answered against agreed service level?
  - Abandoned calls
  - Outbound calls
  - Contact center staff productivity
- Adjudication/further information to quantity cost:
  - Number of visits
  - Cases insourced to insurer's own inspectors/outsourced to third-party administrator
  - Productivity
  - Within service level
  - Claims rejected
  - Complaints
- Indemnification/fulfilling terms of settlement:
  - Number of cases
  - Time to settle
  - Average cost
  - Closure within service level agreement
- Recovery/capturing legal entitlement from third party:
  - Number of cases referred
  - Amount recovered
  - Time to recover

- • Legal team involved
- • Intercompany agreements
- • Complaint management:
  - • Number of complaints
  - • Compliments
  - • Executive referrals
  - • Time to resolve this
  - • Potential cost to rectify

Prudent organizations will become aware through the monitoring of these performance metrics that there is a correlation among repair cost, duration of repair, and customer (property owner) satisfaction. Generally, the longer the repair process, the more dissatisfied the property owner and the more expensive the total cost of repair. Provided that no short-cuts are taken in terms of the repair—either drying or repair quality—including workmanship, the more likely it is that the property owner will be happy (or less likely to complain) if repairs are completed promptly, and the total cost of repair will usually be less.

The increasing use of supply chains in managing the repair process also requires consideration of the management of standards in that area. Increasingly, key supply chain initiatives focus on customer-facing and internal performance factors such as reliability, responsiveness, as well as more traditional measures such as agility, cost reduction, and asset management. Thus, they provide focus and actionable insight into the most pressing supply chain demands and customer needs.

The use of standardization and scorecarding to highlight supplier delivery and quality performance helps to

- • Track key metrics
- • Manage supplier on-time performance, quality, price, and service levels, and identify supplier issues immediately
- • Share scorecards to reward best performers and allocate workloads

Key performance indicators are often used within the supply chain industry, both to measure direct tactical delivery (that is, time to respond) and to measure the performance of the supplier against strategic objectives (that is, level of insourcing or outsourcing of the repairer's organization).

Performance audit refers to an examination of a program, function, operation, or the management systems and procedures of an entity to assess whether the entity is achieving economy, efficiency, and effectiveness in the employment of available resources. The examination is objective and systematic, generally using structured and professionally adopted methodologies.

It is a prerequisite that, for an effective audit to take place, there must be a set of standards against which to undertake that audit. If audit is the basis on which consistent service and performance are provided, then standards are the bedrock of that audit function. That is, without clear and relatively transparent standards, there is little or no advantage in carrying out audit activity.

10.4 INTERNATIONAL PERSPECTIVE

Therefore, the alignment between standards and quality cannot be overlooked, including from an international perspective. Useful information on the topic of building quality is available from the Building Cost Information Service of RICS, known as BCIS. Established in 1962, the BCIS provides independent cost information about the built environment. The BCIS pioneered elemental cost planning in the early 1960s that is now the basis of early cost advice in the construction industry.

The issue of standardization of property types—and therefore the management of quality in flood repairs—is less clear. In a recent report on international standardization, the BCIS not only highlighted what property standards are in place, but also indicated the best sources for finding accurate information and data in each of the countries surveyed (BCIS, 2009).

Other significant report findings include

- Respondents from 19 out of the 40 countries did not claim any published standard elemental classification of building parts for costing purposes.
- Many countries where such standards exist tend to be linked historically with the United Kingdom.
- In the absence of locally agreed standards, professionals commonly adopt "foreign" standards or ad hoc in-house developed standards.
- There are said to be no classifications of standard building types in use in 11 of the 40 countries, and the respondents from a further 19 countries did not identify the title of a published third-party standard.

The indications are therefore that, at least conceptually, flood repair standards developed in the United Kingdom can be adopted, at least in part, to meet the needs of local situations. With a large proportion of flood damage repairs being funded by insurance companies, insurers are also increasingly seeking to optimize their supply chains by entering into Pan-European and in some cases global supply relationships. For this reason, insurers acting as key stakeholders increasingly need to recognize that there are different codes of practice for property repair from country to country, and that inter-country benchmarking is problematic, albeit that European directives are increasingly leading to convergence of standards, as far as is practically possible.

In considering the topic of international flooding, it is impossible to overlook the challenge of catastrophe events happening, where large tracts of land and properties are affected resulting in widespread damage and abject misery, as represented by the tragic events of 2005 in New Orleans and more recently in Pakistan and Australia. Detailed consideration of such major events is beyond the scope of this book but it is reasonable to suggest that, in such circumstances, relatively traditional considerations of quality become obsolete, and service becomes in many cases a matter of life and death.

Notwithstanding, there are lessons to be learned from a disaster approach that adopts standardization at its core. Agreed processes and methodology, as far as practically possible, improve the transparency of response activities and allow all key components of the wider solution to plan and interact more efficiently. Standardization provides a framework for event planning, adapted for local environments, with the overall impacts being those of improved speed of response, lower cost, and—ultimately—better service.

10.5 SUMMARY

Standardization and quality in the topic of flooding is a complex topic. It accommodates the issues of location, quality, repair cost, and property owner expectation, to name just a few elements. Even the definition is open to some degree of ambiguity: By *standardization*, do we refer to legal standards, that is, codes of practice or health and safety requirements, or do we refer to personal standards (those expectations of the property owner or their representative)?

Does compliance with legal standards imply that the work is of sufficient quality? Probably not, if one accepts that quality is to some degree a subjective requirement or an aspirational target linked perhaps to the nature of the damaged building itself, that is, basic or luxury.

Clearly there are some elements of standardization that can be codified: compliance with basic minimum standards of materials and workmanship at the very least but also evidence of professional competency to handle flood cases, and some level of agreement as to what comprises the basic criteria for agreeing when a property is dry or not dry. To some degree, the "proverbial glass" is at least one-quarter full, rather than three-quarters empty. The continued threat of flooding due to climate change and flash flooding, whatever the cause, must encourage the wider flood industry to continue to collaborate on standards, yet in doing so still leaving room for competitive differentiation where appropriate.

If quality remains a key objective, then it is against accepted standards that quality is measured. Standards underpin the audit process by which quality is gauged, and against which continuous improvement is driven. These standards relate not only to process but also to the suitability of materials for purpose, the performance of the supply chain, and the competence of flood professionals. Increasingly, technology has a part to play in benchmarking quality against minimum and internal standards, and will continue to be a key enabler. The UK model continues to mature, with individual organizations such as the BDMA providing a catalyst for improvement but, more importantly, offers a vision of a coordinated response to improvement of the flood response.

In the case of catastrophic events even in Europe, leveraging the knowledge and resources of the wider flood industry across boundaries as an industry is an essential element of progress if the true benefits of scale are to be fully realized. Knowledge transfer and, to some degree, standardization remain at the heart of this ambition, adapted for local conditions. Finally, standardization has a role to play in the global response to tragic flood events, helping to create order from chaos by improving the planning process and simplifying the response mode.

REFERENCES

BDMA (British Damage Management Association) (2010). BDMA Education [online]. Available at <http://www.bdma.org.uk/education> [Accessed September 2010].

BCIS (Building Cost Information Service) (2009). 2009 BCIS International Cost Elements Enquiry Report. London: BCIS.

BRE (Building Research Establishment) (2010). BRE. Available at <http://www.bre.co.uk> [Accessed September 2010].

Constructing Excellence (2010). Constructing Excellence in the Built Environment [online]. Available at <http:// www.constructingexcellence.org.uk> [Accessed September 2010].

Egan, J. (1998). Rethinking Construction—The Report of the Construction Taskforce. London: Department of Trade and Industry.

Garvin, S., Reid, J., and Scott, M. (2005). Standards for the Repair of Buildings Following Flooding. London: Construction Industry Research and Information Association.

Health and Safety Executive (2002). Control of Substances Hazardous to Health (COSHH) Regulations. Available at <http://www.hse.gov.uk/coshh>.

National Archives (1992). Personal Protective Equipment at Work Regulations. Available at <http://www.legislation.gov.uk/uksi/1992/2966/contents/made>.

National Archives (1997). Confined Space Regulations, UK. Available at <http://legislation. gov.uk/uksi/1997/1713/contents/made>.

National Archives (1999). Management of Health and Safety at Work Regulations. Available at <http://www.legislation.gov.uk/uksi/1999/3242/contents/made>.

National Flood School (n.d.). National Flood School—About Us. Available at <http://www. nationalfloodschool.co.uk/about_us.htm> [Accessed September 2010].

Office of Public Sector Information (1974). Health and Safety at Work Act, UK. Available at <http://www.hse.gov.uk/legislation/hswa.htm>.

Office of Public Sector Information (1984). Building Act 1984, United Kingdom. Available at <http://www.opsi.gov.uk/RevisedStatutes/Acts/ukpga/1984/cukpga_19840055_en_1>.

Office of Public Sector Information (2003). Building (Scotland) Act, United Kingdom. Available at <http://www.opsi.gov.uk/legislation/Scotland/acts2003>.

Richardson, D. (1996). *Insuring to Value*. Los Angeles, CA: Marshall and Swift.

Tagg, A., Escarameia, M., Von Christierson, B., Lamond, J., Proverbs, D., and Kidd, B. (2010). CLG Building Regulations Research Programme—Lot 13 Water Management, Signposting of Current Guidance. London: Communities and Local Government.

11 Resilient Repair Strategy

Richard Ayton-Robinson

CONTENTS

11.1 INTRODUCTION

Government research has shown that it is only really cost-effective to undertake extensive changes to homes to make them more resilient to flooding if the property is likely to flood more often than every 25 years (Thurston et al., 2008). This equates to a 4% annual risk of flooding. Where the likelihood is that extensive repairs would be needed following flooding, the risk factor is based on a frequency of 50 years, or 2% annual risk of flooding. It also concluded that resilient reinstatement should only be considered for homes where there is complete information on the nature and type of flood risk and where the options have been professionally evaluated.

Following the summer 2007 floods in the United Kingdom, the question arose as to whether previous government research was too cautious and whether a stronger case now exists for action to promote property-level resilience measures. In 2009, the Association of British Insurers (ABI) published Research Paper No. 14, which evaluated the cost of flood-resilient reinstatement of domestic properties (ABI, 2009). The research findings that drew on cost data from the 2007 floods were that, if anything, the previous government research was too optimistic and that the cost of providing resilient reinstatement was significantly higher than had been expected.

The 2007 floods resulted in 180,000 home insurance claims and a total bill for insurers of £3 billion (ABI, 2008). Despite the significance of the event and the fact that for some this was a second or even more frequent flood event, most homes were reinstated on a like-for-like basis. The reality of this is that should these homes be flooded again, the level of damage will be similar to that suffered in 2007. In many cases the cost may actually increase, as many homeowners will have taken the opportunity to modernize or upgrade their properties by adding further nonresilient materials.

11.2 RESILIENT VERSUS RESISTANT REINSTATEMENT

An alternative to resilient reinstatement is to consider resistant measures designed to keep floodwater out of the home. Where this is possible, this is likely to be preferable. Resistant products and methods can broadly be divided into two categories:

1. Those offering temporary resistance (devices or products that need to be activated or installed prior to the flooding)
2. Those offering a permanent solution (usually integral or built into the property)

See Table 11.1 for a summary of the cost of a selection of resistant measures.

11.2.1 TEMPORARY RESISTANT MEASURES

Temporary resistant measures at the property level generally attempt to seal openings within the external envelope of the property, such as doors, windows, air vents, and ducts. They can be further categorized as

- Local: Those restricted to within the confines of the grounds of the property and defenses, which may be erected beyond the curtilage of the property, respectively.
- Remote: Defenses that are often deployed by water authorities for the protection of the wider community. These are often sited around housing estates or to contain local watercourses.

11.2.2 PERMANENT RESISTANT MEASURES

These measures are permanently built in to the property and include defenses such as watertight doors and windows and water-sealed airbrick covers. They are usually more costly to incorporate than temporary measures. The success of these resistant measures, however, depends on a number of property-specific considerations:

- Expected maximum depth of flooding
- Level of detachment of the property (i.e., detached, terraced, semi-detached)
- Age and design layout of the property
- Material the property is built from that may come into contact with floodwater (porosity, rigidity, and jointing are relevant factors)

Where a property is attached to another property, the relative success will be limited by the extent of similar measures being adopted by the adjoining property owners, due to the vulnerability posed by the party wall and continuous cavities.

Where the depth of flooding is greater than 900 millimeters, this can have structural implications caused by the impact from floating debris such as furniture, trees, and vehicles. The movement of such a volume of water or the unequal hydrostatic pressure exerted by retained floodwater can result in the collapse of walls (USACE, 1998). Measures designed to utilize the structure to retain water at depths greater than 900 millimeters are inadvisable unless the structure can be reinforced to resist the hydrostatic pressures exerted. In most cases, this will not be economically viable.

Temporary resistant measures at the property level tend to offer a more cost-effective alternative compared to permanent measures and can be particularly effective against shallow flooding and where sustained resistance is not required. The economic benefit from property-level resistant measures is largely limited to detached properties where flood depths will not exceed 900 millimeters. This would equate to about 5% of the UK housing stock at risk of flooding (ABI, 2009).

11.2.3 Resilient Reinstatement

Resilient repair measures offer additional robustness in the event of future flooding. In some cases these measures will involve substituting a nonresilient material, such as plasterboard, with more resilient alternatives, such as cement or lime-based render (Soetanto et al., 2008). In other cases, an alternative method of construction could be adopted, for example, raising electrical sockets and apparatus to above the likely flood level. These alternative methods, combined with the resilient materials, would serve to minimize the extent of actual damage and reduce the drying time in the event of future flooding. As a consequence, fewer items would need to be stripped out and reinstated, reducing both the future repair cost and also the resultant time spent on repairs. This enables occupants to get back into their homes sooner, thereby minimizing the cost of alternative accommodations.

There is no definitive specification or set of standards for flood-resilient upgrading, although some repairs will fall under the control of building regulations. Various publications have proposed a set of benchmark standards but none have been universally adopted by industry (Proverbs and Soetanto, 2004). When considering resilient repair options, it is necessary to undertake a detailed review of the structure—element by element. The key elements of the building are considered further for resilient reinstatement potential.

External walls: By their very nature, most materials used in the construction of the external envelope and façade offer a high degree of water resistance to weather. Variations will inevitably occur due to the wide variety of materials and construction techniques that have been employed in domestic properties over the years. With age, weaknesses in the integrity of the façade may develop, making the structure susceptible to penetration by the pressure exerted by floodwaters. There are few flood-resilient options available for treatment of the external façade beyond the application of a waterproof render. In many situations this may not be acceptable aesthetically.

TABLE 11.1

The Cost and Relative Advantages of Resistant Measures (Comparative Costs at 2008 Prices)

Type	Approx. Cost to Retrofit (£K exc. VAT)	Advantages	Disadvantages
Temporary			
Free-standing barriers	6–12	Prevents floodwater from reaching property	Predominantly designed to offer protection to communities rather than individual houses
			Requires advance warning for timely mobilization
			Storage
Flood skirts (in situ flexible rubber barrier raised and fixed to the perimeter of the property)	10–35	Prevents floodwater from reaching property	Requires advance warning for timely mobilization
			Storage
Door barriers/guards	0.75 (DIY type) 1–2 (those requiring specialist installation)	Inexpensive, particularly for DIY type	Varying degrees of effectiveness: some now kite-marked, others less proven
		Easily and quickly installed by owner	Storage required
Air brick/vent covers	0.5 (DIY type)	Relatively inexpensive	Requires deployment by owner
		Rapid deployment	
Permanent			
Waterproof external doors (incorporating auto-sensing pneumatic seals)	8–12	Inconspicuous	Not a solution in isolation
		No user intervention required	Relies on being complemented by other flood protection products or methods
			Expensive
Low-level bunds	Depends on a multitude of factors	Prevents floodwater from reaching property	Only suitable or practical for certain property types and locations
			Aesthetically detracting
			Application more suited to community protection

Airbricks/vents (built in)	0.5–1	No intervention by occupant required Completely passive	Relatively unproven in real flood scenarios Relies on sound workmanship
Raised thresholds	1.5–2 (inc. replacement doors)	No intervention by occupant required Completely passive	Access problems for occupants, particularly the nonambulatory Door barriers offer more practical/economical alternative Only nominal increases can be achieved Relies on being complemented by other flood protection methods Any ramps required for easy access will contribute toward costs
Storm porch	5–7.5	No intervention by occupant required, completely passive	Access problems for occupants, particularly the nonambulatory Expensive for level of additional resistance achieved Must be further supplemented by similar resistance measures to door to ensure effectiveness Practicalities of implementing to all external doors Any ramps required for easy access will contribute toward costs
Boundary walls	3–6 (depending on size of curtilage)	No intervention by occupant required Completely passive	Relies on continuity of barrier, otherwise weaknesses in line of defense may be exploited
External wall treatment (such as render)	1.5–2	Improves performance of poorly performing substrates	Expensive for level of additional resistance achieved Must be supplemented by same resistance measures as original entrance to ensure effectiveness Visually alters the external façade of the property
Basement or cellar tanking	10–15	Offers protection from groundwater ingress	Needs supplementing with a sump and pump Vulnerable to penetration caused by later repairs or alterations No additional protection offered to upper floors No protection for surface-level flooding
Anti-flood valves	0.1–0.5	No intervention Completely passive	Not a solution in isolation Relies on being complemented by other flood protection products/methods

Source: Adapted from ABI (Association of British Insurers) (2009). Resilient Reinstatement—The Cost of Flood Resilient Reinstatement of Domestic Properties. Research Paper No. 14. London: Association of British Insurers.

Cavity walls can present a particular challenge, particularly where retrofit insulation has been incorporated (Bowker, 2007). Both the assessment of and the resultant repair may require the involvement of a specialist. In general, the most susceptible insulation is blown fiber and bead type that readily absorb water, often resulting in clumping and thereby compromising the thermal efficiency of the external envelope and presenting a risk of moisture transfer in exposed locations. Where cavity wall insulation needs to be removed or is installed to achieve compliance with building regulations thermal element upgrading, a closed cell type should always be specified that will be resilient to future flood damage.

Internal walls and partitions: For shorter periods of duration and shallow-depth flooding, the wall components will require little if any repair, relying on a controlled drying process instead. Any finishes such as plasterboard must be removed where they cannot be successfully retained by early restoration, perhaps using rapid drying methods. Materials react differently to flooding in different situations and environments. Where left in a saturated state, such finishes may need to be removed to an arbitrary height above the flood line, as their porosity combined with capillary attraction will allow water absorption above the level of the floodwaters. Where internal walls are built of masonry, a resilient repair option is to replace finishes with sand and cement backing and a lime-based finish. For partitions constructed of wood or metal studs, complete removal and replacement might be the economic option. Modern construction commonly incorporates partitions supported by internal floor structures built within a shell. A significant issue in repairing such properties is the need to remove the sheet floor finishes to gain access for drying or replacing saturated insulation. When reinstating this type of construction, a revised support detail for the partitions and the use of closed cell insulation will provide resilience against future damage.

Floor finishes: There are few, if any, floor finishes that can provide complete resilience across the full range of flood depths and durations. Often the need for removal of floor finishes will have been dictated by the need to gain access to the supporting structure, as discussed previously for internal walls and partitions. Where the floors are of solid concrete construction and access to services or insulation is not an issue, ceramic finishes could be considered as a resilient reinstatement option that will survive most scenarios.

Staircases: The construction of staircases makes repair difficult when flooding has damaged the lower levels of the structure. In some instances the most economic method of repair may be to replace the entire flight. Because domestic staircases are almost exclusively constructed from wood, a repair providing resilience against future flooding would necessitate the use of resilient materials such as concrete or steel. It might also be more cost effective to replace the entire staircase with the resilient material rather than trying to repair it. The most appropriate way of dealing with a wooden staircase may in fact be to avoid stripping out water-damaged components at an early stage.

Resilient kitchens: One of the most significant repair costs in any domestic flood claim is the reinstatement of the kitchen. There are a number of possible flood-resilient alternatives—not only for the materials, but also the type and design of the units used. Traditional kitchen units constructed from chipboard offer little if any resilience against flood damage. The introduction of free-standing kitchen units can make them easier to remove (providing access for drying) and, because the units are free standing, they do not need to be deconstructed and thus are less likely to suffer damage when removed. Kitchen units constructed of stainless steel are commonplace in the commercial kitchen environment but have yet to be widely adopted in a domestic context. Steel kitchens cost, on average, between 90% and 200% more than their wooden counterparts, depending on the size and quality required. Replacement kitchens are generally considered to involve the most sacrifice (in terms of design, style, and materials) on behalf of homeowners if they are to adopt a resilient approach.

11.3 COST CONSIDERATIONS

The cost-effectiveness of any resilient reinstatement will always need to consider an assessment of the risk of future flooding and the likely depth. An overestimate could result in an uneconomical approach, whereas an underestimate could leave areas vulnerable and therefore increase the extent of future repairs should flooding recur. Although more difficult to predict, the likely depth of future flooding risk will have a significant impact on the cost effectiveness of a resilient repair option.

- *Flood depth greater than 900 millimeters:* Beyond a flood depth of 900 millimeters, resilient repair methods become costly to provide and of limited benefit. This is because beyond this depth, damage to the fabric and finishes will become widespread. For example, raised sockets and services will again be within the flood level and suffer damage.
- *Flood depth less than 100 millimeters:* Below this low flood depth, kitchens, electrics, and plumbing would not be replaced or subject to significant repair cost if future flooding occurred. For such scenarios, the estimated additional cost over traditional reinstatement for providing resilient measures would average 12%.
- *Flood depth between 100 and 900 milliliters:* Deep floods cause damage to kitchen units and built-in closets, thereby requiring replacement. Units are commonly constructed from a form of particleboard that offers little resilience to future flooding. Fitted kitchens represent high-value items that significantly affect reinstatement costs. To offer resilience, units should be constructed of resilient materials (e.g., stainless steel). Other alternatives might include plastic kitchen units; however, domestic standard components are not widely available. If they were available, the cost is likely to be comparable to stainless steel. For this greater depth of flooding, the estimated additional cost over traditional reinstatement for providing resilient measures would average 34%.
- *Flooding greater than 900 millimeters:* There are associated complications with flooding over this depth—that is, structural complications and increased water pressures. For most property types, resilient repairs could not be justified by economic considerations. Other factors such as increased risk frequency or minimizing disruption periods might have a bearing on the decision.

11.3.1 OTHER COST CONSIDERATIONS

Although the adoption of flood-resilient construction may have a long-term positive cost benefit by reducing the cost of any future physical reinstatement, this only forms part of the total spend on flood claims, and there should be consideration given to the potential for this to impact on claims spend in other areas.

It must also be borne in mind that these financial considerations would not normally be the main driver for determining methods of construction adopted in repairing a property. Consideration must be given to other drivers, such as live-ability and, most important, the acceptability of any change in construction methods to the average homeowner.

The main issues for consideration by insurers in the adoption of flood-resistant and -resilient construction are what impact this would have on claims costs and on the claimant experience, the key factors being

- Cost of repairs
- Speed of repairs
- Requirement for alternative accommodation

It is necessary to establish if the adoption of the flood-resistant and -resilient construction methods would have an impact on claim lifespan, as this has a direct bearing on alternative accommodation spend and, as a result, on customer satisfaction (Samwinga, 2009). There is also evidence that longer claim duration increases claim costs, in that the policyholder's attitudes toward the level of indemnity harden if insurers appear to, in the policyholder's opinion, drag their feet.

11.4   IMPACT OF RESILIENT REPAIRS ON REINSTATEMENT DURATIONS

The reinstatement durations of resilient repair alternatives are considered comparable to similar-value traditional repair scenarios. However, because some of the resilient work can commence at slightly higher moisture levels, the extent and duration of drying may be reduced slightly and particularly where a speed-drying method is used. In the case of repeat flooding of a resiliently reinstated property, strip-out, drying, and reinstatement would, however, be significantly reduced. Time savings will accrue due to

- *Reduced or eliminated strip-out:* Resilient materials' ability to withstand water inundation, or design features enabling elements to be removed in the early stages of flooding (e.g., doors).
- *Reduced drying time:* There are reduced opportunities for water to be trapped within the building structure.
- *Reduced reinstatement phase:* The extent of reinstatement is minimized due to the strip-out being reduced.

The actual phase durations that would be incurred with the repair of second-event resiliently reinstated properties will be significantly influenced by actual flood depth, duration, and water type.

11.5 HOMEOWNER ATTITUDE AND ACCEPTANCE OF RESILIENT REINSTATEMENT

It is important to consider the acceptability of flood-resilient and -resistant construction methods to homeowners. A great degree of resistance will result in confrontation during the claim handling process, especially if the homeowner will likely have to make a financial contribution toward the costs of the works.

Following the 2007 floods, a sample of homeowners (ABI, 2009) were asked a number of questions concerning the financial aspect of the reinstatement costs. They were asked whether they would be prepared to bear these costs themselves and, if so, how much they would be willing to spend.

They were also asked what their attitude would be to a number of resilient construction items that may be perceived to have a detrimental effect on the aesthetics of their homes. This was to ascertain what level of intrusion they would be prepared to accept to mitigate future damage and claims.

11.5.1 RISK AWARENESS

Homeowners' attitudes toward the adoption of flood-resistant and flood-resilient construction will be influenced by their assessment of the risk of their property re-flooding. In the sample of homeowners surveyed, the largest proportion of respondents (i.e., 47%) did not have any firm assessment of the likely risk of future flooding, while the next largest group (33%) believed that their property would never flood again. Without a reasonably accurate understanding of the likely risk of future flooding, homeowners are unlikely to want to carry out resilient repairs, particularly if they are responsible for funding the cost.

A clear understanding of the actual risk that homeowners face would allow them to take a more educated view of the benefits of the use of flood-resilient and -resistant construction methods when compared to any trade-offs—be they aesthetic or financial.

Of those homeowners surveyed who did believe that there was a significant risk of their home being flooded again in the future, 95% said they would be willing to adopt one or more items from a menu of resilient construction methods offered:

- The most popular measure was the re-siting of services to above waist height; 74% of the respondents were willing to consider this. This measure appears to be generally attractive, as its purpose is easily understood.
- 64% of respondents would consider replacing suspended timber floors with solid concrete slabs.
- 51% of respondents would consider tiled floors as an alternative to carpets.

- 49% of respondents would be prepared to replace chipboard or timber carcass kitchen base units with units made of plastic or a similar (e.g., stainless steel) impervious material.
- 41% of respondents would consider replacing baseboards with tiled backsplashes.
- 33% of respondents would consider replacing wooden doors, doorframes, with plastic units, but generally this measure is not popular, as it was believed to have a detrimental effect on the appearance of the home.

The re-siting of services was seen to be something that could be achieved with minimal disruption and at a reasonable cost. The benefit produced if the property were to reflood was easily understood, and the homeowners also saw some benefits from improved normal usability (i.e., no need to bend down to plug in appliances). The aesthetic impact was deemed minimal although there were still a number of respondents who said they would not like to have "plugs in the middle of their walls."

The use of solid floor slabs and tiled floor coverings was also easily understood but respondents showed concern in substituting carpeting or laminate for tiles. A substantial minority of the respondents had misgivings, because it was felt that tiling was "cold" compared to carpeting or that it would not fit in with their interior design schemes.

The respondents were interested in the adoption of flood-resistant kitchen units. While kitchens may form a large part of the claims spent on reinstatement following the recent flooding, it is not certain, given that the public is now being encouraged to upgrade and replace kitchens on a regular basis, whether "robust" kitchens may be seen as desirable when homeowners reflect further on this.

The final suggestion was the use of plastic substitutes for doorframes, architraves, and casings. This was not, on the whole, felt to be something that homeowners would like, as the potential visual impact would be too great. The perception was that the use of plastic would limit the number of potential decorative finishes available to homeowners.

It is clear that the voluntary adoption of flood-resilient construction methods by homeowners is linked to the perceived benefit as compared to the perceived obtrusiveness and its potential limiting effect on the decorative finishes available to homeowners. It may be possible to overcome many of the objections raised through education, and as more acceptable substitute products become more widely available.

11.5.2 INCENTIVES

Of the respondents surveyed who stated that they were unwilling to consider flood-resilient construction if their premises were to flood again, 75% said that they would be willing to consider the matter further if the incentive of a premium discount at insurance renewal was offered.

11.6 FLOOD-RESISTANT CONSTRUCTION

All respondents to the survey were asked whether they would be interested in carrying out work that would prevent floodwater from entering their homes in the future (i.e., flood-resistant construction). Of the respondents, 58% said that they would be interested in flood-resistant measures. Of those who were unwilling to carry out any work, the most popular responses were that they were not sure what it would involve, it would cost too much, or that it was not possible to actually prevent water from getting in. The lattermost is a valid point for many semi-detached and terraced properties where a homeowner may waterproof their own property but if their neighbors do not have any work carried out, the floodwaters are able to circumvent the measures utilized via an adjoining property.

While many homeowners suffered disturbance as a result of the 2007 flooding, the survey undertaken showed that less than 50% of them had taken steps to mitigate the impact of future flooding, although 58% of them would consider such work if further flooding was experienced. In another survey conducted by a major insurer after the floods of 2007, it was revealed that of the people interviewed—not all who had been flooded—95% had taken no steps to prevent future damage (Norwich Union, 2008).

11.7 CONCLUSIONS

The following summarizes the evidence presented:

1. On average, resilient reinstatement costs 30% to 40% more than traditional reinstatement.
2. Some resilient measures can be introduced on a cost-neutral basis, for example, setting power points further up the wall where the electricity supply drops down from the ceiling.
3. It is possible to reinstate some homes that flood to a depth of less than 100 millimeters at a lower increased cost average of 12%. This would make economic sense where there is reasonable evidence that the risk of deeper flooding is very unlikely.
4. At most, resistant measures could be appropriate for no more than 5% of the UK housing stock that is at risk of flooding, unless similar measures can also be taken for adjoining houses. This is because resistant measures are less likely to be appropriate for semi-detached or terraced properties where it may not be possible to ensure that all elevations are protected from floodwater.
5. Resistant measures are only considered appropriate for flood depths up to 900 millimeters.

REFERENCES

ABI (Association of British Insurers) (2008). The Summer Floods 2007: One Year On and Beyond, 2008. London: Association of British Insurers.

ABI (Association of British Insurers) (2009). Resilient Reinstatement—The Cost of Flood Resilient Reinstatement of Domestic Properties. Research Paper No. 14. London: Association of British Insurers.

Bowker, P. (2007). Flood Resistance and Resilience Solutions. An R&D Scoping Study. R&D Technical Report. Department for Food and Rural Affairs.

Norwich Union (2008). Homeowners Fear Future Flooding but Fail to Take Measures to Protect Their Property. Norwich, UK: Norwich Union.

Proverbs, D., and Soetanto, R. (2004). *Flood Damaged Property: A Guide to Repair.* London: Blackwell.

Samwinga, V. (2009). Homeowner Satisfaction and Service Quality in the Repair of UK Flood-Damaged Domestic Property. School of Engineering and the Built Environment. Wolverhampton, University of Wolverhampton.

Soetanto, R., Proverbs, D., Lamond, J., and Samwinga, V. (2008). Residential properties in England and Wales: An evaluation of repair strategies towards attaining flood resilience. In Boscher, L. (Ed.), *Hazards and the Built Environment: Attaining Built-In Resilience.* London, UK: Taylor and Francis.

Thurston, N., Finlinson, B., Breakspear, R., Williams, N., Shaw, J., and Chatterton, J. (2008). Developing the Evidence Base for Flood Resistance and Resilience. Bristol, Environment Agency.

USACE (U.S. Army Corps of Engineers) (1998). Flood Proofing Performance: Successes and Failures. Tulsa, OK: U.S. Army Corps of Engineers, Omaha District.

Section III

Mitigation and Adaptation to Flood Risk

12 International Historical, Political, Economic, Social, and Engineering Responses to Flood Risk

David Crichton

CONTENTS

12.1 INTRODUCTION AND HISTORICAL CONTEXT

According to the United Nations, more than 90% of all deaths from natural disasters are water related, and 99% of deaths from flood from 1975 to 2001 (i.e., over 250,000 people) were from low-income groups (UN, 2004). In richer countries, total disaster losses are generally less than 2% of the GDP (gross domestic product), while in poorer countries the figure is nearly 14%.

12.1.1 SOME MAJOR FLOOD EVENTS

If men could learn from history, what lessons it might teach us! (Coleridge)

- The Yangtze River floodplain covers 70,000 square miles. This area produces 45% of China's rice. In 1931, the Yangtze River flooded and more than 3.7 million Chinese died, mainly from famine. It flooded again in 1954 when 30,000 people died, and since then the government has redoubled its efforts to build dykes and reservoirs.
- For 20 years after World War II, as a proportion of the GDP, Japan spent three times more on flood defenses each year than England. Despite this, high population densities in flood hazard areas meant that during this period, more than 1,000 people were killed every year by floods in Japan, with two of the floods each killing more than 5,000 people. The worst event was the Isewan Typhoon flood in 1959 that killed 6,000.
- The Lynmouth flood of August 15, 1952, resulted in 34 deaths and 130 cars washed out to sea. An unprecedented 250 millimeters (10 inches) of rain fell on Exmoor in 22 hours. There were subsequent allegations that the severity of the flood was due to government cloud-seeding tests. This has never been proven, but secret information declassified in 2002 shows that the government was indeed carrying out such tests in the area just before the flood event.
- The River Arno flood in Florence on November 3, 1966, killed 35 and left 5,000 homeless. Many priceless paintings, frescoes, books, and sculptures were destroyed or damaged.
- Paris has a contingency plan to evacuate works of art (but not people) away from the Seine in case of a repeat of the 1910 floods (Mellot, 2003) (which have a 100-year return period) when the Seine rose by 8 meters. Nothing has been done since then to contain a rise of more than a few centimeters. Meanwhile, the Seine's width has been restricted in the center of Paris. Some 90% of the floodplain is urbanized in the City of Paris and the three surrounding départements: the Val de Marne, the Seine-Saint-Denis, and

the Hauts-dé-Seine. Since 1910, in addition to several major schools and hospitals built in the floodplain and a huge dependence on electricity, over 600,000 underground car parking spaces have been created and 158 kilometers of underground railway constructed. As a consequence, a major flood today would cause many losses in this area and about 17 billion euros of damage (excluding infrastructure). Approximately 1 million inhabitants and more than 100,000 companies, factories, and networks would be impacted for several months (Floodresilientcity, 2010).

- On Saturday, July 31, 1976, 3,500 people were camping in Estes Park in Colorado (United States) when 12 inches of rain fell in only 4 hours. The Weather Service issued a flood warning but this was not broadcast by the radio services, which did not want to interrupt a football game with a weather bulletin. Many campers were washed away.
- A contractor at a famous art gallery in London set up a temporary water supply by attaching a hose to a unused water main. In April 2000, the work was stopped for the long Easter weekend and during that time the hose became detached and water accumulated in the building to a depth of 1.4 meters, causing £5 million worth of damage. The judge reviewed the previous case law at a hearing in 2007 and concluded that because there had not been a rainstorm or other natural event, and because no pipes had "burst" in the legal definition of the word (because the pipes were undamaged), there had not been any "flood" in the legal sense. The judge said that in this case the law was "patently absurd" and decided there had been a flood after all, thus creating for the first time in English law, a generally applicable legal definition of the word (see *Board of Trustees of the Tate Gallery v. Duffy Construction Ltd.* and another [2007] EWHC 361).
- Heavy rains in Pakistan in summer 2010 are estimated to have left some 20 million people homeless and caused many deaths, making this catastrophe worse than the 2004 tsunami.

What lessons can we learn from these events?

12.1.2   CHINA: HUMAN SACRIFICES FOR FLOOD CONTROL

Around 500 BC, China used rituals and dyke building to control floods. In one region, priests required the annual sacrifice of a maiden to the river god. The victim was thrown into the river wearing heavy adornments. As the years passed, Chinese historians recorded that those farmers who had eligible daughters moved away from the area in increasing numbers. This policy was at least successful in encouraging people to move away from the floodplain (Clark, 1983).

Eventually, around 400 BC, a magistrate called Hsimen Pao put an end to the practice by having all the priests and officials thrown into the river instead. He introduced a policy of structural engineered flood defenses that lasted well into the second half of the 20th century.

12.1.3   Gilbert White: The Father of Floodplain Management

Gilbert White's doctoral dissertation, "Human Adjustment to Floods" (White, 1942), has been called the most influential ever written by an American geographer (YouTube, 2010a). "Floods are 'acts of God,'" he wrote, "but flood losses are largely acts of man." White championed the holistic management of floodplains, suggesting that we live with Nature instead of fighting it with "structural" solutions. White advocated sustainable uses of natural resources. His ideas are now generally known as "sustainable flood management."

Japan introduced the "human adjustment to floods" concept in 2000, based on Gilbert White's work. The basic idea is to adjust the way humans live with floods by land use zoning, flood-proof buildings, and flood insurance. The role of flood insurance is important in that it provides direct economic incentives to individuals to relocate or take their own precautions against flood while at the same time facilitating rapid economic recovery after a flood.

China also now relies much more on natural flood management, encouraging reforestation upstream to slow the rainfall runoff so the rivers can cope. A sustainable flood management approach has now been adopted in many other countries, including France, Germany, Switzerland, Scotland, as well as the United States and Canada. The latest country to move away from the structural defenses approach to more sustainable flood management is Wales with its "New Approaches Programme" (NAP) in 2007 (Welsh Assembly Government, 2007). For example, in the Conwy Valley, the Welsh government now pays farmers to let their fields flood to protect the village of Trefriw.

12.2   UNDERSTANDING FLOOD RISK

12.2.1   The Crichton Risk Triangle

The Crichton Risk Triangle (Crichton, 1999) was designed for use by the insurance industry for catastrophe modeling (see Figure 12.1). Catastrophe models work on the basis that risk is a function of hazard, exposure, and vulnerability, and each factor needs to be considered independently. Risk is represented by the area of an acute-angled triangle.

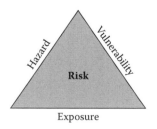

FIGURE 12.1   The Crichton risk triangle. (From Crichton, D. (1999). The risk triangle. In Ingleton, J., Ed., *Natural Disaster Management*. London: Tudor Rose.

- *Hazard:* In the case of flooding, Hazard represents the frequency and severity of rainfall events or storms. Climate change predictions indicate an increasing hazard over which society has little immediate control other than to clean watercourses, provide adequate drainage (Lindholm, Schilling, and Crichton, 2007), and adopt natural flood management practices (WWF, 2007).
- *Exposure:* Exposure represents the density and value of property located in flood hazard areas, such as near rivers, the coast, below dams, or on low-lying land, especially at the foot of a slope. Typical household and business premises contents are more valuable and more vulnerable than ever before. Population growth, migration, and smaller, often single-person households have meant a huge demand for land for development in many countries. In England, new build has reached an average of around 40 dwellings per hectare (Hall, 2008). In the Thames Gateway floodplain, a density of 200 dwellings per hectare is planned. Climate change and the "heat island" effect of densely populated areas can increase the frequency and severity of localized rainstorms in urban areas leading to more flash floods (Munich Re, 2010).
- *Vulnerability:* Vulnerability refers to the resilience of the properties insured and depends on the design and construction. Many countries concentrate nowadays on managing hazard and exposure (Crichton, 2010) instead. Increasingly, they work together; for example, Germany is working with France on the Moselle catchment (CIPMS, 2006) and with Scotland on sustainable flood management techniques developed under the SAFER project (EU-Safer, 2010). However, in its comprehensive approaches to flood risk management, Scotland has carried out extensive materials testing for flood resilience and introduced much stricter building regulations on matters such as water resilience and drainage to take climate change into account (The Building (Scotland) Regulations, 2004). In a current EU-funded research project, Scotland has been held up as an example to the rest of Europe in work on making cities more resilient (Floodresilientcity, 2010).

The risk matrix appears in Appendix 12.A at the end of this chapter and shows how different factors can affect hazard, exposure, and vulnerability.

12.2.2 PERCEPTION OF RISK AND PSYCHOLOGY OF RISK

Since 1970, when Paul Slovic (2000) met Gilbert White, much of Slovic's work has concentrated on natural and technological hazards and public responses. Extensive psychometric surveys have been carried out all over the world as part of this research, and these surveys have found that perceptions of risk vary considerably and consistently by age, gender, and nationality. Slovic concludes that risk is perceived as higher if the hazard is unknown and uncontrollable, and if the effects are dreaded and visible. Much depends on the level of trust the public has in the authorities. If people trust the authorities, they might perceive the flood risk as low, for example.

After all, the Council would not allow the development unless it was safe, would it? The first people who need to understand the risk are local planners, politicians, engineers, and architects.

Without efforts by the authorities to prevent floodplain development and to raise flood risk awareness, the public often does not perceive the flood risk at all, and is happy to live next to a river or the sea. They do not see the need for insurance unless their mortgage lender insists on it. Retired people who have paid off their mortgages may let their policies lapse and often move to the coast.

12.2.3   WHAT LEVEL OF FLOOD RISK IS ACCEPTABLE?

A detailed report on this has been conducted in the United States (Galloway et al., 2006) that uses a 1% standard (100-year return period) for its National Flood Insurance Program (NFIP). It concluded that

> Flood standards in many developed countries far exceed the NFIP 1 percent standard. Japan and the Netherlands use 0.01 percent (10,000-year) protection for coastal works and 0.5 percent (200-year) to 0.05 percent (2,000-year) protection for riverine systems.
>
> Reports by the U.S. Army Corps of Engineers (USACE) indicate that the riverine 0.2 percent flood is expected to cause 2.24 times as much damage as the 1 percent flood and that the 0.5 percent flood is expected to cause 1.5 times as much damage as the 1 percent flood.

The report recommended that

> FEMA should ensure that NFIP guidance and program activities clearly indicate that critical facilities should be located outside the 0.2 percent (500-year return period) floodplain.

12.2.4   THE INSURANCE TEMPLATE

The insurance template (an extract appears in Appendix 12.B at the end of this chapter) does not refer to insurability standards as no maximum risk levels for acceptance have been agreed to by any insurers anywhere in the world. As the template makes clear, it simply gives a guide as to the maximum levels of risk for which insurance should be available somewhere in the market *at normal terms*. This is based on mathematical calculations, not market agreements. Nevertheless, many of the more responsible planning authorities have found it a useful benchmark for their planning policies.

12.3   RESPONSES TO FLOOD RISK

Throughout history and all over the world, flood responses seem to be much the same, just like the laws of physics.

12.3.1 POLITICAL RESPONSES

An object at rest tends to stay at rest, or if it is in motion tends to stay in motion with the same speed and in the same direction unless acted upon by another force. (Newton's First Law of Motion)

Politicians tend to favor the status quo unless pushed by outside forces. There is a term for this in the United States (Kunreuther, 2010): "not in my term of office"—NIMTOF. If we perceive the likelihood of a disaster as below some arbitrary threshold of concern, we then assume that "It won't happen to me—at least not on my watch."

However, if there is a sudden outside force such as a major flood, politicians are also prone to knee jerk reactions. For example, after the Hull floods in England in 2007, the recently appointed minister in charge had previously been in charge of the department responsible for providing aid to developing countries after disasters. He immediately promised government money to assist those who had no insurance (DCLG, 2007). This was the first (and only) time this has ever been done in the United Kingdom. It ignored the inequity for those who had bought insurance coverage or the long-term effects on demand for insurance.

Some pressures are less sudden or obvious but just as powerful, as described below.

12.3.1.1 Property Developer Interests

Property developers have a strong financial interest in being allowed to build in flood hazard areas. Once the property is built and sold, they can walk away from the problem. In England, for example, property developers are given great freedom to build in flood hazard areas, as evidenced by the fact that such a high percentage of properties have been built in flood hazard areas (Table 12.1). Property developer profits far exceed the money allocated for flood defenses.

TABLE 12.1

Percentage of All New Dwellings Built in Flood Risk Areas in England, by Region, 1996–2005

Region	1996	1997	1998	1999	2000	2001	2002	2003	2004	2005[a]
Northeast	6	5	2	3	1	2	2	3	2	2
Northwest	5	5	7	5	6	9	6	8	5	4
Yorkshire and the Humber	11	12	7	10	13	12	11	15	10	13
East Midlands	10	12	6	7	9	11	13	13	11	9
West Midlands	7	4	6	6	2	4	5	4	5	3
East of England	6	7	8	7	7	6	7	8	7	13
London	27	25	26	24	23	20	21	28	26	18
Southeast	6	8	9	10	9	10	8	10	7	7
Southwest	5	6	6	8	7	8	10	7	8	7
England	9	9	9	9	9	9	10	11	10	9

Source: DCLG, Land Use Change Statistics, Department of Communities and Local Government, 2007 press notice 2464.

[a] Provisional figures.

Selected English housebuilders' profits in 2005 before tax (taken from companies' annual reports and accounts) include:

- Slough Estates: £582.3 million
- Persimmon Homes: £495.4 million
- Barratt Developments: £39 4.3 million
- Wilson Bowden plc: £216.4 million
- Total: £1,688.4 million

The government minister responsible for English flood defenses in 2006 said:

I fully understand the concern among insurers about reduction in the Environment Agency's 2006–07 flood resource allocation from £428 million to £413 million. (Milliband, 2006)

It is not clear why the government reduced this expenditure. However, the government later announced (Burnham, 2008) that it would spend £40 million to provide free swimming lessons for children and people who are 60 years of age or older. (Incidentally, around the time of the £15 million reduction in flood defense spending, the UK government's economic development aid given to China was more than £40 million per year.)

The amount of profit which can be made by property developers from building in cheap floodplain land must raise the spectre of political corruption in some countries. Even in England, the amount of new build allowed under the central government's planning policy PPS25 and by local authorities in flood hazard areas (see Table 12.1) must raise concerns about the motives for allowing it, given the distress caused by flood events. It would be interesting to know how much property developers contribute to political party funds. More than £200 billion worth of property and assets in England and Wales are now located in flood plains. It certainly demonstrates the differences in planning rules in England compared with the rest of the UK (see Appendix 12.C at the end of this chapter.

12.3.1.2 Nuclear Energy

There is still the potential for critical infrastructure such as nuclear power stations to be built on the floodplain due to political decisions:

We believe the best locations [for new nuclear power stations] . . . are adjacent to existing nuclear power station sites. (British Energy, 2008)

The most likely sites are all coastal—Sizewell, Bradwell, Hinkley Point, and Dungeness—and therefore vulnerable to sea-level rise from climate change. It is quite possible that radioactive wastes will remain on these sites indefinitely. All of these sites are owned by British Energy, which is owned by EDF (Électricité de France). EDF was a major financial donor to the UK Labour Party from 2003 to 2005. This is not unusual for commercial companies, but EDF at the time was 70% owned by the French government. Decision-making criteria have been laid down to ensure that sites cannot be excluded by planners on the grounds of flood risk.

12.3.1.3 Ecology Interests

Government legislation throughout Europe now gives priority to EU Directives on water quality, habitats, etc. These can conflict with sustainable flood management. For example, the Water Framework Directive prevents the modification of rivers to cope with increased flows due to climate change, and the Habitats Directive prevents the clearance of weeds and silt from watercourses, leading to blockages. Such conflicts should not be necessary. Organizations such as WWF Scotland and RSPB Scotland should be praised for showing the way forward with their active support of sustainable flood management (Johnstonova, 2010). In the whole of the European Union, it is only in Scotland where the government states that sustainable flood management is more important than ecology (HMSO, 2003).

12.3.1.4 Unintended Consequences

A focus on just one element can lead to unintended consequences. Sustainable drainage systems (SUDS) sound good, but in the wrong place or poorly maintained, can make the flood hazard worse. Hidden infrastructure projects such as sewers and culverts can be the victims of cost cutting. In the United Kingdom, the Disability Discrimination Act is resulting in new houses being more vulnerable to flooding. Controls on waste disposal are leading to fly tipping* into watercourses and sewers blocked by sanitary products. In England, reductions in the frequency of waste collection are attracting pigeons and seagulls, and leading to increases in the population of urban rats and foxes and an increase in public health risks.

Comprehensive approaches are needed involving all stakeholders.

12.3.2 ECONOMIC RESPONSES

> An object will accelerate with acceleration proportional to the force and inversely proportional to the mass. (Newton's Second Law of Motion)

Economic forces can lead to a vicious circle that can spiral out of control. Especially in a time of economic recession, a flood hazard can lead to properties being blighted, and this can be contagious. Small and medium-sized enterprises will be especially vulnerable (Crichton, 2006), and economic meltdown is conceivable. A major element in the economic response is, of course, insurance.

12.3.2.1 Insurance versus Gambling

Other things being satisfactory, and unless the market is distorted in some way—for example, by government regulation—insurance could theoretically always be possible at a price, as long as there is uncertainty about occurrence, timing, or quantum, and as long as the event is outside the control of the proposer. There is always the risk that insured people may take less care of their property, but this is less of a risk with natural hazards than with many other perils. The key factor about insurance is that if the market is allowed to operate naturally, then the pricing mechanism will ensure that developments in high-risk locations are not subsidized by

---

* *Fly tipping* is a British term for illegally dumping waste somewhere other than at an authorized site.

lower-risk developments. The problem with artificial regulation, such as the United Kingdom's "Statement of Principles" (see below) or the United States' flood insurance program (see below) is that subsidized flood insurance can enable continued high-risk developments.

In practice, most mainstream insurers will be reluctant to quote for very high risks and there comes a point where the risk becomes so great that the proposer may find it cheaper to self-insure or to visit a betting shop. Insurance helps to maintain businesses after a flood, thus safeguarding livelihoods (AXA Insurance, 2004).

In the past 25 years, the average UK house price has increased from £23,644 to £194,362 but the average household insurance premium is still less than £300. One insurer has calculated that its policyholders in safe areas are subsidizing its policyholders in flood hazard areas by more than £40 million per year. This is not sustainable, and substantial changes in availability and affordability can be anticipated.

12.3.2.2   Insurance Claims Costs

After a major flood event, insurance claims costs can be much higher than expected. There are a number of reasons for this:

- Shortage of repair personnel and dehumidifying equipment can mean a shoddy repair that must be redone later.
- Shortage of materials and labor leads to higher costs. This is called *claims surge*.
- Local flooding can attract predatory "cowboy builders" from other areas.
- Lack of experienced loss adjusters can lead to loss adjusters from other countries being drafted in. Such loss adjusters are often not accustomed to dealing with flood claims, as they are not so widely insured in other countries.
- Delays in carrying out the drying-out and repair process lead to more damage.
- Autumn and winter floods are more expensive because natural drying-out takes longer.
- High volumes of claims mean that there is more potential for fraud to escape undetected, and more temptation for bandwagon fraud.
- Political pressures on insurers can force them to pay out more than they need to.

This is one reason why the British National Flood Insurance Claims Database is so important—because it reflects the true costs of flood claims and helps insurers detect excessive claims.

12.3.2.3   Eleven Insurance Scenarios

- *Scenario 1—Belgium, Italy, France, Germany:* General exclusion of flood coverage. Flood coverage can be offered as an option, and this gives insurers the opportunity to underwrite and price or decline each risk on its own merits. Optional coverage is very subject to adverse selection. Only those who perceive a risk will buy it, and the insurer is likely to end up with an unbalanced book of high-risk business. On the other hand, people on high ground may not buy flood insurance but are still at risk, as the residents of

Baiersdorf in Bavaria found out in July 2007. Rainfall caused €70 million of flooding in more than a thousand houses and a business park (Munich Re, 2010). Most were uninsured.

- *Scenario 2—Australia/Canada: General exclusion of coverage for flooding from rivers and the sea.* Coverage is still provided for flooding due to surface water, burst pipes, or sewage backup on the basis that this could happen anywhere, whereas flooding is considered inevitable in zones close to rivers or the coast.
- *Scenario 3—Sweden/Norway: Litigation.* Insurers can recover the costs of flood damage due to planning decisions or inadequate drainage (Lindholm, Schilling, and Crichton, 2007). This can be an effective way to influence the behavior of planning authorities and water utilities.
- *Scenario 4—United Kingdom: "Statement of Principles."* Under the "Statement of Principles," the industry has agreed to maintain flood coverage for existing customers (for properties built before 2010) until 2013 but say this deadline will not be extended. Thereafter, there could be problems with the availability or affordability of insurance, especially where the flood hazard is greater than the 100-year return period. Where coverage is given at all, there is also likely to be much greater use of substantial excesses or deductibles. One insurer already applies an excess equivalent to 10% of the previous flood claim. A variation of this is to reduce the excess if the policyholder can demonstrate that an effective, insurance-approved, temporary demountable flood defense was deployed in time to minimize flood damage. It should be noted that if the excess exceeds £2,500, the policy is no longer adequate security for a mortgage in the United Kingdom, and this could result in widespread foreclosures.
- *Scenario 5—Scotland: A partnership.* The Scottish Planning Policy SPP7 was produced in consultation with the insurance industry. It has a number of features, including
  - A general presumption against building where the flood hazard is greater than the 1 in 200-years flood event.
  - A risk matrix similar to the insurance template so that vulnerable properties are not permitted where the risk exceeds the levels in the insurance template.

As a result, there has been almost no new building in flood hazard areas since 1995 and the proportion of properties in flood hazard areas of Scotland (other than Moray) is much lower than in any other part of Britain (see Appendix 12.A at the end of this chapter). The Scottish government has pledged to protect all properties against the 100-year river or coastal flood by the end of 2008, and apart from Moray it is well on the way to achieving this. Scottish achievements have passed apparently unnoticed by English-based insurers, but two Scottish-based insurers have reduced their premiums accordingly.

- *Scenario 6—Australian Northern Territories: Government subsidies.* While river and coastal floods are not generally insured in Australia, the government has introduced subsidies for wider flood insurance in the

Northern Territories to encourage economic development in, and migration to, this underpopulated state following the devastation in Darwin from cyclone Tracy on December 25, 1974.

- *Scenario 7—United States: Government intervention.* The government has mapped areas with a flood hazard exceeding the 100-year return period. Flood insurance in those areas is effectively underwritten by the government, which collects the premiums from insurers and pays insurers for the claims they settle. The effect has been to promote economic development and settlement in hazardous areas.

  To stimulate take-up, insurance has been made compulsory for those with mortgages. However, after a flooding event in northern Vermont in 1998, of the more than 1,500 victims of the disaster, FEMA (Federal Emergency Management Agency) found that 84% of the homeowners in flood-hazard areas did not have insurance—even though 45% had been required to purchase such coverage according to Kunreuther (Kunreuther, 2010).

  After huge losses to the program from Hurricane Katrina, flood zones and premiums were changed from January 19, 2010, resulting in substantial premium increases for those in the highest hazard zones (YouTube, 2010b).

- *Scenario 8—Ontario, Canada: Enforced relocation.* Following Hurricane Hazel in 1964, it was decided that no new building or flood defense work should take place in flood hazard areas and to ban the sale of any property in such areas to anyone other than the local authority that would buy the property, demolish it, and turn the site into parkland. This method has since been copied in parts of the United States.

- *Scenario 9—Ireland: In transition?* Ireland has the highest household insurance penetration in the world, with some 98% of households having flood insurance coverage (Irish Insurance Federation, 2010). Until 2009, Ireland allowed unrestricted building in flood hazard areas, but following a series of major floods in November 2000, February 2002, November 2002, and October 2004, they issued their first planning guidelines (Irish government, 2009) for flood in November 2009, the same month as the most serious Irish floods to date, costing insurers €244 million. These planning guidelines totally ignored the pleas from the insurance industry to government to resist pressure from property developers, which in effect could allow floodplain development to continue almost as before. Insurers are considering their position.

- *Scenario 10—The Netherlands: Private flood insurance is illegal.* This avoids the social problems caused by problems with availability and affordability of insurance, but limits the country's access to global reinsurance facilities.

- *Scenario 11—Japan: Standard rates for all. In transition.* Japan has had a market tariff agreement under which every insurer charged the same rates for flood regardless of the risk. Following the Japanese change to sustainable flood management, it has been realized that insurance has a major role to play in providing a disincentive to living in flood hazard areas. The insurance market is now in transition to more selective underwriting but is hampered by a lack of statistical data. In particular, they do not have the equivalent of the British National Flood Insurance Claims Database.

This list is not exhaustive; there are other variations and many different approaches around the world (Crichton, 2008). Sustainable flood management demands that insurance premiums reflect the true risk, and in that way provide a disincentive to living in flood hazard areas.

12.3.2.4   Insurance in Less-Developed Countries

There are a number of ways in which insurance companies are helping less-developed countries:

- *Microinsurance:* Microinsurance schemes have been developed along with microfinance projects mainly aimed at improving the livelihoods of individuals. For example, a woman might obtain a low-interest loan to buy a sewing machine and be able to arrange insurance on the machine so that if it is lost in a storm or flood, she can obtain a new one and still earn a living.
- *Capital projects:* The World Bank will insist on insurance for capital infrastructure projects on which it lends money. For some reason, the insurance is only required during the construction phase, but it would be logical to require the insurance to be maintained thereafter.
- *Education and awareness:* Multinational companies such as hotel chains often arrange a global insurance program that includes their branches in less-developed countries. The global insurers will roll out education and awareness programs to the staff in these branches and often make them available to the wider local community.

12.3.3   SOCIAL RESPONSES

For every action there is a reaction equal in magnitude and opposite in direction. (Newton's Third Law of Motion)

In a time of climate change and increasing rainfall, if flood risk management is neglected, Gaia can soon change into Grendel.*

This can lead to major social problems. Loss of lives, livelihoods, and property could lead to a breakdown in social cohesion.

- This may lead to civil unrest and disobedience.
- People will be looking for someone to blame.
- Litigation may increase.

It could be argued that how society prepares for and reacts to flooding is a measure of its compassion and understanding. The most vulnerable members of society—the poor, sick, elderly, disabled, children, and immigrants—may be those most likely to live on cheaper floodplain land (perhaps the modern-day equivalent of the human sacrifices in 500 BC?).

---

* Gaia is the goddess of the Earth; Grendel was a monster descended from Cain who was defeated by Beowulf in an epic poem written around 1100 AD. The story of Beowulf was made into a film in 2007 starring Anthony Hopkins.

Infrastructure problems with hospitals, electricity, and sanitation could lead to loss of life. Mental and physical health issues such as increases in suicide rates and the effects of water-borne pathogens are vital aspects that have thus far been relatively neglected in research work.

The influence of religion and faith communities after a natural disaster is often neglected in academic studies. There is considerable, little recognized potential from such communities for providing resources and comfort to flood survivors, especially among ethnic minorities or in countries with limited disaster recovery infrastructure (Gaillard, 2010).

Voluntary organizations generally can also play a major role, for example, the Red Cross in the United States, or WRVS or the Samaritans in the United Kingdom. This is even more important in countries where the official emergency services cannot cope. According to a November 2006 report by the Chief Fire Officers Association (Hayden, 2006) in England and Wales, "The UK [sic] simply does not currently have the capability to respond to a major flood event." *(Note:* The report actually only applies to England and Wales, not the United Kingdom. There is a statutory requirement on fire officers in Scotland to provide such capability. This does not apply in England and Wales.)

There has been very little research on the effects of pet ownership on evacuation rates. Research in the United States shows that 30.5% of human evacuation failures could be attributed to dog ownership, and 26.4% to cat ownership (Heath, 1999).

The psychology of risk means that it can be very difficult to accurately convey flood risk to people unless they have experienced the effects themselves. Flood awareness programs are important to help to prepare people at risk.

12.3.4 ENGINEERING RESPONSES

It is easier and more sustainable to work with Nature, not against it. Many countries still try to fight Nature (and the laws of physics) using structural defenses.

12.3.4.1 Structural Flood Defenses to Reduce Hazard

There is, of course, an important role for civil engineers in designing structural flood defenses such as walls, culverts, or reservoirs to reduce vulnerability, especially for critical infrastructure (Crichton, 2002). However, there are dangers to a one-dimensional approach, including

- Some research suggests that structural defenses can actually increase the risk (Tobin, 1997; Criss and Shock, 2001; Takeuchi, 2002; Kelman, 2007).
- Flood defenses can give a false sense of security. When a flood defense fails, the results can be more catastrophic than if the defense had not been built at all because the failure can be sudden and more people may be in the danger zone (FEMA, 2007). In the Grafton incident (Pfister, 2001) in Australia in 2001, residents refused to leave because they were convinced the defenses would protect them (in the event they did, but only just). Often the perception of risk can differ dramatically from the real risk (Slovic, 2000).
- Structural defenses transfer the risk onto future generations, along with the maintenance and repair costs (Etkin, 1999).

- New defense programs cannot keep pace with current floodplain developments in countries such as England and take resources away from other capital projects.
- Flood defenses can themselves be damaged by floods. The Japanese government calculated that after the Naka River and Kokubu River floods of 1998 and the Fukuoka floods of 1999, half of the costs of the floods consisted of repairing damage to the flood defenses themselves (Takeuchi, 2002).
- Structural flood defenses need constant maintenance. The more the defenses, the greater the cost of maintenance, taking much-needed public spending away from schools and hospitals and other essential services.
- If a river is walled in and not allowed to flood onto its banks, it is more likely to deposit sediment in the bed of the river. This caused the Mississippi River floods in 1903, 1912, 1913, 1922, and 1927. Each time the levees were raised, but the sediment simply raised the height of the river.
- Defenses can displace the problem upstream or downstream. This is a particular problem in continental Europe where one country's flood management solution is another country's increased problem.
- Flood walls can act as a barrier to stop the flood from draining back into the river or sea. After the 1953 coastal flood in England, for example, many defenses had to be demolished to let the water drain away.
- Flood walls will have to be repeatedly raised as climate change impacts are felt.

Research by Munich Re (2010) has revealed that storms resulting in floods occur much more frequently than "classic" cases of river flooding, and are likely to increase even more in the future. These can produce overland flows even outside floodplains—for example, the Madeira floods in February 2010. While Madeira, Singapore, and the United States, for example, have emergency flood channels for extreme events, normal structural defenses are of little effect because such floods can happen on any sloping ground, not just near rivers or coastlines. There is still a place for structural solutions to defend essential infrastructure. However, the high costs of maintenance and construction mean that often a more cost-effective solution is simply to abandon the high hazard areas.

12.3.4.2 Temporary Demountables to Reduce Vulnerability

There are many different types of *demountable* flood defenses, but some can make matters worse and some can lead to building collapse (Crichton, 2004). Insurers therefore tend to be careful about which products they approve. Because there is no guarantee that a temporary demountable will be deployed in time, the prudent insurer will not give discounts, but a few are prepared to offer incentives such as low-interest loans to purchase approved systems and may offer a reduced excess if an approved system was deployed in time.

In Scotland, some local authorities provide temporary demountables free of charge or at a discounted cost to property owners in hazard zones. The Scottish Water company provides free demountables in areas at risk of sewage overflows, pending remedial work.

12.3.4.3　Architecture to Reduce Vulnerability

Venice and the Netherlands prove that architecture can adapt to flooding. Floating buildings, buildings on stilts, buildings with waterproofed basements and ground floors, buildings on made-up ground—there are many solutions that architects can offer to make buildings more resilient to flooding. Existing buildings can also be adapted; for example, by moving all vulnerable equipment such as kitchens, electrical switchgear, etc. to upper floors and reserving ground floors for car parking. This is a solution that has been used in Shrewsbury in England, for example, where parking space is at a premium.

Architects could potentially be the vanguards in the struggle to adapt society to climate change. Despite this, adaptation methods are not taught in most architecture schools, and there is only one standard textbook (Roaf, Crichton, and Nicol, 2009) on the subject anywhere in the world.

12.3.4.4　Building Regulations

An easy way for government to reduce vulnerability would be to apply more resilient building regulations covering the design and materials used. Even better would be to make these regulations retrospective should the building need repairs after a storm or flood. In this way, insurers would be obliged to introduce resilient reinstatement, and over a few years the building stock would become more resilient generally at no cost to the taxpayer.

12.3.4.5　Nonstructural Solutions

This topic is discussed later in the book. Such solutions form a component of sustainable flood management.

12.4　CONCLUSIONS

Sustainable flood management can offer cost-effective solutions by working with Nature instead of fighting it. At the end of the day, Nature will always win such battles. The way to work with Nature is to use sustainable flood management methods. If Gaia is indeed changing into Grendel, the answer may be another BEOWULF:

- **B** Ban building in flood hazard areas through planning controls
- **E** Educate the public and raise awareness
- **O** Organize resilience measures and recovery procedures;
- **W** Water management, including
  - Sustainable drainage methods
  - River restoration
  - Maintenance of watercourses
  - Removal of culverts
- **U** Urgent warning systems and evacuation procedures
- **L** Land management practices (natural flood management)
  - Remove land drains
  - Plant trees and hedges to retain surface water
  - Encourage farmers to store water during heavy rainfall
- **F** Financial: insurance costs can discourage living in hazardous areas

These are the main components of sustainable flood management. All of them have been introduced in Scotland, which has also introduced specific actions such as the following:

- *Local planners must set up Flood Liaison and Advice Groups (FLAGs).* FLAGs bring together all relevant stakeholders, including insurers and property developers, to advise on planning applications and flood management schemes and to make collective decisions about risks in an informal, nonconfrontational setting. Property developers often do not appreciate the gravity of flood hazard development problems until they attend FLAGs. Unfortunately, there are no FLAGs outside Scotland.
- *Biennial reports.* Compulsory in Scotland since 1997, these reports record all flood events in the local authority's area, as well as the actions taken to prevent a reoccurrence. Because every local authority must produce these reports, commercial software for standard geographic information systems (GIS) has been developed to assist Scottish local authorities in providing these reports in a consistent format that can be readily accessed by insurers using their own GIS software. For example, "FloodVu" developed by Kaya Consulting Ltd.
- *Money is available to local authorities* not only for structural projects, but also for nonstructural initiatives, such as natural flood management, which are much more cost effective.

These have been very successful as can be seen from Appendix 12.C. In addition, the Scottish Government has promoted the take-up of insurance with rent schemes to help low-income families obtain insurance protection (Vestri, 2007). The take-up rate in Scotland is now 57%, compared with 39% in England.

APPENDIX 12.A: THE RISK MATRIX

Different factors can affect different sides of the risk triangle. The main effects are on the sides indicated by the tick (✓) in the boxes below. The effects may be positive or negative. Only political and nonstructural measures affect all three sides of the triangle directly, although the other factors can have an indirect negative effect on all three sides.

Factor	Hazard Frequency and Severity of Event	Exposure Number of People or Value of Property in Hazard Zone	Vulnerability Resilience of People, Livelihoods, or Property
Political	✓*	✓**	✓***
Economic	*	✓	***
Social	*	✓	✓***
Structural	✓	**	✓***
Insurance	*	✓**	✓***
Non Structural	✓	✓	✓

* Can create additional hazards, such as a breakdown in social cohesion and the rule of law.

** Can create additional exposure if actions result in increased building in flood hazard areas.

*** Can result in less resilient livelihoods and less resilient buildings being constructed.

APPENDIX 12.B: THE INSURANCE TEMPLATE

Extracts from the residential property section of the "insurance template" showing the levels of risk that may be insurable at normal terms. Higher risks may be accepted with premium loadings.

Type of Property	Return Period (Years)	Annual Probability of Flood (%)
Housing for the disabled or elderly	1,000	0.10
Basements	750	0.15
Ground-floor flats	500	0.20
Touring mobile homes for seasonal occupancy only	50	2.00
Other	200	0.50

Source: From Crichton, D. (1998). Flood appraisal groups, NPPG 7, and insurance. In *Proceedings of the "Flood Issues in Scotland" Seminar*, Perth, Australia, December 1998. © Crichton 1998.

APPENDIX 12.C: FLOOD EXPOSURE IN BRITAIN IN 2009

Country	Proportion of Existing Properties at Risk (%)	Proportion of New Build in Flood Hazard Areas
England (100-year return period)	10	11% (2009)
Wales (100-year return period)	12	Negligible (since 2004)
Scotland (100-year return period)	Negligible except for Moray[a]	Negligible except for Moray (since 1995)
Scotland (200-year return period)	3.9	

Source: Environment Agency, Department for Communities and Local Government, Welsh
    Assembly, Scottish government. Figures for Northern Ireland not available.

[a] Due to the Scottish government's commitment to defend all properties against the 100-year
return period river or coastal flood by the end of 2008.

REFERENCES

AXA Insurance (2004). Business Continuity Guide for Small Businesses. London: AXA
    Insurance. Available at <http://www.axa4business.co.uk/bc/guide.asp> [Accessed
    November 14, 2006].
Building (Scotland) Regulations (2004).
Burnham, A. (2008). Andy Burnham, MP, Secretary of State for Culture, Media and Sport,
    Press release, June 2008.
CIPMS (Commissions Internationales pour la Protection de la Moselle et de la Sarre (2006).
    Plan D'action Contre les Inondations dans le Bassin de la Moselle et de la Sarre Bilan
    2001–2005. Secrétariat der CIPMS (French and German language versions only). Trier,
    Germany 2006.
Clark, C. (1983). *Flood*. New York: Time-Life Books Inc.
Crichton, D. (1998). Flood appraisal groups, NPPG 7, and insurance. In *Proceedings of the
    "Flood Issues in Scotland" Seminar*, Perth, Australia, December 1998. Stirling: Scottish
    Environment Protection Agency. pp. 37–40
Crichton, D. (1999). The risk triangle. In Ingleton, J., Ed. *Natural Disaster Management*.
    London: Tudor Rose.
Crichton, D. (2002). UK and global insurance responses to flood hazard. *Water International*,
    27(1), 119–131.
Crichton, D. (2004). Temporary Local Flood Protection in the United Kingdom. — An
    Independent Assessment. London: University College London.
Crichton, D. (2006). Climate Change and its Effects on Small Businesses in the UK. London:
    AXA Insurance.
Crichton, D. (2008). Towards a comparison of public and private insurance responses to flood-
    ing risks. *International Journal of Water Resources Development*, 24(4), 583–592.
Crichton, D. (2010). Towards a comparison of public and private insurance responses to flood-
    ing risks. *International Journal of Water Resources Development*, in press.
Criss, R. E., and Shock, E. L. (2001). Flood enhancement through flood control. *Geology*,
    29(10), 875–878
DCLG (Department of Communities and Local Government) (2007). Department of
    Communities and Local Government press notice 2464, July 2007.
Etkin, D. (1999). Risk transference and related trends: Driving forces towards more mega-
    disasters. *Environmental Hazards*, 1, 69–75.

EU-Safer (Strategies and Actions for Flood Emergency Risk Management) (2010). Available at <www.eu-safer.de>.

FEMA (Federal Emergency Management Association) (2007). Interagency policy review committee. September 2006. The National Levee Challenge: Levees and the FEMA Flood Map Modernization Initiative. Federal Emergency Management Agency. (published June 27, 2007).

Foley, N. (2007). Hidden Infrastructure" Contingency Today, May 2007. Available at <http://www.contingencytoday.com/online_article/Hidden-Infrastructure-/277> [Accessed November 2007].

Floodresilientcity (2010). Available at <www.floodresiliencity.eu> [Accessed September 7, 2010].

Gaillard, J. C. (2010). Religions, natural hazards, and disasters. *Religion,* 40(2), 82–131. Available at <http://www.sciencedirect.com/science/journal/0048721X> [Accessed May 27, 2010].

Galloway, G. E., Baecher, G. B., Plasencia, D., Coulton, K. G., Louthain, J., Bagha, M., and Levy, A. R. (2006). Evaluation of the National Flood Insurance Program: Assessing the Adequacy of the National Flood Insurance Program's 1 Percent Flood Standard. Water Policy Collaborative, University of Maryland for FEMA. Released for publication in May 2007.

Hall, P. (2008). *The siren call to 'densify or die' poses a false choice.* Town and Country Planning, April 2008, pp. 176–179.

Hayden, P. (2006). Management of Major Flood Events: Fire and Rescue Services Contribution to the Emergency Phase. Report for the Chief Fire Officers Association Board. Chief Fire Officers Association, England and Wales. Available at <http://www.hwfire.org.uk/PDF%20Files/CFOA%20Board%20Approved%20Document.pdf> [Accessed November 25, 2006].

Heath, S. (1999). *Animal Management in Disasters.* St. Louis, MO: Mosby Inc. 320 pp.

HMSO (Her Majesty's Stationery Office) (2003). Water Environment and Water Services (Scotland) Act 2003. 2003 asp 3. London: Her Majesty's Stationery Office.

Irish Government (2009). Planning Guideline 20. November 2009, Dublin.

Irish Insurance Federation (2010). Personal communication. April 2010.

Johnstonova, A. (2010). Meeting the Challenges of Implementing the Flood Risk Management (Scotland) Act 2009. RSPB Scotland, Edinburgh.

Kelman, I. (2007). Reliance on Structural Approaches Increases Disaster Risk. Version 1. July 26, 2007. Web-only publication available at <http://www.ilankelman.org/miscellany/StructuralDefences.rtf> [Accessed May 25, 2007].

Kunreuther, H. (2010). Not in my term of office. *Washington Post,* April 14, 2010. Available at <http://views.washingtonpost.com/leadership/panelists/2010/04/not-in-my-term-of-office.html> [Accessed May 28, 2010].

Lindholm, O. G., Schilling, W., and Crichton, D. (2007). Urban water management before the court: Flooding in Freidrikstad, Norway. *Journal of Water Law,* 17(5), 204–209.

Mellot, P. (2003). *Paris inonde* 208 pp. (French language version only). Paris: Editions de Lodi.

Milliband, D. (2006). Speech by Rt Hon David Miliband MP. Secretary of State for Environment, Food and Rural Affairs, at *The Association of British Insurers Annual Conference,* November 7, 2006.

Munich Re (2010). Flash Floods. Topics 01 2010. Munich Re, Munich. Available at <http://www.munichre.com/en/reinsurance/magazine/publications/default.aspx> [Accessed May 28, 2010].

Pfister, N. (2001). Community response to flood warnings: The case of an evacuation from Grafton, March 2001. *The Australian Journal of Emergency Management,* 17(2), 19–29.

Roaf, S., Crichton, D., and Nicol, F., (2009). *Adapting buildings and cities for climate change (second edition).* 38 pp. Oxford: Architectural Press.

Slovic, P. (2000). *Perception of Risk* (Risk, Society and Policy Series), Ragnar Lofstedt, Ed. 473 pp. London: Earthscan Publications Ltd.

Takeuchi, K. (2002). Flood management in Japan—From rivers to basins. *Water International,* 27(1), 20–26.

Tobin, G. A. (1997). The levee love affair: A stormy relationship. *Water Resources Bulletin,* 31(3), 359–367.

United Nations (2004). Guidelines for Reducing Flood Losses, United Nations, Geneva, Switzerland, <http://www.unisdr.org>

Vestri, P. (2007). Exploring the Take-Up of Home Contents Insurance, Hexagon Research and Consulting, Scottish Executive Social Research. (web-only publication; available at <www.scotland.gov.uk/socialresearch>).

Welsh Assembly Government (2007). New Approaches Programme 2007. Available at <http://wales.gov.uk/topics/environmentcountryside/epq/waterflooding/flooding/newapproaches/?lang=en> [Accessed May 24, 2010].

White, G. (1942). Human Adjustment to Floods. Research Paper No. 29, 1942, published 1945. Chicago: University of Chicago Department of Geography.

WWF (2007). *Slowing the Flow.* 16 pp. Dunkeld: WWF Scotland.

YouTube (2010a). Gilbert White in person speaking in a 1994 buyout and relocation seminar. Available at <http://www.youtube.com/watch?v=FTkHcH2yVqk> [Accessed May 25, 2010].

YouTube (2010b). Available at <http://www.youtube.com/watch?v=yHtL7uuaBFk&feature=related> [Accessed May 25, 2010].

13 Risk Management, Adaptation, and Monetary Aspects

Annegret H. Thieken, Holger Cammerer, and Clemens Pfurtscheller

CONTENTS

13.1 INTRODUCTION

Floods are responsible for about 20% to 30% of the economic losses caused by natural hazards worldwide (Douben and Ratnayake, 2005). It is expected that flood risk will rise in many parts of the world as a response to a combination of changing climate and, thus, altered flood frequencies (e.g., Kundzewicz et al., 2005) and an increase in vulnerability, for example, due to increasing floodplain occupancy, landcover changes, susceptible building materials, etc. Thus, there is an enhanced demand for better and reliable flood damage prevention. This can be interpreted as part of climate change adaptation that aims, according to the European Environmental Agency (EEA, 2008), at increasing the resilience of natural and human systems to current and future impacts of climate change such as flooding.

13.2 RISK MANAGEMENT AND ADAPTATION

In recent years, there has been a shift from technical flood protection to an integrated risk management that follows a structured approach that can be separated into four phases (e.g., Kienholz et al., 2004; Figure 13.1): (1) disaster response during a flood event, (2) recovery, (3) risk analysis and assessment, as well as (4) disaster risk reduction, which is primarily aimed at preventing and mitigating damage.

When a hazardous event occurs, immediate measures will be undertaken with the priority to limit both the adverse effects and the duration of the event (response phase). During recovery, the affected society starts to repair damage and tries to regain the pre-disaster standard of living. This phase is setting the stage for the society's next "disaster" (Olson, 2000): If the affected society is willing to learn from a disaster, there will be a period of disaster risk reduction in which measures that are aimed at minimizing the vulnerability of exposed people and their assets will be implemented. These measures can include structural (i.e., technical) as well as non-structural measures (e.g., spatial planning, early warning, private precaution, flood insurance, emergency control, etc.) (see Figure 13.1).

To enhance risk reduction, the disastrous event, society's response, as well as the performance of existing preventive and precautionary measures should be analyzed in the aftermath of an event in the framework of a risk analysis (Kienholz et al., 2004).

This cycle of disaster management has increasingly been used by international and national organizations and various versions have been published (e.g., PLANAT, 1998; Silver, 2001; DKKV, 2003; FEMA, 2004; Kienholz et al., 2004). A common understanding is that a thorough analysis and a subsequent assessment of risks is a prerequisite for effective risk reduction. Ideally, risk analysis should also take place without having experienced a severe disaster and the entire management process of risk analysis, assessment, and implementation of reduction measures should be accompanied by monitoring procedures. Especially in the context of global climate

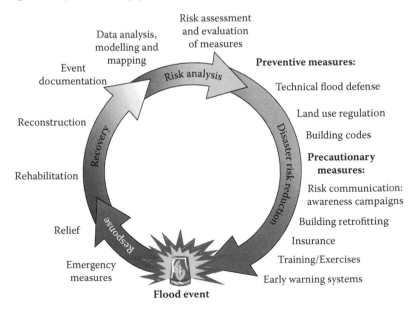

FIGURE 13.1   Flood risk management cycle.

change, updating risk analyses and adapting risk reduction measures are continuous tasks. This has already been adopted in the EU White Paper on Adaptation (EU, 2009), where adaptation is described as a long and continuous process, and in the EU Floods Directive: The preliminary flood risk assessments due in December 2011, the flood hazard and risk maps due in December 2013, as well as the flood risk management plans due in December 2015 shall be reviewed and—if necessary—updated by December 2018, 2019, and 2021, respectively, and every 6 years thereafter (EU, 2007).

At present, the experience of (extreme) flood events still plays a major role in stimulating risk management processes and adaptation measures, in particular for raising public awareness about different climate-related hazards and risks and their mitigation (e.g., Hinkel et al., 2010; Kreibich, Seifert et al., 2010). Adaptation is seen as a social learning process (Hinkel et al., 2010). After flood events there is a window of opportunity in which precautionary measures and preparative behavior at different levels of society—private households, companies, civil protection, administration, and policy— can be effectively strengthened. Examples are given in Thieken et al. (2007), Kreibich and Thieken (2009), Kreibich, Seifert et al. (2010), as well as in the next section.

13.3 CASE STUDY 1: INFLUENCE OF EXPERIENCED FLOODING IN 2005 ON ADAPTATION TO NATURAL HAZARDS OF ENTERPRISES IN AUSTRIA

From an entrepreneurial perspective, the perception of risks, their management, and planning of adaptation strategies are substantial factors of business continuity and success and consecutively for the regional or national welfare. Risk management on the entrepreneurial level should comprise technical prevention (technical or mobile mitigation measures), preparative measures to train adequate behavior in case of a threatening event, and ex-ante risk assessment (Seifert, 2008). Implementation of the management of natural risks is thus closely connected with the overall risk management strategy of a company, its institutional settings (Wittmann, 2000), as well as the perception of risks due to natural hazard processes within the organization.

To analyze the status quo of precaution and preparedness of enterprises against natural hazards as well as the role of flooding experienced in August 2005, a postal survey was carried out in 2009 among 1,300 companies in the Austrian provinces of Tyrol and Vorarlberg. The random sample contains 181 valid cases ($n_{returned}$ = 14%) with 90 variables. These cover the following topics: exposure and perception, adaptation and risk management measures, direct and indirect losses in 2005, drivers of indirect impacts and business interruption, international and national economic linkages, and the company's size and location as well as the economic sector.

Only 22% of the polled enterprises in Tyrol and Vorarlberg, Austria, stated that there was no exposure of the company to natural hazard processes. Nearly 68% estimated their exposure as light or moderate, and 10% evaluated the exposure as severe or highly severe. Because more than 60% of the polled companies have been located directly on or proximate to the waterfront, they are likely to be threatened by floods. Surprisingly, 45 of 175 private businesses (26%) also saw groundwater as a risk for their company. Other perceived threats were logjams, debris flows, landslides, under-cutting or -washing of infrastructure, and erosion.

The status quo of entrepreneurial mitigation and adaptation measures to natural hazard processes in the federal states of Tyrol and Vorarlberg as of 2009 (Figure 13.2) is as follows. Approximately 36% of 181 companies are insured against direct losses of corporate assets, nearly 30% of the polled enterprises are insured against business interruption due to natural hazards, and approximately 20% implemented technical mitigation measures against alpine hazards. To a lesser extent, security agents, emer-gency and evacuation plans, and to a much lesser extent a corporate fire brigade, risk audits of suppliers and customers, comprehensive risk management, and a checklist of threats were part of entrepreneurial prevention (Figure 13.2). However, some of these measures, especially the existence of security agents, emergency plans, and corporate fire brigades, primarily depend on the size of the company, due to indus-trial safety and legal obligations in the Federal Republic of Austria.

In fact, adaptation strategies to natural hazards are biased by institutional settings. Risk awareness and prevention planning depend on existent loss accounting and clear responsibilities. In the poll, approximately 47% of the SMEs (small and medium-sized enterprises with up to 249 employees) were characterized by a total absence of compe-tence and responsibility for natural hazard and risk management and, hence, act upon a "muddling-through" strategy, whereas 100% of the large-scale enterprises (with more than 250 employees) have implemented a risk management system or parts of it.

This status quo has been influenced by the severe flood event that took place in western Austria in August 2005 and the "lessons learned" by the companies. From

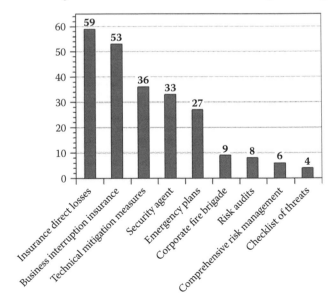

FIGURE 13.2 Number of enterprises in the federal states of Tyrol and Vorarlberg, Austria ($n_{total} = 181$), that implemented adaptation and prevention measures against natural disasters (multiple responses possible).

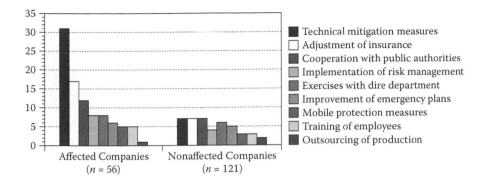

FIGURE 13.3 Number of enterprises in the federal states of Tyrol and Vorarlberg, Austria, that adapted their strategies to cope with natural hazards after the flood in 2005 (multiple responses possible).

177 polled companies, 56 were affected by the floods in 2005; that is, they experienced either direct or indirect losses, or both, and 121 were not affected. After 2005, 40 of 56 of the flood-affected enterprises (i.e., about 71%) adjusted parts or the total risk management strategy, while only 21 of 121 (i.e., approximately 17%) of the non-affected companies changed parts or the total risk management strategy.

In most cases, the affected companies concentrated their efforts on construction or upgrading technical mitigation measures to prevent damage (Figure 13.3). Seventeen companies adjusted their insurance and twelve improved cooperation with the public authorities responsible for flood and alpine risk mitigation, such as the regional departments of water management or the Austrian Torrent and Avalanche Control (TAC). On the contrary, the nonaffected companies only marginally adapted their coping strategies with natural hazards, despite the heavy impacts the floods in 2005 had on the local and regional economy. Very few enterprises implemented a comprehensive risk management strategy.

These results clearly highlight the large influence of flood experience on preventative and precautionary behavior. Siegrist and Gutscher (2008) found that this can be explained by the fact that people without flood experience definitely underestimate the negative effects associated with a flood. They thus concluded that risk communication must not focus solely on technical aspects, but should also help people envisage the negative (emotional) consequences of natural disasters. To further stimulate precautionary behavior of residents in flood-prone areas, Grothmann and Reusswig (2006) emphasized that it is essential to also communicate available measures that can be implemented by companies or homeowners, their costs, and their effectiveness. For this, methods for the quantification of damage and risk are necessary.

13.4 QUANTIFICATION OF FLOOD IMPACTS AND RISK

In their seminal paper, Kaplan and Garrick (1981) suggested a quantitative definition of risk that has found widespread use in risk analyses. The basic idea is that risk analyses should answer the following three questions:

1. What can happen? (What can go wrong?)
2. How likely is it that it will happen?
3. If it does happen, what are the consequences?

Consequently, flood risk analysis not only deals with the flood hazard (e.g., inundation scenarios) and flood frequencies, but also looks at potential flood impacts. Impacts of flooding can be generally classified into direct and indirect damage and losses. Direct damage occurs due to the physical contact of the floodwater with exposed objects or elements at risk, for example, humans, buildings, infrastructure, or any other objects. Indirect damage is induced by the direct impacts and occurs—in space or time—outside the flood event. It mainly results from an interruption of economic and social activities (Parker, Green, and Thompson, 1987). Both types of damages are further classified into tangible and intangible damage, depending on whether or not their quantity can be expressed in monetary values (e.g., Smith and Ward, 1998). Tangible damages are damage dimensions that can be easily specified in monetary terms, whereas intangible damage addresses impacts on goods and services that are not traded in a market and are thus difficult to assess by monetary values.

The quantification of flood impacts and associated risks becomes important when we proceed in the risk management process. During risk assessment it must be decided which kind and extent of risk can be accepted and which cannot (e.g., Petrascheck, 2003). In this context, risk is often defined as a loss or damage that will occur or be exceeded with a given probability—with a strong focus on direct damage. Past disasters demonstrate that absolute safety is neither achievable nor affordable. Thus, an acceptable level of risk must be defined or negotiated within a community at risk. In this process, risk reduction and adaptation measures are increasingly evaluated and compared by multicriteria, cost effectiveness, or cost-benefit analyses (e.g., Hinkel et al., 2010); that is, the benefits of a risk reduction measure—assessed in terms of avoided damage and losses during the lifetime of the measure—must exceed its costs. This idea of optimizing investments is illustrated in Figure 13.4.

Evaluation of measures will gain even more importance with regard to adaptation to climate change. Good adaptation practices should be appropriate, proportionate, and cost effective in the long term (EEA, 2008). According to the EU White Paper on Adaptation (EU, 2009), priority should be given to adaptation measures that would generate net social or economic benefits irrespective of the uncertainty in future scenarios—these are so-called no-regret measures.

Altogether, monetary assessments and methods for the quantification of flood damage and losses are becoming increasingly important because they are needed for a number of tasks in the flood risk management process, particularly for supporting optimal decisions on flood risk reduction measures, flood risk mapping, and financial appraisals during and after flood events and in the (re-)insurance sector. In most cases, flood damage is currently restricted to direct damage.

13.5   MONETARY ASSESSMENT OF DIRECT FLOOD DAMAGE

In principle, the total (direct) costs of a disaster could be calculated on the basis of loss compensation for property owners (residential, commercial, industrial, and

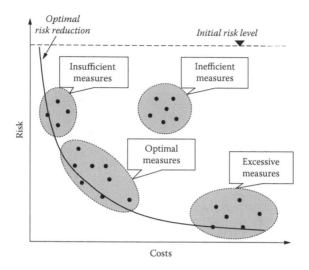

FIGURE 13.4 Risk–Cost-Diagram for assessing the efficiency of risk reduction measures. (Adapted from Einstein, H. H. (1997). Landslide risk—Systematic approaches to assessment and management. In Cruden, D. and Fell, R. (Eds.), *Landslide Risk Assessment.* Rotterdam: Balkema.)

agricultural) and costs for repairing damaged public infrastructure by summing up the repair costs spent on the individual items. However, it depends on the problem or (research) question at hand as to how damage is correctly assessed in monetary terms. Table 13.1 summarizes the two important aspects of the monetary evaluation of flood losses: (1) the asset volume (replacement value or depreciated value) and (2) the price level. Damage compensation and repair work are normally based on the replacement value of the damaged objects; that is, value as new, and represent the financial costs of a disaster that are relevant for, for example, insurance. For economic assessments (e.g., for damage evaluation in the context of cost–benefit analyses of mitigation projects), damage must, however, be assessed by depreciated value, which takes into account the current state of maintenance of the object under study and its present value (e.g., Penning-Rowsell et al., 2005; Van der Veen and Logtmeijer, 2005; Messner et al., 2006). Depreciated values are often roughly only half the replacement values (Penning-Rowsell et al., 2005).

TABLE 13.1

Monetary Evaluation of Direct Tangible Evidence

Asset/Stock ⟍ Prices	Past	Present
Gross (past)	Acquisition costs (historic costs)	Replacement value in current prices
Net (present)		Depreciated value in current prices (present value)

Moreover, prices are changing in time. Hence, the reference year of costs must be reported so that it is possible to correct data from different years by accounting for inflation. Data and assessments from different years can be referenced to the price level of one particular year by prices indices published by statistical agencies.

Due to the basic differences in the monetary evaluation and their different application purposes, it is a must that details about the underlying monetary assessment are documented when actual flood losses are collected or when potential flood losses are estimated. Despite these methodological difficulties, quantitative estimates of direct damage allow us—among other things—to analyze and quantify the amount of damage reduction due to different mitigation measures. This is illustrated by the following case study.

13.6 CASE STUDY 2: THE INFLUENCE OF FLOOD EXPERIENCE, PRECAUTIONARY ADAPTIVE BEHAVIOR, AND EARLY WARNING ON FLOOD LOSSES OF PRIVATE HOUSEHOLDS

Flood experience and consecutive adaptive (i.e., precautionary) behavior of private households can strongly influence the extent of building and content damage in the case of recurring floods. Several investigations have indicated that damage mitigation due to private precaution and appropriate response to flood warnings can be considerable (Smith, 1981, 1994; Wind et al., 1999; ICPR, 2002; Kreibich et al., 2007a,b). For example, Wind et al. (1999) analyzed the 1993 and 1995 Meuse floods in the Netherlands and detected that flood losses in the residential sector were 35% lower in 1995 than in 1993 although flood volumes and inundated areas were of comparable magnitude. This reduction was explained by a marginal increase in forecast lead times and experience gained from the flood in 1993. Similarly, Smith (1981) stated that the actual damages in private households from the 1974 flood in Lismore (Australia) were only 52.4% of the potential flood losses due to frequently experienced flooding and longer forecast lead times. According to Brilly and Polic (2005) as well as Grothmann and Reusswig (2006), households with recent flood experiences are more aware of flood risks and tend to better invest in precautionary measures than those who have never been struck by a flood or only a long time ago. To further quantify risk reduction effects, this case study analyzes the precautionary behavior and responses to flood warning within different flood-experienced private households and the effects on the amount of damage.

The analysis is based on four campaigns with computer-aided telephone interviews undertaken in private households in Austria and Germany in the aftermath of the severe flood events in Central Europe in 2002, 2005, and 2006, as shown in Figure 13.5. The most comprehensive telephone survey was conducted in Germany in 2003 among 1,697 private households in Bavaria, Saxony, and Saxony-Anhalt that were affected by flooding in August 2002. This survey focused on flood losses on buildings and contents and also on potential loss influencing factors like water level, flood duration, precautionary behavior, contamination, etc. (for details, see Thieken et al., 2005, 2007). Another poll with 605 interviewees aimed at analyzing loss-influencing factors in the case of high groundwater levels in Dresden, Germany

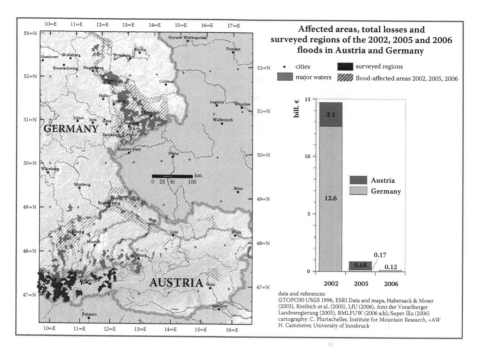

FIGURE 13.5 Affected areas, total losses, and surveyed regions of the 2002, 2005, and 2006 floods in Austria and Germany.

(Kreibich et al., 2009). Further, 461 households were interviewed after they were hit by floods in Bavaria in August 2005 or along the Elbe in March/April 2006 (Kreibich and Thieken, 2009). The poll in Austria (219 interviews) investigated the insurance structure in private households in Tyrol and Vorarlberg after the severe floods in August 2005 and also contained questions with regard to building and content losses and precautionary behavior (Raschky et al., 2009). Because all four of these surveys contained building and content damages and questions on potential flood loss influencing factors (i.e., hydrological parameters like water depth and flood duration, early warning, precautionary and emergency measures, contamination, flood experience, building characteristics), it was possible to join these surveys to make one consistent database (a total of 2,982 interviews). Additionally, the flood losses as well as the loss ratios for buildings and contents were scaled to the reference year 2007 and were derived for various flood-experienced groups (i.e., no flood experience before; once, twice, and more, or past flood experience less than 2 years ago, 2 to 6 years, longer than 6 years). Significant differences between two or more independent samples were tested by the Mann-Whitney U-test or the Kruskal-Wallis H-test, respectively.

In general, households without any flood experience have the largest flood losses, in contrast to those who experienced one or more damaging event in the past. The loss ratio on buildings of the inexperienced group is significantly higher (1.7 times, or €12,782) in comparison to the households being affected already once before and 2.8 times higher (€15,756) in comparison to those with two or more prior damaging events.

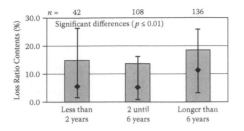 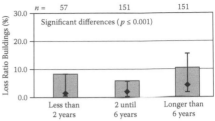

FIGURE 13.6   Differences between the loss ratios of contents (left) and buildings (right) dependent on the time since the last flood event occurred (bars = means, points = medians, and 25th to 75th percentiles).

Data from households with at least one previously experienced flood were further subdivided considering the different time periods since the last flood occurred (Figure 13.6). This temporal classification shows that both building losses and content losses rise significantly with increasing time gap. While the monetary difference between the groups "less than 2 years" and "2 to 6 years" is relatively low, the flood damage in the third group ("longer than 6 years") jumps up remarkably. The loss ratios of the latter are 2.8 times (€11,115) higher for buildings and 2.2 times higher (€1,512) for contents than the respective figures in the class "less than 2 years."

To find out why households with no or lapsed flood experience have higher flood losses, all subgroups were further divided by different forecast lead times and several precautionary measures (e.g., flood-adapted building use, protection of the oil heating).

The results of the analysis of various forecast lead times were very contradictory. When affected households received the warning 2 to 48 hours before inundation, building losses decreased significantly in all subgroups in comparison with cases with only one-hour forecast lead time. However, when the warning time amounts to more than 48 hours, the building losses rose surprisingly in all classes (by a factor 1.7 to 5; see Table 13.2).

The answers to the question, "Do you know how to protect your household and yourself due to the official flood warning?" may provide some reasoning for these unexpected results. In general, people who knew how to behave in case of inundation after having received the warning had significantly lower building damage ratios (by approximately 60 to 70%). For example, the subgroup that experienced the last flood more than 6 years ago had, in fact, 81% (€28,898) lower building loss ratios when they were informed on how to protect themselves compared to the uninformed subgroup. These outcomes illustrate that flood loss mitigation due to flood warning does not only depend on longer forecast lead times but also on a great sharing of the knowledge of the people who have to respond appropriately. Furthermore, the last result supports the statement of the International Commission for the Protection of the Rhine (ICPR, 2002) that, after 7 years without any flood event, the preparedness of people is almost faded so that they do not know how to behave in case of an emergency (with high damage being the consequence). For this reason it is absolutely essential to release flood warnings together with information about adequate—that is, damage-reducing—behavior.

Permanently installed precautionary measures can be crucial in mitigating flood losses and can compensate for missing or fading flood experience to a large degree.

TABLE 13.2

Loss Ratios and Absolute Building Damage Distinguished by Differences in Flood Experience and Forecast Lead Times

#		Loss Ratio Buildings (in %)			Total Building Losses (in €)		
Never	Warning time	Up to 1 h	2 to 48 h	Over 48 h	Up to 1 h	2 to 48 h	Over 48 h
	Sample size	66	298	118	78	345	133
	Mean value	11.9	12.2	20.2	50,911	46,623	63,375
	Median	6.2	5.4	16.6	27,481	22,229	55,573
	Significance		0.000			0.000	
Once	Warning time	Up to 1 h	2 to 48 h	Over 48 h	Up to 1 h	2 to 48 h	Over 48 h
	Sample size	19	94	18	25	115	23
	Mean value	13.6	6.2	15.9	45,620	24,946	64,671
	Median	4.5	2.4	7.6	32,610	11,115	43,480
	Significance		0.022			0.009	
≥ Twice	Warning time	Up to 1 h	2 to 48 h	Over 48 h	Up to 1 h	2 to 48 h	Over 48 h
	Sample size	27	96	30	30	111	33
	Mean value	5.8	5.1	15.0	21,798	22,489	45,895
	Median	2.3	1.7	11.3	10,670	7,467	33,344
	Significance	0.001				0.019	

Note: € = Euro; # = number of previously experienced floods.

In our data set, people with no flood experience had the highest loss mitigation potential for buildings (85%, or €24,452) and for contents (94%, or €5,040) when they undertook private precautions before the flood event when compared to those with no or only marginally precautionary behavior. Even if the last flood occurred more than six years ago, well-installed precautionary measures reduced building damage by 74% (€22,229) and content damage by 70% (€7,056; see Figure 13.7). Among the single precautionary measures, flood-adapted use (i.e., no use as office or living room) and flood-adapted interior fitting (i.e., the usage of water-repellent materials such as a tile floor instead of carpet or parquet) of the flood-exposed stories were the most effective. This illustrates those private precautionary measures that do not depend so much on the behavior of the affected residents in case of the event,

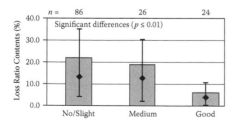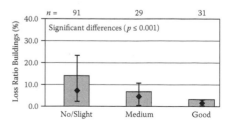

FIGURE 13.7 Differences between the loss ratios of contents (left) and buildings (right) dependent on the precautionary level in the group "longer than 6 years" (bars = means, points = medians, and 25th to 75th percentiles).

that help to reduce flood damage in the longer term. In the future, such analyses can be used to evaluate the cost effectiveness of precautionary measures at the household level as started by Kreibich, Schwarze, and Christenberger (2010).

13.7 CONCLUSIONS

Adaptation as an effort to better cope with floods on the levels of companies and households is greatly influenced by personally experienced impacts. The case study in the Austrian provinces of Tyrol and Vorarlberg showed that enterprises tend to follow a minimalism strategy in terms of technical mitigation measures and to neglect comprehensive risk management. One of the reasons for this could be the kind of public risk transfer system in Austria, which provides nearly no incentives for private (and entrepreneurial) prevention measures and adaptation (Raschky and Weck-Hannemann, 2008). Despite the awareness of risks and threatening processes and the fact that nearly every second company in western Austria is located near a waterfront, risk management strategies to optimally reduce direct and indirect losses at the entrepreneurial level are still missing.

Case Study 2 showed that private households always had huge damages when they had no or only lapsed flood experience before. Despite all that, the results revealed that it is possible to keep damages low by building in precaution and response to flood warnings as long it is clear how to behave adequately. Therefore, it is essential to propagate long-term precaution and inform people well about appropriate emergency measures, especially in regions where flood risk awareness seems to have faded.

REFERENCES

Amt der Vorarlberger Landesregierung (2005). Das Starkregen- und Hochwasserereignis des August 2005 in Vorarlberg, Feldkirch.

BMLFUW (Bundesministerium fur Land-und Forstwirtschaft, Umwelt und Wasserwirtschaft) (2006a). Hochwasser 2005—Ereignisdokumentation, Teilbericht der Wildbach- und Lawinenverbauung, Vienna. 18

BMLFUW (Bundesministerium fur Land-und Forstwirtschaft, Umwelt und Wasserwirtschaft) (2006b). Hochwasser 2005 -Ereignisdokumentation der Bundeswasserbauverwaltung, des Forsttechnischen Dienstes für Wildbach- und Lawinenverbauung und des Hydrographischen Dienstes, Vienna.

Brilly, M., and Polic, M. (2005). Public perception of flood risk, flood forecasting and mitigation. *Natural Hazards and Earth System Sciences,* 5, 345–355.

DKKV (Deutsches Komitee für Katastrophenvorsorge) (2003). Hochwasservorsorge in Deutschland—Lernen aus der Katastrophe 2002 im Elbegebiet, Lessons learned. Schriftenreihe des DKKV 29, Bonn.

Douben, N., and Ratnayake, R. M. W. (2005). Characteristic data on river floods and flooding: facts and figures. In J. van Alphen, E. van Beek, and M. Taal (Eds.), Floods, from defence to management. London: Taylor and Francis Group, pp. 11–27.

EEA (European Environmental Agency) (2008). Impacts of Europe's Changing Climate— 2008 Indicator-Based Assessment. EEA-Report 4/2008, Chapter 6, pp. 161–166.

Einstein, H. H. (1997). Landslide risk — Systematic approaches to assessment and management. In Cruden, D., and Fell, R. (Eds.), *Landslide risk assessment.* Rotterdam: Balkema, Rotterdam. pp. 25–50.

EU – Commission of the European Communities (2007). Directive 2007/60/EC of the European Parliament and of the Council of 23 October 2007 on the assessment and management of flood risks. *Official Journal of the European Union*, L288, 27–34.

EU – Commission of the European Communities (2009). White Paper—Adapting to climate change: Towards a European framework for action. Brussels 1.04.2009, 147, 16 pp.

FEMA (Federal Emergency Management Agency) (2004). About FEMA—What We Do. Available at <http://www.fema.gov/about/what.shtm> [Accessed on October 28, 2004].

Grothmann, T., and Reusswig, F. (2006). People at risk of flooding: Why some residents take precautionary action while others do not. *Natural Hazards*, 38, 101–120.

Habersack, H., and Moser, A., Eds. (2003). Ereignisdokumentation Hochwasser August 2002. Vienna: Plattform Hochwasser.

Hinkel, J., Bisaro, S., Downing, T. E., Hofmann, M E., Lonsdale, K., McEvoy, D., and Tabara, J. D. (2010). Learning to adapt: Re-framing climate change adaptation. In Hulme, M., and Neufeldt, H., Eds., *Making Climate Change Work for Us*. Cambridge: Cambridge University Press. pp. 113–134.

ICPR (International Commission for the Protection of the Rhine) (2002). Non-structural Flood Plain Management—Measures and Their Effectiveness. Koblenz: ICPR.

Kaplan, S., and Garrick, B. J. (1981). On the quantitative definition of risk. *Risk Analysis*, 1(1), 11–27.

Kienholz, H., Krummenacher, B., Kipfer, A., and Perret, S. (2004). Aspects of integral risk management in practice—Considerations with respect to mountain hazards in Switzerland. *Österreichische Wasser- und Abfallwirtschaft*, 56, 43–50.

Kreibich, H., Merz, B., and Grünewald, U. (2007a). Lessons learned from the Elbe River floods in August 2002 — With a special focus on flood warning. In Vasiliev, O., Gelder, P., Plate, E., and Bolgov, M. (Eds.), *Extreme Hydrological Events: New Concepts for Security*. Berlin: Springer.

Kreibich, H, Müller, H., Thieken, A., and Merz, B. (2007b). Flood precaution of companies and their ability to cope with the flood in August 2002 in Saxony, Germany. *Water Resources Research*, 43, doi: 10.1029/2005WR004691, W03408.

Kreibich, H., Schwarze, R., and Christenberger, S. (2010). Nutzen und Kosten privater Hochwasservorsorge. In Thieken, A. Seifert, I., and B. Merz, B., Eds., *Hochwasserschäden Erfassung, Abschätzung und Vermeidung*. Munich: Oekom-Verlag. pp. 263–270.

Kreibich, H., Seifert, I., Thieken, A. H., Lindquist, E., Wagner, K., and Merz, B. (2010). Recent changes in flood preparedness of private households and businesses in Germany. *Regional Environmental Change*, doi 10.1007/s10113-010-0119-3.

Kreibich, H., and Thieken, A. (2009). Coping with floods in the city of Dresden, Germany. *Natural Hazards*, 51(3), 423–436, doi: 10.1007/s11069-007-9200-8.

Kreibich, H., Thieken, A., Grunenberg, H., Ullrich, K., and Sommer, T. (2009). Extent, perception and mitigation of damage due to high groundwater levels in the city of Dresden, Germany. *Natural Hazards and Earth System Sciences*, 9, 1247–1258.

Kreibich, H., Thieken, A., Petrow, T. Müller, M., and Merz, B. (2005). Flood loss reduction of private households due to building precautionary measures. Lessons learned from the Elbe floods in August 2002. *Natural Hazards and Earth System Sciences*, 5, 117–126.

Kundzewicz, Z. W., Ulbrich, U., Brücher, T., Graczyk, D., Krüger, A., Leckebusch, G. C., Menzel, L., Pinskwar, I., Radziejewski, M., and Szwed, M. (2005). Summer floods in central Europe—Climate change track? *Natural Hazards*, 36, 165–189.

LfU (Landesamt für Umwelt) (2006). Endbericht August-Hochwasser 2005 in Südbayern. Available at <http://www.hnd.bayern.de/ereignisse/hw220805/hw200508_endbericht.pdf>.

Messner, F., Penning-Rowsell, E., Green, C., Meyer, V., Tunstall, S., and Van der Veen, A. (2006). Evaluating flood damages: Guidance and recommendations on principles and methods. Project Floodsite. Available at <http://www.floodsite.net/html/partner_area/search_results3b.asp?docID=293>, [Accessed April 2010].

Olson, R. S. (2000). Toward a politics of disaster: Losses, values, agendas, and blame. *International Journal of Mass Emergencies and Disasters*, 18, 265–287.

Parker, D. J., Green, C. H., and Thompson, P. M. (1987). *Urban Flood Protection Benefits: A Project Appraisal Guide*. Aldershot: Gower Technical Press. 284 pp.

Penning-Rowsell, E., Johnson, C., Tunstall, S., Tapsell, S., Morris, J., Chatterton, J., and Green, C. (2005). *The Benefits of Flood and Coastal Risk Management: A Manual of Assessment Techniques*. Middlesex: Middlesex University Press. 238 pp.

Petrascheck, A. (2003). Brauchen wir den Paradigmenwechsel vom Hochwasserschutz zum Risikomanagement? In Grünewald, U. (Hrsg.): Hochwasservorsorge in Deutschland— Stand, Defizite, Konzepte, Tagungsband zum Erfahrungsaustausch 8./9.7.2003 in Potsdam, Cottbus, S. pp. 38–41.

PLANAT (1998). Plattform Naturgefahren: Von der Gefahrenabwehr zur Risikokultur. Broschüre zur nationalen Plattform Naturgefahren. Bern, Landeshydrologie und -geologie.

Raschky, P., Schwarze, R., Schwindt, M., and Weck-Hannemann, H. (2009). Alternative Financing and Insurance Solutions for Natural Hazards—A Comparison of Different Risk Transfer Systems in Three Countries—Germany, Austria and Switzerland— Affected by the August 2005 Floods. Bern: Public Insurance Companies for Buildings, Prevention Foundation.

Raschky, P., and Weck-Hannemann, H. (2008). Vor- oder Nachsorge? Ökonomische Perspektiven. In Glade, T., and Felgentreff, C. (Eds.), *Naturrisiken und Sozialkatastrophen,* Berlin, pp. 269–280.

Seifert, I. (2008). Risikomanagement von Unternehmen bei Hochwasser. http://digbib.ubka. uni-karlsruhe.de/volltexte/1000009285, Ph.D. thesis, University of Karlsruhe.

Siegrist M., and Gutscher H. (2008). Natural hazards and motivation for mitigation behavior: People cannot predict the affect evoked by a severe flood. *Risk Analysis*, 28(3), 771–778.

Silver, M. L. (2001). International best practices in disaster mitigation and management recommended for Mongolia. UNDP. *Disaster Management Conference.* December 2001, Mongolia, pp. 1–9.

Smith, D. (1981). Actual and potential flood damage: A case study for urban Lismore, NSW, Australia. *Applied Geography,* 1, 31–39.

Smith, D. (1994). Flood damage estimation—A review of urban stage-damage curves and loss functions. *Water SA,* 20, 231–238.

Smith, K., and Ward, R. (1998). *Floods: Physical Processes and Human Impact.* Chichester: John Wiley & Sons.

Super Illu (2006). Ausgabe vom 12.04.2006. Berlin: Superillu Verlag GmbH and Co. KG.

Thieken, A., Kreibich, H., Müller, M., and Merz, B. (2005). Flood damage and influencing factors: New insights from the August 2002 flood in Germany. *Water Resources Research*, 41, W12430, doi: 10.1029/2005WR004177.

Thieken, A., Kreibich, H., Müller, M., and Merz, B. (2007). Coping with floods: Preparedness, response and recovery of flood-affected residents in Germany in 2002. *Hydrological Sciences Journal*, 52, 1016–1037.

Van der Veen, A., and Logtmeijer, Ch. (2005). Economic hotspots: Visualizing vulnerability to flooding. *Natural Hazards*, 36, 65–80.

Wind, H., Nierop, T., de Blois, C., and de Kok, J. (1999). Analysis of flood damages from the 1993 and 1995 Meuse flood. *Water Resources Research,* 35, 3459–3465.

Wittmann, E. (2000). Risikomanagement in internationalen Konzernen. In D. Dörner, P. Horváth, and H. Kagermann (Eds.), *Praxis des Risikomanagements: Grundlagen, Kategorien, branchenspezifische und strukturelle* Stuttgart: Aspekte. pp. 789–820.

14 Blue Space Thinking

Robert Barker

CONTENTS

14.1   INTRODUCTION

Flood risk is set to be an increasing problem throughout the world. Urbanization of rainfall catchment areas has increased surface water runoff rates and resulting river flows far beyond the rates of the natural environment. Climate change is expected to increase peak rainfall, river flows, winter groundwater levels, mean sea levels, and peak wave heights—all of which will put increased strain on the ability of existing infrastructure to cope with storms.

The percentage of the world population living in urban areas has risen from 14% in 1900 to just over 50% in 2008 and is expected to rise to 70% by 2050 (Population Reference Bureau, 2010). In developed countries, where the urban population is already at 74%, the cities and towns have been designed to cope with less extreme weather occurrences than predicted for the future. Therefore, the infrastructure is likely to be overwhelmed more frequently, resulting in increased flooding.

14.2   FLOOD MANAGEMENT

Typically, local and national governments are responsible for designing the infrastructure to manage flood risk. In turn, it has been the role of engineers to develop the solutions, which for the most part have been civil engineering solutions. These works have dramatically altered the natural environment to create an artificial, anthropogenic landscape. This landscape changes throughout countries and counties but there are many recognizable features associated with or contributing to flood risk management.

In rural areas, regularized field patterns, sometimes created by engineered drainage networks typical of the Netherlands, combined with agricultural land use, affect the runoff rates from rainfall. Embanked river edges prevent the river from overtopping its banks and spilling onto the land. They also prevent the natural meander of river courses across the floodplain. River channels have often been rerouted or straightened to remove the natural winding meander and decrease the navigation time, which equally increases the rate at which floodwater passes down the channel. In some cases, bypass channels have been created to avoid using the river altogether. In 2005, the storm surge from Hurricane Katrina also used the river bypass (Mississippi River-Gulf Outlet, or MR-GO) to flow directly from the Gulf of Mexico into New Orleans (Figure 14.1). Artificial flood storage areas and wetlands, such as the Nene Washes, have been created to store floodwater away from urban areas.

In urban areas, brick, concrete, and sheet piling embankments have been used to regularize flows and protect land from flooding. Over time and with subsequently higher floods, these embankments have been raised bit by bit, often canalizing the river channel meters below the top of the embankments or obscuring the view of the river behind the embankments. This has disconnected the land from the river and diminished the relationship between local inhabitants and the river.

In urbanized areas, underground drainage systems are typically relied upon to carry rainwater away from properties. Although systems have been designed to incorporate built-in redundancy (particularly the London sewer system, which with

FIGURE 14.1 MR-GO river bypass canal and view toward the breach location and Lower 9th Ward.

some modifications has served Londoners for almost 150 years), many of the original systems were planned for smaller populations than they cater to now and cannot always be readily upgraded for future growth projections. With growth that has already occurred, pockets of permeable green land within the city have contracted or been developed and covered by impermeable tile roofs, concrete pavement, and tarmac roads. Furthermore, drainage systems have been designed to cope with projected 1 in 30-years storms and not the more extreme storms that make newspaper headlines when they overwhelm our river courses. In these big storms, sewers have often been overwhelmed by the inundation of rainwater.

14.3   PRECEDENT CHANGE

This medley of civil engineering work has transformed our landscape; it has shaped the environment we live in and, for the most part, created a safer environment—until now. The advent of climate change has led policymakers to question whether this hard-engineered approach to manage flooding is the most appropriate (Defra, 2005). The uncertainty of the scale of future of flooding, combined with the difficulty in managing or controlling floods, has resulted in a shift in emphasis toward reducing and managing risk rather than eliminating it altogether. This leads us to question the traditional civil engineering approach:

- Should we keep raising the height of river embankments? If so, to what extent? And how will this affect the residual risk posed by failure or breach of the embankments?
- Is it cost effective to increase the capacity of our underground drainage systems?
- Can we increase the capacity of rivers, roads, and landscape to absorb water and reduce the risk of flooding in a more naturalistic way?

And if we do create more flood storage areas, do we have sufficient space to still enable growth, produce food, and grow fuel?

Architects, planners, and engineers have also asked questions:

- Do we want to, or have to, live behind concrete walls?
- Can we live with occasional flooding?
- Can we design our landscape and our buildings to cope with flooding without significant disruption to our daily lives?
- Can we reshape our towns and cities to cope with flooding and at the same time create more beautiful and sustainable places to live?

14.4   THE UNITED KINGDOM

Baca Architects (headquartered in London, England) identified that the construction and planning industries have a role to play in reducing flood risk. New development and urbanization have tended to increase the amount of impermeable surfaces and increase drainage into our sewer networks, thereby increasing runoff rates and putting greater pressure on our sewage infrastructure, respectively. UK planning policy states that 60% of all new homes should be built on previously developed (Brownfield) land. While this provides an opportunity to clean up and reuse redundant sites, many previously developed land areas are located on low-lying land close to river courses, the sea, or other areas susceptible to flooding.

New development can be designed to improve the ability of urban areas to absorb water, decrease existing runoff rates (particularly on Brownfield land), separate and recycle waste water treatment, and, through redevelopment, simultaneously pay for flood risk management measures that would not occur otherwise.

14.5   INTEGRATED APPROACH

The main drivers for change in the way development is planned in the twenty-first century are going to be sustainability and climate change. These are manifest in the need to make developments zero carbon, conserve water, reduce unsustainable transport use (namely, cars, including electric or biodiesel cars unless powered from renewable sources), reduce overheating, and manage flood risk. There are potential benefits from measures introduced to respond to these issues but understanding which to prioritize is fundamental to good planning. Among these issues, flooding is the main risk to life and therefore a high priority. Simultaneous assessment of the other requirements helps to develop integrated proposals.

Baca Architects has been developing spatial planning in the United Kingdom and the Netherlands to implement these improvements and enhance our urban environments to cope with future extreme weather resulting from climate change. In parallel with adaptation to climate change, Baca has sought to identify opportunities to introduce renewable energy, decrease water usage, and encourage more sustainable transport to minimize carbon emissions and mitigate climate change. Some of the measures and principles are described in the following projects.

14.6   LifE PROJECT

In 2005, Baca Architects developed the LifE (Long-term initiatives for flood-risk Environments) Project. The aim of this project was to create a more integrated approach to development that sought to reduce the cause of climate change (carbon emissions), while at the same time reducing the effects of climate change (particularly flood risk) through the creation of responsible sustainable development. The work sought to find appropriate development solutions (whether it be for recreation, farming, industry, or housing) that respond to the flood risk and likely impacts of climate change, rather than prohibiting all development. This was to be a positive approach, enabling development and growth to occur in accordance with planning policy but also to create better place making and planning, using flood risk and climate change as a driver.

The three key principles of this approach (illustrated in Figure 14.2) are

1. *Living with water:* Establishing safe development in water environments by managing risk. Adapting to increased flood frequency and severity, likely to happen with climate change.
2. *Making space for water:* Working with natural processes to provide room for rain to fall, the river to grow, and the sea to expand in times of flood, and reduce reliance on defenses, where possible.
3. *Zero carbon:* Providing all energy needs from renewable sources on site, such as wind, tidal, and solar power.

Emphasis is placed on nondefensive flood risk management measures on the basis that they may reestablish floodplains and the natural capacity of the land to store floodwater. This has the benefit of re-creating a connection with the natural water cycle, and coping with flooding in exceedance of engineering design standards. This may be particularly important when the future effects of climate change are still only estimated. Instead of keeping water out of sites, space for water is provided within developments and water permitted on to sites in a controlled and predetermined manner. This may not be appropriate if the floodwater is deep or fast flowing.

Funding by Defra was used to evolve this approach, test it through master planning three sites in the United Kingdom, and establish design principles for other sites at risk of flooding. Baca teamed up with the BRE, Halcrow, Cyril Sweett, Fulcrum Consulting, and LDA Design to form a multidisciplinary team that could test all aspects of planning for the three sites. The three sites were chosen on the basis of their location within the water catchment. Site 1 was located in the upper catchment of the River Wandle, Site 2 was located in the middle catchment of the River Nene, and Site 3 was located in the lower catchment of the River Arun.

A detailed study of all three sites was carried out, including current and future flood risk. The team carried out sensitivity testing of the future flood risk to examine

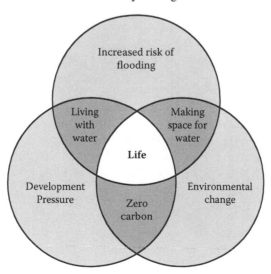

FIGURE 14.2 Integrated approach to design.

the impacts of future flooding if they were worse than the guidance given by government. For example, government guidance when looking at 100 years in the future is to examine the effect of a 30% increase in peak rainfall, a 20% increase in peak river flows, and approximately one-meter rise in sea level (DCLG, 2006). On each of the sites, measures in excess of these figures were explored to identify if and how that might change the development approach to be carried out in the immediate future.

Each site was master planned to establish the potential use, mix, layout, and scale to determine the spatial and infrastructure demands to create a sustainable development plan. At each site, the area required for amenity space, renewable power sources, and space for water far exceeded the site area, even at relatively low development densities. This highlighted the difficulty in creating sustainable master plans within urban areas even before making additional space for water. The solution at each site was to establish relatively dense areas of development, within or divided by a series of multifunctional landscape areas.

The heavily urbanized catchment of Site 1 meant that the rainfall would run off the land quickly, causing localized flooding and potentially overwhelming the sewers and river. Development was organized around rain courtyards, green rainfall storage areas surrounded by buildings. Figure 14.3 shows how a local play area would be set

FIGURE 14.3 Multifunctional rain courtyards (from the LifE Project Site 1). (From Baca Architects with BRE (2009). *The LifE Handbook*. Watford, UK: IHS BRE Press. With permission.)

within the rain courtyard and slightly raised above the surrounding area to keep dry during more frequent storms.

At Site 2, the river was the main source of flooding, with predicted increased river flows potentially affecting almost the entire site area (approximately 50 hectares) during an extreme flood (Baca Architects and BRE, 2009c). Conveyance corridors and green drainage channels were introduced between compact safe development zones (Figure 14.4) to allow water to pass through the site during a flood. At other times, these corridors would provide public amenity space, private gardens, allotments, and space to locate small-scale wind turbines, set back from the development to avoid issues of noise or vibration.

In Site 3, the main source of flooding would be from the tidal river. The volume of water would be to make space for water within the development site (Baca Architects and BRE, 2009d). Therefore, the proposal was to divert water into large flood storage areas, such as wetlands or meadows, which would create a fantastic opportunity for habitat improvement. These storage areas would also provide the opportunity to locate an array of highly efficient 2 megawatt wind turbines away from development. Development set behind river flood defenses could still be affected by pluvial or groundwater flooding. Within the urbanized development area, public squares, an amphitheatre, a skate park, and other play areas (Figure 14.5) are designed to store

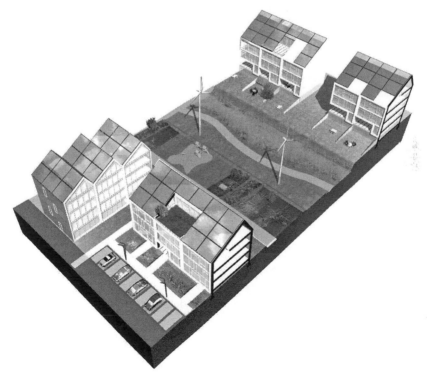

FIGURE 14.4 Multifunctional floodplain corridors (from the LifE Project Site 2). (From Baca Architects with BRE (2009). LifE Report, Appendix A. Watford, UK: IHS BRE Press. With permission.)

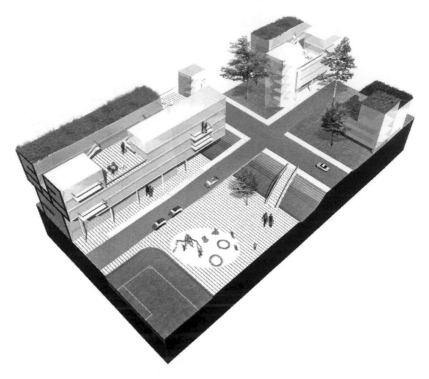

FIGURE 14.5 Multifunctional urban square (from the LifE Project Site 3). (From Baca Architects with BRE (2009). LifE Report, Appendix B. Watford, UK: IHS BRE Press. With permission.)

rainwater. The river defenses were also designed to provide space for intertidal habitat, promenading, and planting to create an attractive riverside feature.

The findings from the LifE Project have been published in the *LifE Handbook* and *LifE Report*, both available from BRE Press. They are also influencing a number of projects that Baca Architects are working on in the United Kingdom and the Netherlands.

14.7   EILAND VEUR LENT AND NIJMEGEN, THE NETHERLANDS

The land to the north of Nijmegen is the site for one of forty major projects in the Netherlands to make "Room for the River." Extensive dyke relocation work is programmed along the River Waal to reduce peak water levels along the river course by approximately 270 millimeters. Approximately one kilometer of dykes will be realigned to create a secondary flood channel, leaving the existing dyke as a raised peninsula between the river and the new flood channel. Behind the new dyke, a development of some 30,000 homes has been planned. Figure 14.6 indicates the flood relief channel in the context of the City of Nijmegen to the south and Lent to the north.

The engineered banks of the flood channel could have been designed using rock revetments, as is common to many dykes in the Netherlands. However, the Nijmegen

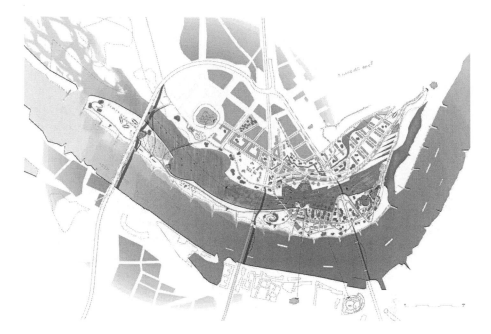

FIGURE 14.6 Plan of the flood relief channel in Nijmegen and Lent indicating Baca Architects' proposals to enhance the channel and surrounding landscape.

Municipality invited a number of international designers to prepare ideas for the new peninsula and surrounding landscape.

Baca Architects identified the potential to create a new protected water arena for nature and recreation and to exploit the significant changes in river level from 6 meters Normaal Amsterdams Peil (NAP) (during the summer) to 14 meters NAP (during flood). The new channel will be excavated from low-lying land behind the existing dyke and separated from the River Waal by a low dyke, which allows water to pass over during annual peak river levels. The channel will be approximately 300 meters wide and 1 kilometer long. Baca proposed developing greater variety in the channel depth and width to create a range of seasonal habitat at different levels and opportunity for differing forms of recreation (Figure 14.7). A cascade in land levels will be created from the low dyke down to the new water channel (Figure 14.8).

A strict grazing regime is required to reduce the summer vegetation and to create a smooth channel surface that does not impede the flood flows during the 5 to 10 days of annual flooding.

The waste excavated material is used to form a "hill dyke" to the north, providing a safe platform for development and protection for the new town planned in Lent. The hill dyke incorporates low and high areas for parkland, playgrounds, and development. The banks of the dyke are designed with a gentle gradient that extends down into the new flood channel, creating a naturalistic water's edge, complete with mud banks and riparian habitat.

Ironically, the network of dykes that has been used to reclaim land in the Netherlands from within the river deltas has disconnected the people and the natural

FIGURE 14.7 An amphitheater and park are integrated into a superdyke, overlooking the new flood channel and recreation lake.

FIGURE 14.8 The cascade, water meadows, and marginal river stands.

water environment. In combination with the naturalized environment, the reconnection of the river course with part of the former floodplain creates an opportunity to explore the architectural relationship with water and flood risk management.

A range of flood-resilient buildings will showcase the best of twenty-first-century architectural and technological innovation, as the Weissenhof Estate did in the twentieth century. A mix of resilient (Figure 14.9), amphibious, and floating buildings will

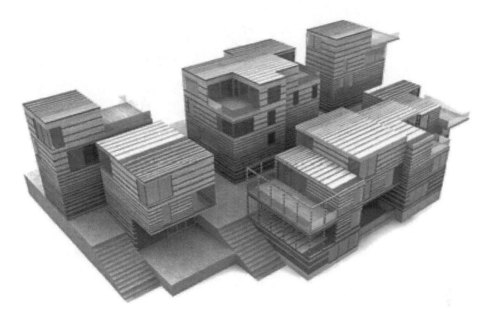

FIGURE 14.9   Clustered resilient homes provide security and safety from floodwaters.

cascade down toward the flood relief channel to provide access to the water and a new floating quay. Multifunctional amenity space between buildings is designed to naturally drain water back into the flood channel and be easily cleaned for fast recovery. The designs will provide a safe and secure community that works in harmony with the River Waal's natural processes and contributes to wider flood risk reduction.

14.8   LITTLEHAMPTON WEST BANK

The West Bank is a thin strip of land on the edge of the River Arun, extending down to the coast. It comprises a leisure marina built on a former landfill, yacht clubs, private houses, a declining shipyard, and farmland (Baca Architects, 2010).

The challenges to redevelop this site are typical of Brownfield sites in coastal areas: flood risk, access, contamination, and ecology. Climate change will only exacerbate these challenges, with rising sea level placing more of the land at risk of flooding and increasing the risk to those already living on the West Bank. The government has identified the need to improve the existing flood defenses; however, there is no national funding currently allocated and it is unlikely that funding will be allocated in the near future. Redevelopment could provide the opportunity to help deliver much-needed defenses, deal with the contamination issues, and revitalize this part of the town. Although the site has been repeatedly identified for regeneration over the past twenty years, redevelopment has not progressed due to the technical challenges.

Baca Architects proposed turning around conventional constraints mapping to focus on the opportunity that substantial development could bring to overcome the challenges and deliver improvement. Baca, together with Knight Frank, Haskoning, and Cyril Sweett, established a financial model to determine the quantum of

FIGURE 14.10   The West Bank development and multifunctional tidal lagoon.

development required to fund new flood defenses and additionally provide wider flood management improvements. This was used to inform a vision for an "exceptional new waterside quarter on the West Bank, combining the best in design, environmental protection, job creation and enhancing the existing character of the area" (Carnegie, Littlehampton Harbor).

The major new development would be raised out of flood risk using soil excavated as part of the creation of a 100 hundred-hectare multi-use tidal lagoon (Figure 14.10), providing flood storage, recreation, floating properties, salt marsh habitat, and tidal energy generation.

Where flood defenses were to be improved, they were conceived as a superdyke, integrated within the urban realm and providing a high-level riverside promenade, with public space, play areas, and retail facilities. The tidal lagoon has the benefit of creating space for hundreds of new berths for the various marinas, providing the capacity for additional support facilities and a large controlled water body. It will also be an exemplar of positive coastal flood risk management in the United Kingdom.

14.9   DORDRECHT FLOODPROOF PILOT PROJECT

A major Brownfield site on the edge of the historic city of Dordrecht unusually does not benefit from flood defenses and is located within the floodplain of the Oude Maas

River. The site is susceptible to flooding during extreme river levels, when water will overwhelm low-lying areas of the site, particularly along the river edge. The river is tidally influenced at this point, and climate change will put further areas of the site at risk of flooding.

The concept for the pilot project was to create a floodproof development of homes that work with water rather than defend against it. The scheme demonstrates how different building types can be used to manage different levels of flood risk across the site. These buildings were organized around access roads, extending from the high ground down to the water. At higher levels, the buildings were linked across the road via communal decks, which provided safe access to emergency escape routes. The streets, houses, and decks were conceived as development piers. The piers accommodated four floodproof building types: floating units, amphibious units, wet-proof resilient units, and dryproof resistant units (Figure 14.11).

- *Floating units:* Deep-section floating platforms provide buoyancy for light-weight timber frame houses. Secondary floating platforms can be docked against the houses and used for private gardens and terraces. Units are linked together to provide added stability and are secured with steel and concrete guideposts. Each unit has a private mooring within the curtilege of the property. The bedrooms are located on the water level, allowing the living areas to take advantage of the roof space and views.
- *Amphibious units:* Three-and-a-half-story timber frame units are paired together on concrete buoyancy decks that rest within a dry dock on the ground. This allows units to enjoy access to the gardens and parking during normal conditions but then rise out of the dock on the buoyancy deck during high-level floods.
- *Resilient units:* At higher levels, a mix of dryproof and wet-proof construction techniques is used to resist and cope with water inundation, respectively. Secondary access decks and escape routes are provided above ground floor during times of flooding to allow continued use of the property and access for the emergency services.

Between the piers, flood-resilient landscaped gardens and play areas (referred to as *water garden courtyards*) provided public amenity for the houses. Street furniture

FIGURE 14.11   Different floodproof construction types. From left to right: floating, amphibious, and dryproof.

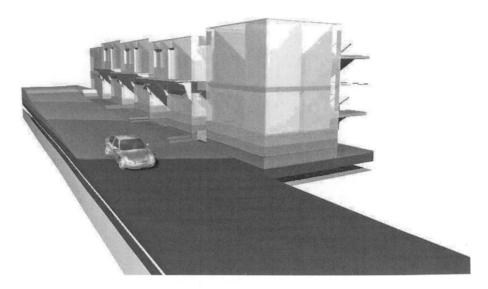

FIGURE 14.12   Intuitive landscapes highlight flood hazard areas within the development.

and materials are designed to withstand floodwater, protect the development from floating debris, and prevent impeding access for emergency services.

The entire development will be linked to a district heating system, elevated above the flood level. Additionally, the development will utilize an evacuated waste system that will take waste directly from each unit to a central disposal area, located above the flood level, thereby reducing the risk of contamination of the watercourse during a flood.

Diagrammatic coding of street surfaces is incorporated into the public realm to aid intuitive evacuation during flooding and to indicate hazard (similar to tactile paving used to aid people with visual impairments). Smooth, colorful surfaces highlight safe routes. Rougher textures are used closer to the watercourse and source of flooding. Corresponding materials are used on the elevation of buildings to highlight the level variations across the site (Figure 14.12).

14.10   CONCLUSIONS

Traditional engineering solutions should no longer be relied upon solely to tackle flood risk. Space for water must be considered in every aspect of our environment and particularly in new development or when redevelopment occurs. If space for water will be incorporated, it should be a prime consideration and used to inform design from the outset. It should be seen as an asset, at the heart of good design. It should be multifunctional, to help deliver other environmental design aspirations and create better place making.

This type of multifunctionality can help achieve the higher densities required for the current and future populations, without losing the quality of space and environment that makes for successful communities. It may also help overcome the economic and social objections to implementing zero carbon and nondefensive flood risk management measures.

The design of flood risk management measures must be seen as an opportunity to provide other functions. Likewise, when considering the design or improvement to facilities such as infrastructure and landscaping, the opportunity to integrate flood-risk management measures should be seen.

Integrated spatial planning requires the simultaneous assessment of multiple issues and criteria, which can be time consuming and, therefore, expensive. However, the sums associated with design time during the initial stages of a project are small, relative to the overall cost of development. This initial investment will have a significant return in the overall reduction in risk and improvement in the quality of our environment. The master planning work for the three LifE project sites, each designed to be carbon-neutral, confirmed that the cost of integrated solutions could provide savings of up to 10% of the capital cost of the development, a margin significant enough to help responsible development proceed. On other sites, an integrated approach may be the only cost-effective way to tackle the obstacles to sustainable redevelopment.

REFERENCES

Baca Architects (2010). Littlehampton West Bank Regeneration Study for Arun District Council and West Sussex County Council. London, UK: BACA Architects.

Baca Architects with BRE (2009a). *The LifE Handbook.* Watford, UK: IHS BRE Press.

Baca Architects with BRE (2009b). LifE report, Appendix A. Watford, UK: IHS BRE Press.

Baca Architects with BRE (2009c). LifE report, Appendix B. Watford, UK: IHS BRE Press.

Baca Architects with BRE (2009d). LifE report, Appendix C. Watford, UK: IHS BRE Press.

DCLG (Department of Communities and Local Government) (2006). Planning Policy Statement (PPS)25: Development and flood risk. The Stationery Office. Department of Communities and Local Government (2006a) Planning Policy Statement (PPS)25: development and flood risk. Norwich, UK: The Stationery Office.

Defra (Department for Environment, Food and Rural Affairs) (2005). Making Space for Water: Taking Forward a New Government Strategy for Flood and Coastal Erosion Risk Management in England. First Government Response to the Autumn 2004 Making Space for Water consultation Exercise. London, UK: Department for Environment, Food and Rural Affairs.

Population Reference Bureau (2010). Human Population: Urbanization (online) Available at http://www.prb.org/Educators/TeachersGuides/HumanPopulation/Urbanization.aspx [Accessed September 2010].

15 Adapting to and Mitigating Floods Using Sustainable Urban Drainage Systems

Susanne M. Charlesworth and Frank Warwick

CONTENTS

15.1 INTRODUCTION

In terms of its magnitude, occurrence, geographical distribution, loss of life and property, displacement of populations, and socioeconomic impacts, flooding is the most common global environmental hazard (Loster, 1999). It accounts for at least a third of global natural disasters per annum with approximately 20,000 lives lost and 20 million people displaced (Loster, 1999; Smith, 2004). Approximately one third of the world's land area is prone to flooding and with 82% of the world's population inhabiting these areas, it represents a significant hazard (Dilley et al., 2005). According to Jonkman (2005), during the period from 1975 to 2002, 176,864 people were killed in the course of 1,883 freshwater flooding events, and a further 2.27 billion people were affected.

There are many reasons for the flooding that caused these mortalities. Jonkman (2005) lists six: coastal flooding, flash floods, fluvial flooding, drainage problems, tsunamis, and tidal waves or tidal bores. This chapter concentrates on two of these that are unique to cities: (1) flooding due to drainage problems (wherein the drainage system is blocked or overwhelmed) and (2) the flooding typical of urban areas—that of flash flooding. According to the Parliamentary Office of Science and Technology (POST, 2007), urban flooding such as this is due to endogenic factors within the drainage basin. These factors include drainage infrastructure that is aging and therefore unable to cope with excess storm water; additional construction that seals the surface, making it impermeable, together with the addition of the building's drainage to a sewerage infrastructure that is already at capacity; and changes to river hydrology and the encroachment of impermeable surfaces from buildings already in place as homeowners pave over their front gardens (Perry and Nawaz, 2008). Finally, the impacts of global climate change are an exogenic factor that will change the characteristics of urban drainage basins according to their geographic location (White and Howe, 2004; Evans et al., 2008).

During the summer storms of 2007 in the United Kingdom, rainfall totals were extremely high (Blackburn, Methven, and Roberts, 2008) due to persistent rain. Flooding resulted from the inability of the aging storm sewer infrastructure to cope; this scenario repeated itself in 2008 and 2009 (e.g., Met Office, n.d.). During the 2007 storms in the United Kingdom, 13 people died, 48,000 homes and 7,300 businesses were flooded, and the total cost was £4 billion, including £1 billion for cleaning up the damage (ABI, 2007; EA, 2010). If a nation's surface water infrastructure cannot cope with rainstorms now, with the likelihood of increased storminess in the future due to climate change, such flooding and devastation will become increasingly frequent (White and Howe, 2004).

This chapter examines the mechanisms involved in urban flooding, suggesting a sustainable approach to the management of flood risk. It introduces the concept of sustainable urban drainage systems (SUDS), detailing the way in which individual devices and their combinations can be used to construct a sustainable drainage system. Cities need to make space for water, not only by utilizing SUDS but also by the reinstatement of functional floodplains by creating blue and green corridors as ribbons of flood-resilient infrastructure reconnecting fractured urban catchments. Because "around two thirds of the building stock that will still be standing in 2050 has already been built" (DBERR, 2008), a fact that is probably true for much of the developed world, the emphasis in this chapter will therefore not be on utilizing SUDS in new developments, but rather of retrofitting SUDS devices to the existing built environment. For developing countries, using SUDS is subject to additional specific problems related mainly to climate and disease. Hence, a separate section considers the use of SUDS technologies focusing on the tropics. Regardless of the location of the city in question, flooding is generally due to the hard infrastructure installed to support its inhabitants, and the following section explores the impact of this infrastructure on the hydrology of urban areas.

15.2   FLOODING OF URBAN AREAS

Flooding is a natural hydrological event (Marsalek et al., 2000; Jeb and Aggarwal, 2008); according to Wheater and Evans (2009), "the effects of urbanization are well understood." However, the characteristics that are unique to urban areas lead to unique impacts during flooding events. These impacts are best illustrated using the storm hydrograph (Figure 15.1), which shows the increased storm peak and decreased lag time associated with smooth, straight urban channels and the increased drainage density brought about by the storm sewer network collecting surface water flows from housing and industrial estates (Campana and Tucci, 2001; Charlesworth, Harker, and Rickard, 2003; WMO, 2008). Increased impermeable areas and removal of vegetation increase surface runoff and decrease evapotranspiration (Butler and Davies, 2004; Jones and Macdonald, 2007), all of which exacerbate the flashy hydrology of urban environments.

Such hard drainage infrastructure could lead to an unsustainable situation where, for example in the United Kingdom, a Foresight Report (Evans et al., 2004) predicted that by the 2080s, the cost of flooding could rise to between £1 and £10 billion if no action was taken to tackle the problem. In the update of this report published in 2008 (Evans et al., 2008), it was estimated that the risk due to intra-urban flooding (i.e., that caused by excess surface water) would be of the same order of magnitude as that of fluvial and coastal flooding. Globally, climate change will change rainfall regimes

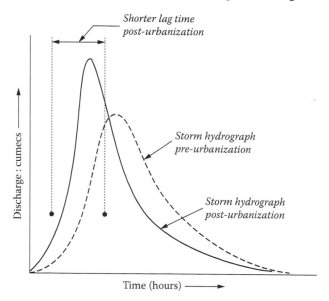

FIGURE 15.1   Storm hydrograph showing the increased storm peak and decreased lag time associated with smooth, straight urban channels and the increased drainage density brought about by the storm sewer network collecting surface water flows from housing and industrial estates. (Adapted from Leopold, L.B., *Hydrology for Urban Planning—A Guidebook on the Hydrologic Effects of Urban Land Use.* Geological Survey Circular 554. U.S. Dept. of the Interior, 1968.)

(Wheater and Evans, 2009), leading to some regions suffering increased storminess and hence increased flooding incidents. Looking to the future, cities must adopt flood-resilient strategies that are sustainable in the long term if they are to avoid the socioeconomic impacts of multiple flooding events. Some still rely on hard infrastructure to provide flood resilience, such as Barcelona's (Spain) detention tanks and flow diversion measures (Cembrano et al., 2004) and the Lower Thames (London) flood alleviation scheme utilizing flood diversion channels (EA, 2008). However, the water industry representative body, Water UK (2008, p. 5), states: "Bigger pipes are not the solution to bigger storms"; hard infrastructure is inflexible and has few benefits apart from its ability to remove water from urban areas as quickly as possible. Water UK (2008) furthermore suggests that SUDS and "sacrificial areas" are the ways forward.

The remainder of this chapter is therefore devoted to considering the way in which SUDS in cities can aid in the mitigation of surface water flooding, and also considers how Water UK's (2008) sacrificial areas can benefit urban drainage in a sustainable way.

15.3   SUSTAINABLE DRAINAGE

Conventional hard drainage tends to concentrate on managing water quantity (Figure 15.2a) by gathering all the runoff water from impervious streets and pavement into storm sewer systems that pass via gullypots (the first entry point of road runoff into an urban drainage network), pipes, and water treatment facilities into the receiving watercourse. These systems can either be combined or separate, whereby foul and surface waters are either combined in the same system or separated and flow separately. The majority of problems with raw sewage finding its way into dwellings during the UK's summer floods were due to the combined sewerage systems' capacity being exceeded by surface water flow, which then backed up to the surface and into people's living rooms. Water quality is of less concern to conventional drainage, and biodiversity and amenity are hardly considered; hence, urban streams have become "neglected, abused, or modified" (Keller and Hoffman, 1977, p. 237).

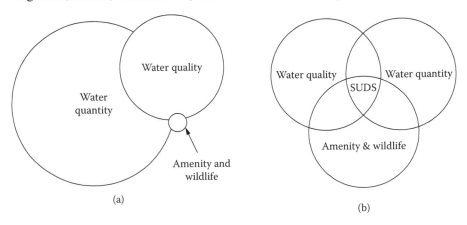

FIGURE 15.2   Balance of benefits offered by (a) conventional drainage and (b) SUDS triangle.

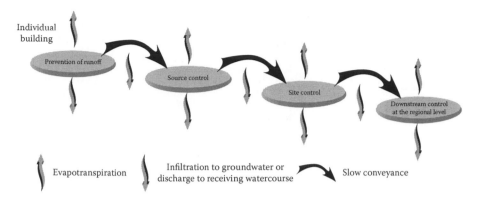

FIGURE 15.3   SUDS management train (Based on CIRIA (Construction Industry Research and Information Association) (2000). Sustainable Urban Drainage Systems: A Design Manual for England and Wales. Report C522.)

SUDS are a suite of measures that treat water in a different way than that of conventional drainage. Instead of constraining surface water into pipes and conduits, forcing it to leave a city as quickly as possible, SUDS encourage infiltration and detention of surface water on site. Figure 15.3 (CIRIA, 2000) shows the way in which the SUDS Surface Water Management Train begins with prevention for individual premises involving good housekeeping measures, then controlling as much of the water at source as possible, followed by water management at the site scale, and then regionally. At each stage, water is allowed to percolate either to the receiving watercourse, or to groundwater through devices that provide water treatment capabilities and amenity and biodiversity benefits.

Figure 15.2b illustrates this in terms of the oft-quoted SUDS triangle in which there is an equal balance between water quantity, water quality, and biodiversity or amenity, in contrast to that of conventional drainage. It is beyond the scope of this chapter to cover the water-quality benefits of SUDS; however, there is much evidence of such improvements in the literature (e.g., Charlesworth et al., 2003). SUDS represent a different way of managing water; rather than treating it as an embarrassment, to be hidden from sight and forgotten, instead it should be treated as a "liquid asset" (Semadeni-Davies et al., 2008) in which society takes account of the behavior of water, rather than water's behavior having to change for the sake of society.

15.3.1   SUDS DEVICES AND FLOOD MITIGATION

It is beyond the scope of this chapter to provide a detailed discussion of the design and construction of SUDS devices; there are many sources of information that provide this (e.g., US EPA, 2000; Charlesworth et al., 2003; SEPA, 2005; GDSDS, 2005; DTI, 2006; Woods Ballard, Kellagher, Martin, Jefferies, Bray, and Shaffer, 2007; EA, n.d.). However, SUDS have been used to mitigate surface water flooding for several decades in the United States; Scandinavia, notably Sweden; and continental Europe, in particular Germany and France. As a result, the flood attenuation benefits of SUDS devices are clear and unequivocal, and there is considerable literature

TABLE 15.1

Typical Sustainable Drainage Devices

Vegetated Devices	Hard Devices
Green roofs and walls	Porous paving (PPS)
Rain gardens	Concrete built street rain garden
Constructed wetlands	Rainwater harvesting
Filter strips	Other: Using existing urban green infrastructure: front gardens,
Swales	school playing fields, traffic islands, grass verges, parks
Vegetated PPS	
Individual householder's rain garden	
Street trees	

detailing their performance. This section therefore outlines individual SUDS devices and the ways in which they achieve these benefits.

Table 15.1 lists some of the individual devices or stormwater best management practices (BMPs in the United States) utilized in the SUDS approach, dividing them into those that rely on vegetation, or "soft" techniques—such as green walls and roofs, constructed wetlands, and swales—and those that are constructed of more engineered materials—such as asphalt, blocks, and concrete; and "hard" infrastructure such as porous paving systems (PPS) and rainwater harvesting systems (RWHS). Other classifications include that of Jefferies et al. (1999), whereby the devices are divided into those that are installed above or below ground (Charlesworth et al., 2003).

However, while useful, these classifications give little information on the *processes* involved in mitigating or adapting to flood risk; the following lists four main processes:

1. Filter strips and swales are long, shallow ditches or channels that make use of vegetation and gentle slopes to convey, infiltrate, store, and treat runoff by slowing the passage of stormwater (see Table 15.2). There are designs that incorporate permeable under-drainage, which have been shown to be effective for all but the heaviest storms.

2. Permeable surfaces, or PPS, comprise porous asphalt, porous concrete, or porous blocks and can be vegetated. The surfaces overlie various grades of aggregate that may also incorporate the use of a geotextile (a permeable fabric that, when used in association with soil, has the ability to separate, filter, reinforce, protect, or drain). They provide a large volume through which the water slowly percolates, and can be stored before entering groundwater. If the soil or aquifer is unsuitable for infiltration—for example, if the soil is contaminated—an impermeable layer can isolate the water from the ground, effectively forming a tank where the water can either be stored or conveyed elsewhere in the catchment where infiltration is more suitable or else to the receiving watercourse. In a review of the performance of PPS, Scholz and Grabowiecki (2007) discuss many of the available structures, stating that "high peak flow ... (is) ... effectively controlled" (p. 3833). Booth (2000) further states that, regardless of PPS structure, they "dramatically reduce surface

TABLE 15.2

Design Criteria for Incorporation of a Swale into SUDS

Side Slope (vertical/ horizontal)	Long Slope	Depth (mm)	Grass Height (mm)	Design Storm Event	Velocity/ HRT	Minimum Length (m)
<1:3	<1:50	300–500	100–200	5 year/24 hour Check for: 10 year/24 hour	<0.25 m sec^{-1} 8–10 min	60

Source: Based on Escarameia, M., Todd, A. J., and Watts, G. R. A. (2006). Pollutant Removal Ability of Grassed Surface Water Channels and Swales: Literature Review and Identification of Potential Monitoring Sites. Highways Agency. 46 pp.

Note: HRT= hydraulic residence time.

runoff volume and attenuate peak discharge." By infiltrating through a PPS structure, water quality is also improved by trapping particulate associated pollutants (PAPs) and by treatment in the biofilm, which is associated either with the geotextile and subsurface aggregates in PPS (Coupe et al., 2003) or which is found naturally within the soil.

3. Infiltration devices such as rain garden street planters (Figure 15.4) can be retrofitted at street level and provide storm peak attenuation; made of stone, they integrate well into the built environment (DTI, 2006). As shown in Figure 15.2b, SUDS devices not only address issues with floodwaters but also reduce the negative impacts of urbanization with respect to water quality. Charlesworth et al. (2003) list the pollutant removal efficiencies of various SUDS devices and found that infiltration devices in particular had either very high or high efficiencies for a range of pollutants from total suspended solids, to individual heavy metals such as zinc and lead. However, the effectiveness of these devices depends very much on the infiltration potential of the soil.

4. Detention devices such as basins, ponds, and wetlands, also rainwater harvesting systems (RWHS) and green roof and walls. Detention and retention ponds rely on capturing water and both allowing its infiltration into the ground and detaining it for subsequent slow release or dissipation. The use of vegetation in SUDS devices slows water velocity, thus attenuating the storm peak and also allowing time for the water to infiltrate. Installing a green roof can absorb up to 100% of incident rainfall, depending on conditions, and regionally 10% of greened roofs can result in a 2.7% reduction in stormwater runoff, with a 54% average reduction in runoff per individual building (Mentens, Raes, and Hermy, 2006). Kellagher and Maneiro Franco (2005) have demonstrated (in Evans et al., 2008, p. 113) that RWHS can reduce downstream flood risk. In a review of the performance of the Scottish SUDS train at DEX (Dunfermline Eastern Expansion; Figure 15.5), MacDonald and

FIGURE 15.4   Water from the downspout flows into a rain garden in Berlin.

FIGURE 15.5   Detention pond incorporated into a road traffic island, Dunfermline Eastern Expansion (DEX), Scotland. (Urban Water Technology Centre, University of Abertay Dundee).

Jefferies (2003) found that the six ponds, wetland, and associated upstream detention basins and swales yielded significant lag times. In a parallel study of an Environment Agency (EA) demonstration SUDS train at Hopwood Motorway Service Area (MSA) in England, Malcom et al. (2003) reported significant reductions in peak flow for all but the largest events.

These devices can be used individually, or they can be designed in combination as a SUDS management train, examples of which in the United Kingdom include the Environment Agency SUDS demonstration sites at the Hopwood MSA near Bromsgrove (United Kingdom) on the M42, junction two (Heal et al., 2008; Figure 15.6), and also the Wheatley MSA at Oxford on the M40 (Bray, 2000). Charlesworth et al. (2003) describe such trains as a "cascade" that is able to tackle many of the negative impacts of flooding that a single strategy alone would have been unable to address.

Therefore, the advantages of SUDS in reducing flood risk in general include the ability to reduce high-velocity, excess water from entering watercourses and reduce peak flows (Woods Ballard et al., 2007). SUDS can also reduce the volume of stormwater (Wilson, Bray, and Cooper, 2004; Kirby, 2005), dealing with it at source so that it finds its way more naturally to the receiving watercourse, or into groundwater (Ferguson, 2002).

15.3.2   SUDS FLOOD-RESILIENT DESIGN

In the United Kingdom, 80% of the dwellings required by 2040 are already in existence. Incorporating SUDS into urban design is relatively straightforward for new

FIGURE 15.6   Part of the SUDS train at Hopwood Motorway Service Area, Bromsgrove, M42 junction two, UK.

build, in that they can be considered at the earliest stage of planning the development. However, if flooding of existing cities is to be mitigated, then thought needs to be devoted to the installation of SUDS devices in and around buildings that have already been constructed. While this section does concentrate on retrofit SUDS, it begins with a brief consideration of the use of SUDS in new developments because, until recently, this has been the focus for planning and drainage design.

15.3.3 Incorporating SUDS Techniques into New Build Developments

The incorporation of SUDS into new build requires their addition at the planning stage. They need to be *designed* to be fit for purpose, as is conventional drainage. In general, hard drainage infrastructure is designed for the 1 in 30-years storm. A simple example of drainage design using the SUDS approach is the use of a single swale that has design considerations (Table 15.2).

Many developed countries' planning laws (e.g., England's Planning Policy Statement 25 (PPS25) (DCLG, 2010)) stipulate that new build must render the site able to deal with surface water at greenfield runoff rates, that is, the rate at which the site would have infiltrated or stored the water prior to development. This is generally calculated using standard formulae such as a rational method (e.g., IEA, 1987), or a flood estimation formula based on the characteristics of the individual catchment such as described in the Institute of Hydrology (IH) report No. 124 Flood Estimation for Small Catchments (IH, 1994). While not necessarily ideal when applied to small urban catchments or their "truncated portions" as described by Cawley and Cunnane (2003, p. 42), they do admit that these formulae are nonetheless "in widespread use for SUDS applications." Furthermore, it is suggested that an average greenfield runoff rate of between 2 and 5 liters per second per hectare is used because the design should take into account the small areas of urbanization (HR Wallingford, 2008). Specific details of the calculations required for computing greenfield runoff rates are beyond the scope of this chapter but further information can be found in HR Wallingford (2005); Wilson, Bray, and Cooper (2004); and Gibbs (2004). Greenfield runoff estimation methods are also reviewed in Balmforth et al. (2006a). While the implementation of greenfield runoff rates for new build is fairly straightforward, PPS25 does stipulate that redevelopment should not exacerbate any existing flooding problems. The next subsection therefore explores the possibilities of incorporating SUDS into an already existing urbanized environment, and Section 15.3.5 examines the problem of designing SUDS for densely populated cities.

15.3.4 Retrofit SUDS

Planning processes and guidance mainly address changes to the urban landscape. Thus, most SUDS installations result from new development or from redevelopment of an existing site (Mitchell, 2005; MacMullan and Reich, 2007; Stovin, Swan, and Moore, 2007). However, such developments represent a limited component of the built environment. The majority of any urban area comprises existing sites rather than new developments. For example, the average rate of development of urban areas in the United Kingdom is estimated to be approximately 1% per annum (Gordon-

Walker, Harle, and Naismith, 2007; DBERR, 2008). The term *retrofit* in the context of sustainable drainage encompasses modifications to existing drainage infrastructure in order to improve water flow and quality (SNIFFER, 2006). Retrofitting SUDS devices may allow new development where existing sewerage systems are close to full capacity, thereby enabling development within existing urban areas. Consequently, improvements aimed at reducing flood risk will advance more rapidly by retrofitting SUDS into existing locations (British Academy, 2005; Bosher et al., 2007).

The objectives of SUDS retrofits do not differ from their implementation in new developments. However, because retrofit is often constrained by the existing urban fabric and infrastructure, retrofit implementations may be more expensive than new developments, and the set of available techniques may be restricted due to lack of space or public acceptability (Kloss and Calarusse, 2006; POST, 2007; Schueler et al., 2007; Shaver et al., 2007; Shutes, 2008). Despite these limitations, studies have demonstrated that it is not difficult to design retrofit schemes that are technically feasible (Stovin, Swan, and Moore, 2007). Worldwide, cities such as Malmö (Sweden), Portland (Oregon), and Tokyo (Japan) have demonstrated the viability and effectiveness of retrofitting SUDS (DTI, 2006; SNIFFER, 2006; Stahre, 2008).

In Malmö, refurbishment of the suburb of Augustenborg has employed source controls, swales, and dry and wet detention areas to relieve pressure on the combined sewerage system (Villarreal, Semadeni-Davies, and Bengtsson, 2004; Stahre, 2008). In Seattle, Washington, and Portland, Oregon, smaller-scale SUDS techniques have been implemented in individual streets, and these may be more appropriate to dense urban developments (DTI, 2006). Retrofit installations have also been employed to solve specific flooding issues—for example, due to excess overland flow at two schools in Worcestershire, England (Bray, 2004; SNIFFER, 2006).

Retrofit can also take place across wider scales. After the successful implementation of pilot retrofit projects (e.g., Emscher Cooperative, 2009), the 865-square-kilometer Emscher catchment in Germany aims to disconnect 15% of runoff and peak flow from the existing combined sewerage system over 15 years (Sieker et al., 2006), and to create an urban lake to store water in a 10-year flood event (Urban Water Project Partnership, 2008).

The size of an area to be retrofitted influences which techniques are suitable. Schueler et al. (2007) contrasted two approaches to retrofit in the United States, largely differentiated on the spatial scale of the retrofit and land ownership issues. In the United Kingdom, Singh et al. (2005) made a similar distinction based on the availability of open space. For larger areas (greater than two hectares), appropriate techniques include modifying and increasing existing stormwater management facilities, utilizing open spaces such as parks, golf courses, highway and infrastructure rights of way, and large car parks (Claytor, 1998; Schueler et al., 2007). For catchments with a relatively high percentage of undeveloped areas (up to approximately 50% impermeable), a smaller quantity of larger SUDS features may be beneficial, and in such locations assessment of retrofit suitability is based on factors such as stormwater pond density, stream density, available area in stream corridors, and the extent of publicly owned land (Schueler et al., 2007). In contrast, land availability may be limited in high-density urban areas, and available sites are likely to be smaller, so appropriate locations include small car parks, individual streets and roofs, filter strips adjacent to impervious areas, and underground sites (SNIFFER, 2006; Schueler et al., 2007). Feasibility in these more densely built-up sites may be assessed by considering factors such as land cover and ownership, the presence of combined sewers, areas due for redevelopment, and the number of problem locations (SNIFFER, 2006; Schueler et al., 2007).

Given the complexities created by multiple landowners in urban settings, quick wins are more achievable when potential retrofit sites belong to a single landowner, developer, or public authority. Successful retrofits have often been driven by a single organization with the authority to implement solutions (Stovin, Swan, and Moore, 2007). In the United States, public organizations and municipalities are often the driving force behind SUDS retrofit, for example because they own appropriate land (Schueler et al., 2007); or can implement financial incentives for disconnection, for example, Portland, Oregon (DTI, 2006); or for reducing impermeable areas, for example, Maryland (DTI, 2006). Downpipe disconnections in Portland have been implemented by 40% of the homes in the city, and are considered to account for an 80% reduction in runoff volume in a 25-year storm event (DTI, 2006). The Seattle, Washington, Street Edge Alternatives project implemented small infiltration basins along road or pavement boundaries to capture road and roof runoff. It has achieved full attenuation of 19 millimeters of rainfall, and a 98% reduction in runoff (Horner, Lim, and Burges, 2002). A similar project in Portland, Oregon (City of Portland, 2005), reduced the peak flow and runoff volume from a 25-year storm by over 80% (Perry, 2009).

Funding in retrofit situations is more problematic than for new developments (Balmforth et al., 2006b), commonly requiring identification of drivers of change. In both new and retrofit developments, organizations responsible for maintenance, especially where they are different from the developer, may be cautious of incurring commitment to ongoing costs (Balmforth et al., 2006b; Defra, 2007). Furthermore, additional land may be required to retrofit SUDS. The Scottish Environment Protection Agency (SEPA, 2005) has estimated a requirement of 5% to 7% of the contributing area, which may not be readily available in inner-urban locations (Charlesworth et al., 2003; Hyder Consulting, 2004; Woods Ballard et al., 2007), so the cost of land acquisition can add to the outlay for SUDS implementation in highly urbanized environments (RCEP, 2007).

In the United States, Schueler et al. (2007) estimated that retrofit construction costs were between 1.5 and 4 times greater than new developments. The increased costs were associated with a number of factors, including site constraints, greater design complexity, higher excavation and landscaping costs, more construction and engineering contingencies, and the prototype nature of designs. Although cost justification of individual SUDS retrofits initially appears problematic, a large-scale German project aimed at disconnecting domestic properties from the sewerage system has achieved average savings of €22 per annum for each of the 37,500 properties disconnected (Jefferies et al., 2008). A review of the potential for wide-scale retrofit of SUDS in England (Gordon-Walker, Harle, and Naismith, 2007) determined that, if implemented nationally, both permeable paving and water butts would provide economic benefits for property owners, but that the benefits of vegetated SUDS were less clearly identifiable at a national level.

In addition to technical performance, public engagement and education is regarded as a key element in SUDS retrofit success (DTI, 2006; SNIFFER, 2007). The importance of maintenance in the continuing effectiveness of SUDS has also been emphasized (Weinstein, 2002; Doyle, Hennelly, and McEntee, 2003; SNIFFER, 2004; Apostolaki, Jefferies, and Wild, 2006; Hirschman and Kosco, 2008). Key problems include (Hirschman and Kosco, 2008) inadequate funding, insufficient staff, lack of awareness surrounding responsibilities, poor designs that complicate maintenance, inaccurate information about the location of devices and their specific maintenance requirements, and lack of regulatory enforcement guidelines and capabilities.

Overall, although retrofitting sustainable drainage is subject to more constraints than its implementation in new developments, examples worldwide demonstrate that retrofit SUDS can be effective in reducing flood risk in the built environment. However, Defra (2008) states that flooding due to excess surface water in urban areas may only be able to be solved by the complete redevelopment of urban centers and associated housing. It was asserted that this may well take many years before full implementation could be achieved.

15.3.5 Designing SUDS into the Built Environment

Like any other set of drainage infrastructures, SUDS devices and trains must be designed properly in order for them to carry out their purpose efficiently. There is guidance (see Section 15.3.3) on the calculations needed in order to comply with regulations regarding design storms and greenfield runoff rates but, particularly with reference to retrofit, the necessary land area to install larger devices (e.g., wetlands and ponds also swales or filter strips to a certain extent) may not be available and Table 15.2, for example, notes that the optimal length for a swale to be effective is 60 meters and unless this is built in at the planning stage, access to such a length is likely to be limited unless it becomes part of Water UK's (2008) sacrificial areas mentioned in Section 15.2.

Shaw, Colley, and Connell (2007) suggest a range of techniques that can be applied at three different scales: (1) the individual building, (2) neighborhood, and (3) the whole catchment or conurbation. However, this does not take into account the density of construction, types of building, or specific land use. It has been shown, for example, that a central business district (CBD) usually has the least vegetation (e.g., Rose, Akbari, and Taha, 1999; Akbari, Rose, and Taha, 2003), and less affluent residential areas also lack green space (Whitford, Ennos, and Handley, 2001). A simplistic means of suggesting the integration of SUDS devices into cities is presented here using less-than-satisfactory nomenclature in the absence of more suitable terminology. Hence, *city center* describes an area with the highest density of buildings, which is therefore the most impermeable and has the least amount of vegetative cover and highest percentage of built structures.

Any water bodies present in the city center will more than likely be located underground or are bounded on all sides by the built environment, which can include the CBD, dense retail areas, and even residential areas where gardens may have been sealed and the houses built in close proximity to one another, perhaps as terraces or high-rise blocks. SUDS devices can be retrofitted to such existing built-up areas, but some do need space that is not readily available under such circumstances.

Once a development is built, it is not normally possible to allocate space for greening, as suggested by Wilby (2007). However, a simplistic bull's-eye approach (Figure 15.7) allows a hierarchy of suitable devices to be constructed from the urban center to the periphery, much like a combination of the zonal and multiple nuclei models of urban structure (Burgess, 1924; Harris and Ullman, 1945, respectively). Figure 15.7 shows that small-scale patches of retrofit can be installed in densely occupied urban centers—for example, green roofs and walls (e.g., those retrofitted to high-rises on the Ethelred Estate in Lambeth, London, UK) (Figure 15.8), areas of PPS and RWHS.

In the urban center, SUDS can become a supporting mechanism, relieving the pressure on conventional systems, but can also enable further development of a site where the traditional drainage infrastructure is at capacity and where utilization of SUDS can ensure that no additional surface water burden is placed on it. Suburban areas can support larger devices such as roadside swales, ponds incorporated into roundabouts such as the Dunfermline Eastern Expansion Roundabout Detention Basin (Figure 15.5 and Section 15.3.1), larger areas of PPS—for example,

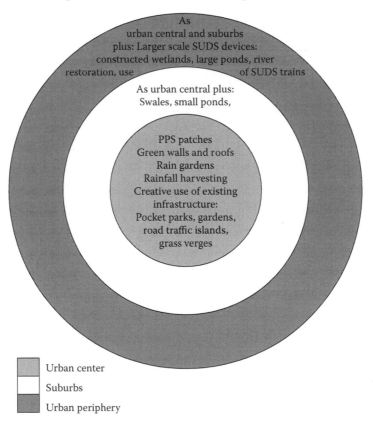

FIGURE 15.7 SUDs bull's-eye.

FIGURE 15.8   Green roof retrofitted to a high-rise block on the Ethelred Estate, Lambeth, London, UK. (From Lambeth Council (n.d.). Available at <http://www.lambeth.gov.uk/Services/HousingPlanning/Planning/EnvironmentalIssues_EXTRA.htm>.)

on supermarket and industrial estate car parks—and the largest devices such as constructed wetlands and ponds could be used in suitable areas on the urban periphery. Here, combinations of devices can be used in trains; examples of these are the Hopwood and Oxford MSAs in England (Figure 15.6), further details of which can be found in Section 15.3.1. New build requires that SUDS be designed into it at the outset, whereby trains of ponds, wetlands, and swales provide the area with the multiple benefits associated with a sustainable drainage system. There is an opportunity to undertake the "smart landscaping" and "smart design" suggested by Antonelli (2008) to make the most use of the ecosystem services that SUDS can provide.

Private frontages can be used as rain gardens, encouraging water to infiltrate and dissipate slowly. The recent trend of "sealing" house frontages using impermeable materials to provide off-road parking for residents' cars has led to the loss of up to two thirds of London's front gardens (London Assembly, 2005). This could amount to a total area of 32 square kilometers (12 square miles) of vegetation removed, potential habitat lost, as well as the loss of permeable surface that would have infiltrated excess rainfall in the event of storms. The Royal Horticultural Society (RHS, 2005) quotes a figure of 10 liters of rainfall per minute as the capacity of an average suburban garden, or 10% of the incident rainfall absorbed. Cumulatively across a city, this could represent thousands of liters of water that does not subsequently contribute to flooding.

To further encourage disposal of surface water on site, roofwater or rainwater harvesting (RWH), using a tanked PPS system as described above, or simple water butts or barrels for later water reuse outside the home in watering a green roof, for example, or inside for toilet flushing, will also reduce the amount of water to be managed as it leaves an individual plot. In areas of the world where droughts are becoming more common (e.g., Australia), RWH enables a resource to be saved when it is available.

A large-scale approach that integrates well into a SUDS approach is that of the River Restoration Centre (RRC, 1999, 2002); in fact, Bray (2006) suggests that both SUDS and river restoration share common objectives. Rivers and streams passing directly through the more built-up areas of the city may have been straightened, channelized, and canalized, and it may not be possible due to land-use constraints to reinstate these. However, those in the urban periphery can be returned to a more natural profile of meanders and riparian vegetation. Upstream and downstream of an urban area, a restored river can slow water and allow it to wash over the restored floodplain, thus protecting the urban area from flooding. The following section explores the concept of floodplain restoration in the context of the sacrificial areas introduced earlier in Section 15.2 (Water UK, 2008).

15.3.6 Sacrificial Areas

Floodplain development disconnects rivers from their floodplain (Wheater and Evans, 2009) whose natural function is to enable water overtopping riverbanks to spread out across the flat land on either side of the river course. Unfortunately, this flat land is seen as ideal to build on, and the risk of flooding is exacerbated by the associated hard surfacing, channelization, and canalization of the river as it passes through the housing or industrial estate. As reported by Shaw, Colley, and Connell (2007, p. 24), "Property on an undefended floodplain is at the highest risk of flooding." The concept of *sacrificial areas* was introduced in Section 15.2; it has been called *managed retreat* or more recently *managed realignment* where it manages flood-prone areas by deliberately realigning river, estuary, or coastal defenses (Defra, 2002). This can be achieved by abandoning topographically lower land and effectively retreating to higher ground, setting back the flood defenses and shortening their overall length, reducing the heights of walls or embankments, and widening the floodplain. These management approaches allow the floodwater to rise and wash over the floodplain, accepting that properties and businesses already built there will be flooded. The UK Environment Agency and Department for Environment, Food and Rural Affairs (Defra) have described this approach as having changed the emphasis from *defending* against floods to *management* of them and have included floodplain restoration within their flood-resilience strategies (Defra, 2002). However, Defra (2002, p. 39) states that "urban watercourses tend to be a neglected resource, are unattractive and inefficient conveyor of floods. Improvements to their environmental status (i.e., reintroduction of natural features in the channel) whilst improving flood conveyance should be sought. This may require space and mean the loss of gardens and even demolition of properties."

Managed retreat has been implemented in coastal areas since 1991 when it was first attempted in the United Kingdom at Northey Island where coasts were eroding rapidly and hard coastal defenses were considered too expensive. Elsewhere, managed realignment has been implemented primarily as part of nature conservation or (re-) creation projects in estuarine locations in the United States (e.g., the Mississippi delta, San Francisco Bay area), the Netherlands (e.g., Friesland, the Scheldt Estuary), and Germany (e.g., the Elbe River) (McCarthy et al., 2001). However, application of this approach to terrestrial areas at risk of pluvial and fluvial flooding is relatively new. Defra's "making space for water" strategy (Defra, 2005, 2008) places greater emphasis on managed realignment, which the Chartered Institution of Water and Environmental Management (CIWEM, 2009) have called "radical" and Myatt-Bell et al. (2002) have called "challenging." However, organizations such as the CIWEM have called for similar approaches in the past—for example, "blue belt" land (CIWEM, 2008). The London Plan (GLA, 2008) refers to "blue ribbons" of "improved river corridors."

This joint approach—of restoring floodplains and river corridors alongside managed realignment—has led to the development of so-called *blue corridors* (BC) (a blue corridor, according to the EA, should be at least 120 meters wide). These may have an associated *green infrastructure* (GI), which can include swales and filter strips, ponds, and wetlands, and which provides what is termed *ecosystem services*—those services provided by the natural environment that benefit people (Bolund and Hunhammar, 1999; Defra, 2010). Pitt (2007) suggests implementing green corridors to make room for floodwaters for the up to 10% of existing UK housing stock considered at risk of flooding. By including blue *and* green corridors, not only would river channels be considered but also overland routes, as for example, in the major summer 2007 floods in the United Kingdom, two thirds of the flooding was due to surface water. Stahre (2008) details the blue and green infrastructure of such an approach in the Swedish city of Malmö, where, from the 1980s, multifunctional regional eco-corridors gradually replaced traditional hard drainage, in order to protect downstream areas from flooding. In this respect, one of the goals of the strategy was that urbanization would not affect the water balance; that is, that the infiltration rate would be as if the site was still greenfield. The concept of a greenfield runoff rate was discussed in more detail in Section 15.3.3.

Countries with temperate climates can countenance the "skylighting" of their urban rivers (opening up a previously sealed watercourse) and the installation of water bodies into the cityscape. However, those cities located in tropical areas have regular heavy downpours associated with their particular climate to deal with, as well as health issues associated with waterborne diseases and vectors of disease, in particular mosquitoes. The next section examines the extent to which SUDS can be used under these conditions and constraints, and finds that SUDS devices have been designed and used successfully. However, the barriers to their use center on political will, lack of knowledge, and lack of funds to explore their use.

15.3.7 TROPICAL SUDS

The built environment in developing and middle-income countries is affected by drainage problems similar to those in the developed world. These relate to increasing urbanization, reduced infiltration capacity, and poor maintenance of stormwater infrastructure, resulting in surface water and flash flooding, with greater impacts on poorer communities (e.g., Tucci, 2002; Parkinson, Tayler, and Mark, 2007; Steinberg,

2007; Douglas et al., 2008; Ogba and Utang, 2008). In poorer communities, the principal responses to flooding are short-term coping strategies performed by individuals. These strategies include relocating people and property above the expected flood level, building short-term barriers to prevent water ingress into individual properties, and constructing temporary ditches to reroute the water (Douglas et al., 2008).

Collective or governmental action to address surface water flooding issues is not widespread. There is limited recognition of the possibilities for SUDS in developing countries, with responses largely focused on improving conventional drainage (e.g., Gwebu, 2003), although awareness of the need for more sustainable approaches to drainage is increasing in middle-income countries (e.g., Malaysia) (Zakaria et al., 2007). Most factors driving the design of SUDS in subtropical and tropical environments are similar to those in temperate environments. An important exception is that aboveground water features risk-creating environments that are favorable to mosquitoes and other insect vectors of disease. In these cases, vegetated techniques that utilize subsurface flow can create an anaerobic environment that is not conducive to mosquitoes, although these devices may be more expensive to install and maintain (Reed, Crites, and Middlebrooks, 1998). This solution has been used to effect in biological–ecological (or Bio-Ecod) drainage systems in Malaysia (Sidek et al., 2002). Bio-Ecods comprise three components: a swale, subsurface biofiltration storage, and a variety of ponds (Zakaria et al., 2007). The excess stormwater is encouraged to pass through the swale and is conveyed in the subsurface storage units to detention and retention ponds where it is allowed to dissipate slowly. Subsurface storage is used throughout the system, being found beneath the dry ponds as well as the swales.

Many cities throughout the world either currently utilize SUDS as a flood-resilient measure, or plan to do so possibly as part of their global climate change (GCC) strategy to adapt to the changes in flooding regime that have already occurred, but also to mitigate any further changes that may occur in the future. The following section examines some of these strategies globally to further strengthen the argument that SUDS are an important technique that can be integrated into a city's drainage infrastructure. SUDS can be used not only to combat flash flooding and address water resource issues but also to provide amenity, health, and a variety of other benefits associated with wider issues linked to GCC.

15.4   CASE STUDIES

Table 15.3 summarizes the SUDS devices specified in the GCC strategies from 13 of the 17 "C40 Participating Cities" that originally met in 2005 to join together in tackling GCC under the auspices of the Clinton Foundation. Only 2 of the 17 cities (Madrid, Spain, and Houston, Texas) make no mention of drainage infrastructure whatsoever, and the strategies of a further two (Bogata, Colombia, and Rome, Italy) were unavailable. However, of the remainder, 7 cities pledge an increase in GI overall, with 11 and 9 specifying green roofs and street trees, respectively. water sensitive urban design (WSUD) is at the heart of Melbourne's strategy, and Chicago uses green urban design to embed a SUDS infrastructure in its GCC document. Many cities identify flooding resilience as the focus of their approach

TABLE 15.3
SUDS Devices Quoted in Climate Change Strategies of Participating Cities in the C40 Clinton Climate Initiative

City	SUDS Devices	Role
Chicago	PPS in "green alleys," rooftop gardens, rain barrels, rain gardens, increasing urban forest canopy, green roofs, bioswales, street trees	Green urban design, runoff reduction, reduce flooding, urban cooling, provide shade, cool individual homes and hence the city, reduce UHIE, reduce greenhouse gases, increase energy efficiency of buildings with green roofs
Hong Kong	Increased green space, street trees, green roofs and walls	CSS, urban cooling, reduced air conditioning usage
London	Urban greening program, street trees, green roofs and walls, river restoration, pocket parks, "SUDS and flood storage in riverside parks," green space connectivity	Surface water flood risk, increase the quality and quantity of greenspace and vegetation as a buffer from floods and hot weather, "increasing green space and vegetation.... Manage and offset rising temperatures (and manage flood risk)," UHIE mitigation, provide biodiversity
Los Angeles	Stormwater capture and reuse, street trees, "skylight" reaches of the Los Angeles River, stormwater infiltration, more parks, green and cool roofs	Water conservation and recycling, recharging of aquifers, reducing the impact of heat waves
Melbourne	Water-sensitive urban design: "all water streams in the urban water cycle are a resource"	WSUD guidelines applied to climate change mitigation and adaptation, including references to, human health, water resource resilience, flooding, UHIE mitigation
Mexico City	Green roofs	Medium-term impact to adapt to climate change
New York	Source controls, green infrastructure for bioretention and biofiltration, low-impact developments, best management practices, blue and green roofs, "bluebelt" areas using open spaces to absorb excess water, cisterns, RWHS, PPS, street trees	Stormwater management and control due to "more intense and frequent rainfall expected from the effects of climate change" Reduction of UHIE: benefits include "cooling and cleansing the air, reducing energy demand, sequestering and reducing emissions of greenhouse gases"
Philadelphia	Increase green space, green infrastructure, PPS, street trees, green roofs, skylight waterways, clean and green vacant lots; increase street trees	Reconnect land and water so that "green infrastructure becomes the City's preferred stormwater management system" Reduce air pollution, manage stormwater, moderate UHIE, sequester carbon, increase property values
Sao Paulo	Permeable areas, water absorption zones	UHIE mitigation
Seoul	Green space, green roofs, stream restoration	"To increase urban climate control ability"

(continued)

TABLE 15.3 (continued)
SUDS Devices Quoted in Climate Change Strategies of Participating Cities in the C40 Clinton Climate Initiative

City	SUDS Devices	Role
Sydney	RWHS, street trees, open space, urban forest, green roofs	Water efficiency; promotes "environmental, health, social, and financial outcomes for the City" Reduction of human health impacts
Tokyo	Promotion of green space, street trees, PPS, urban forests, water-retaining pavement	Reduction of UHIE, recharge of groundwater, reduction in stormwater flow
Toronto	Street trees, rainwater harvesting, PPS, vegetative landscaping, cool/reflective surfaces, "greening projects," green roofs	Increase shade, clean and cool the atmosphere, water reuse, flooding resilience, reduction of the UHIE "To reduce climate change impacts" Reduction in air-conditioning demands, reduce storm runoff

Note: All quotes are taken from those strategies. NB Reports not available: Bogota, Rome. Reports focusing on energy with no mention of water or drainage: Madrid and Houston. Abbreviations are defined in Section 18.7. All the above city climate change strategies are available through the C40 Cities website (C40 Cities, n.d.) apart from that for London (*Source:* GLA (Greater London Authority) (2010). The draft climate change adaptation strategy for London. Available at <http://www.london.gov.uk/climatechange/strategy>.)

as well as mitigation of the "urban heat island effect" (UHIE), while others such as Los Angeles, Melbourne, and Sydney reflect their drier climate in identifying the importance of water resource management, with both Melbourne and Sydney emphasizing the human health benefits of GI. It would seem, therefore, that SUDS have a significant role to play in any strategy implemented to adapt to or mitigate against surface water flooding, and are found in cities globally. However, Table 15.3 also shows the multiple benefits associated with a SUDS strategy, with many cities citing a wide variety of benefits from positive human health attributes to carbon sequestration and storage (CSS) associated with vegetated techniques in particular.

15.5 CONCLUSIONS

The SUDS concept is not new; it was utilized in the Negev Desert up to 4,000 years ago (Kirby, 2005, p. 120) and, as such, it can be thought of as tried and tested practice. Kirby (2004) also states that SUDS can perform better than traditional piped systems, but only if they are designed, installed, and maintained properly, and also afford sustainability—which hard drainage plainly does not. Sustainable drainage can also provide a multiple-benefit approach to flood risk management by providing a flexible flood-resilient infrastructure.

This flexibility is also reflected in the ability of SUDS to combine individual devices into a train as well as being able to integrate with conventional, hard drainage approaches. SUDS need not stand on their own (Mentens, Raes, and Hermy, 2006); they can integrate with a wide range of other strategies being developed for more efficient, sustainable buildings and cityscapes. These devices also provide opportunities for wildlife, amenity for the city's human inhabitants, and aid in providing a sustainable, healthy urban future. SUDS will not be efficient if local conditions are not accounted for. Tropical SUDS, for instance, not only have to take into account the climate with regular torrential downpours being common, but also that devices must be designed to minimize open water health risks, such as attracting mosquitoes to ponds and wetlands. The same is true for overall urban design that must take a holistic approach to flood-proofing individual buildings, streets, and cities. While the characteristics of the physical structures that make up SUDS have been concentrated on here, implementation of any strategy must obviously be "guided and supported by national policies and strategies" (Burton et al., 2006, p. 9) having regard for the fact that these structures are effectively local and will need to be managed and owned locally.

While adaptation and mitigation are important in addressing problems associated with flooding resilience, there is inherent uncertainty in simply identifying the factors driving change and hence in being able to predict the magnitude of possible changes. The benefits of the SUDS approach are clear, however "... we cannot prevent all flooding—there is always the possibility that extreme events will overtop or breach any defences" (Defra, 2007, p. 3) and hence "a portfolio of responses ..." is required (Evans et al., 2008, p. 51), which includes the integration of SUDS devices and trains globally across the built environment.

REFERENCES

ABI (Association of British Insurers) (2007). Summer Floods 2007: Learning the Lessons. 27 pp.

Akbari, H., Rose, L.S., and Taha, H. (2003). Analyzing the land cover of an urban environment using high-resolution orthophotos. *Landscape and Urban Planning*, 63, 1–14.

Antonelli, L. (2008). Alive and well: Bringing nature back into building design. *Construct Ireland*, 4(4), 6.

Apostolaki, S., Jefferies, C., and Wild, T. (2006). The social impacts of stormwater management techniques. *Water Practice & Technology*, 1(1), 1–8.

Balmforth, D., Digman, C., Kellagher, R., and Butler, D. (2006a). *Designing for exceedance in urban drainage—Good practice (C635)*. London: Construction Industry Research and Information Association.

Balmforth, D., Digman, C., Butler, D., and Shaffer, P. (2006b). Integrated Urban Drainage Pilots: Scoping Study. Available at <http://www.defra.gov.uk/environment/flooding/documents/manage/surfacewater/scoperev.pdf> [Accessed April 2010].

Blackburn, M., Methven, J., and Roberts, N. (2008). Large-scale context for the UK floods in Summer 2007. *Weather*, 63(9), 280–288.

Bolund, P., and Hunhammar, S. (1999). Ecosystem services and urban areas. *Ecological Economics*, 29, 293–301.

Booth, D. (2000). Field Evaluation of Permeable Pavements for Stormwater Management: Olympia, Washington. EPA-841-B-00-005B. United States Environmental Protection Agency. 2 pp.

Bosher, L., Carrillo, P., Dainty, A., Glass, J., and Price, A. (2007). Realising a resilient and sustainable built environment: Towards a strategic agenda for the United Kingdom. *Disasters*, 31(3), 236–255.

Bray, B. (2000). Design of the "treatment train" at Hopwood Park MSA, M42. *Standing Conference on Stormwater Source Control*. Coventry University.

Bray, B. (2004). Sustainable drainage retrofit at Waseley Hills School, Rubery, Worcs. In *Atkins Water: Scottish Water SUDS Retrofit Research Project*. Available at <http://www.scotland.gov.uk/Topics/Environment/Water/bathingwaters/RetrofittingSUDS:Appendix D> [Accessed April 2010].

Bray, B. (2006). Integrating SUDS and river restoration to protect and enhance urban watercourses. *River Restoration News*, 24, 4–5.

British Academy (2005). Submission to the Royal Commission on Environmental Pollution for its Study of the Urban Environment. Available at <http://www.britac.ac.uk/templates/asset-relay.cfm?frmAssetFileID=6437> [Accessed April 2010].

Burgess, E. W. (1924). The growth of the city: An introduction to a research project. *Publications of the American Sociological Society*, 18, 85–97.

Burton, I., Diringer, E., and Smith, J. (2006). Adaptation to Climate Change: International Policy Options. Pew Center on Global Climate Change. 24 pp.

Butler, D., and Davies, J. W. (2004). *Urban Drainage*. London: E & FN Spon.

C40 Cities (n.d.). Available at <http://www.c40cities.org/cities/> [Accessed April 2010].

Campana, N. A., and Tucci, C. E. M. (2001). Predicting floods from urban development scenarios: Case study of the Dilúvio Basin, Porto Alegre, Brazil. *Urban Water*, 2, 113–124.

Cawley, A. M., and Cunnane, C. (2003). Comment on estimation of greenfield runoff rates. *National Hydrology Seminar*, pp. 29–43.

Cembrano, G., Quevedo, J., Salamero, M., Puig, V., Figueras, J., and Martí, J. (2004). Optimal control of urban drainage systems. A case study. *Control Engineering Practice*, 12, 1–9.

Charlesworth, S. M., Harker, E., and Rickard, S. (2003). A review of sustainable drainage systems (SuDS): A soft option for hard drainage questions? *Geography*, 88(2), 99–107

CIRIA (Construction Industry Research and Information Association) (2000). Sustainable Urban Drainage Systems: A design manual for England and Wales. Report C522. London: CIRIA.

City of Portland (2005). NE Siskiyou Green Street Project. Available at <http://www.portlandonline.com/shared/cfm/image.cfm?id=78299> [Accessed April 2010].

CIWEM (The Chartered Institution of Water and Environmental Management) (2009). Blue Corridors Make Space for Water. Available at <www.ciwem.org/press/20090223_ACBlueCorridors1.doc> [Accessed April 2010].

CIWEM (The Chartered Institution of Water and Environmental Management) (2008). Flood and Coastal Erosion Risk Management. Policy Position Statement. Available at <http://www.ciwem.org/policy/policies/flood_risk_management.asp> [Accessed April 2010].

Claytor Jr., R. A. (1998). An eight-step approach to implementing stormwater retrofitting. In *National Conference on Retrofit Opportunities for Water Resource Protection in Urban Environments: Proceedings*. Chicago, Illinois February 9–12, pp. 212–218. Available at <http://www.epa.gov/nrmrl/pubs/625r99002/625r99002.pdf> [Accessed April 2010].

Coupe, S. J., Smith, H. G., Newman, A. P., and Puehmeier, T. (2003). Biodegradation and microbial diversity within permeable pavements. *European Journal of Protistology*, 39, 495–498.

DBERR (Department for Business, Enterprise and Regulatory Reform) (2008). Strategy for Sustainable Construction. Available at <http://www.berr.gov.uk/files/file46535.pdf> [Accessed April 2010].

DCLG (Department for Communities and Local Government) (2009). Planning Policy Statement 25: Development and Flood Risk. London: The Stationery Office.

Defra (Department for Environment, Food and Rural Affairs) (2002). Managed Realignment Review Project Report. Policy Research Project FD 2008. 324 pp.

Defra (Department for Environment, Food and Rural Affairs) (2005). Making space for water. Available at <http://www.defra.gov.uk/environment/flooding/documents/policy/strategy/strategy-response1.pdf> [Accessed April 2010].

Defra (Department for Environment, Food and Rural Affairs) (2007). Funding and Charging Arrangements for Sustainable Urban Drainage Systems. Available at <http://www.defra.gov.uk/environment/quality/water/waterquality/diffuse/non-agri/documents/suds-report.pdf> [Accessed April 2010].

Defra (Department for Environment, Food and Rural Affairs) (2008). Making Space for Water Urban Flood Risk and Integrated Urban Drainage Pilot Summary Report. Available at http://www.defra.gov.uk/environment/flooding/documents/manage/surfacewater/urban drainagereport.pdf [Accessed April 2010]

Defra (Department for Environment, Food and Rural Affairs) (2010). Natural Environment Adapting to Climate Change. 52 pp. Available at <http://www.defra.gov.uk/environment/climate/documents/natural-environment-adaptation.pdf> [Accessed April 2010].

Dilley, M. J., Chen, R. S., Deichmann, U., Lerner-Lam, A. L., Arnold, M., with Agwe, J., Buys, P., Kjekstad, O., Lyon, B., and Yetman, G. (2005). *Natural Disaster Hotspots: A Global Risk Analysis*. Disaster Risk Management Series No. 5. Washington, DC: The World Bank, Hazard Management Unit. 132 pp.

DTI (2006). Sustainable drainage systems: a mission to the USA. Global Watch Mission Report. 148 pp. Available at <http://www.britishwater.co.uk/Document/Download.aspx?uid=b0e0f905-9ffc-4a99-a057-38e7feff93a1> [Accessed April 2010].

Douglas, I., Alam, K., Maghenda, M., McDonnell, Y., McLean, L., and Campbell, J., (2008). Unjust waters: climate change, flooding and the urban poor in Africa. *Environment and Urbanization*, 20(1), 187–205.

Doyle, P., Hennelly, B., and McEntee, D. (2003). SUDS in the Greater Dublin area. In *Irish National Hydrology Seminar.* Available at <http://www.opw.ie/hydrology/data/speeches/I_Padraig%20Doyle%20et%20al.pdf> [Accessed April 2010].

EA (Environment Agency) (no date). Sustainable Drainage Systems (SUDS): An Introduction. Bristol: Environment Agency. 24 pp.

EA (Environment Agency) (2008). Lower Thames Flood Risk Management Strategy. Available at <https://consult.environment-agency.gov.uk/portal/re/flood/thames/lts?pointId=9094 89#document-909489> [Accessed April 2010].

EA (Environment Agency) (2010) The Costs of the Summer 2007 Floods in England. Project: SC070039/R1. Available at <http://evidence.environment-agency.gov.uk/FCERM/Libraries/FCERM_Project_Documents/The_cost_of_summer_2007_floods_in_England.sflb.ashx> [Accessed April 2010].

Emscher Cooperative (2009). Rain Water Projects. Available at <http://www.eglv.de/wasser-portal/regenwassernutzung.html> [Accessed April 2010].

Escarameia, M., Todd, A. J., and Watts, G. R. A. (2006). Pollutant Removal Ability of Grassed Surface Water Channels and Swales: Literature Review and Identification of Potential Monitoring Sites. Highways Agency. 46 pp.

Evans, E., Ashley, R., Hall, J., Penning-Rowsell, E., Sayers, P., Thorne, C., and Watkinson, A. (2004). *Foresight Future Flooding*, Volume I and Volume II. London: Office of Science and Technology.

Evans, E. P., Simm, J. D., Thorne, C. R., Arnell, N. W., Ashley, R. M., Hess, T. M., Lane, S. N., Morris, J., Nicholls, R. J., Penning-Rowsell, E. C., Reynard, N. S., Saul, A. J., Tapsell, S. M., Watkinson, A. R., and Wheater, H. S. (2008). *An Update of the Foresight Future Flooding 2004 Qualitative Risk Assessment.* London: Cabinet Office.

Ferguson, B. K. (2002). Stormwater management and stormwater restoration. In France, R. L. (Ed.), *Handbook of Water Sensitive Planning and Designs.* New York: Lewis Publishing.

GDSDS (Greater Dublin Strategic Drainage Study) (2005). *Sustainable drainage systems. Regional drainage policies—Volume 3, Environmental management,* Chapter 6. 25 pp. Available at <http://www.dublincity.ie/WATERWASTEENVIRONMENT/WASTEWATER/DRAINAGE/GREATERDUBLINSTRATEGICDRAINAGESTUDY/Pages/ClimateChange.aspx. [Accessed April 2010].

Gibbs, G. (2004). Preliminary rainfall runoff management for developments (EA/DEFRA W5-074/A) Summary information. Available at <http://www.ciria.org.uk/suds/pdf/preliminary_rainfall_runoff_mgt_for_development.pdf> [Accessed April 2010].

GLA (Greater London Authority) (2010). The draft climate change adaptation strategy for London. Available at <http://www.london.gov.uk/climatechange/strategy> [Accessed April 2010].

GLA (Greater London Authority) (2008). The London Plan: Spatial Development Strategy for Greater London. 508 pp.

Gordon-Walker, S., Harle, T., and Naismith, I. (2007). Cost-Benefit of SUDS Retrofit in Urban Areas. Science Report SC060024 . Bristol: Environment Agency. Available at <http://publications.environment-agency.gov.uk/pdf/SCHO0408BNXZ-e-e.pdf> [Accessed April 2010].

Gwebu, T. D. (2003). Environmental problems among low income urban residents: An empirical analysis of old Naledi-Gaborone, Botswana. *Habitat International,* 27, 407–427.

Harris, C. D., and Ullman, E. L. (1945). The nature of cities. *Annals of the American Academy of Political and Social Science,* 242, 7–17.

Heal, K. V., Bray, R., Willingale, S. A. J., Briers, M., Napier, F., Jefferies, C., and Fogg, P. (2008). Medium-term performance and maintenance of SUDS: A case study of Hopwood Park Motorway Service Area, UK. *11th International Conference on Urban Drainage,* Edinburgh, Scotland, UK.

Hirschman, D. J., and Kosco, J. (2008). *Managing Stormwater in Your Community: A Guide for Building an Effective Post-Construction Program.* Ellicott City, MD: Center for Watershed Protection. Available at <http://www.cwp.org/Resource_Library/Controlling_Runoff_and_Discharges/sm.htm> [Accessed April 2010].

Horner, R. R., Lim, H., and Burges, S. J. (2002). Hydrologic Monitoring of the Seattle Ultra-Urban Stormwater Management Projects. Available at <http://www.ci.seattle.wa.us/util/stellent/groups/public/@spu/@usm/documents/webcontent/spu02_020016.pdf> [Accessed April 2010].

HR Wallingford (2005). Preliminary Rainfall Runoff Management for Developments. EA/Defra Technical Report W5-074/A.

HR Wallingford (2008). UK Stormwater Drainage: Guidance and Tools. Available at <http://gamma.hrwallingford.co.uk/UKStormwaterDrainage/index.htm> [Accessed April 2010].

Hyder Consulting (UK) Ltd. (2004). Retrofitting Sustainable Urban Drainage Systems: Case Study—Dunfermline. Report No: NE02351/D1. Available at <http://www.scotland.gov.uk/Topics/Environment/Water/bathingwaters/RetrofittingSUDS> [Accessed April 2010].

IEA (Institution of Engineers, Australia) (1987). *Australian Rainfall and Runoff—A Guide to Flood Estimation.* Vol. 1.

IH (Institute of Hydrology) (1994). *Flood Estimation for Small Catchments.* Wallingford, UK: Institution of Engineers, Australia.

Jeb, D., and Aggarwal, S., (2008). Flood inundation hazard modelling of the River Kaduna using remote sensing and geographic information systems. *Applied Sciences Research,* 4 (12), 1822–1833. Available at <http://insipub.net/jasr/2008/1822-1833.pdf> [Accessed April 2010].

Jefferies, C., Aitken, A., McLean, N., Macdonald, K., and McKissock, G. (1999). Assessing the importance of urban BMPs in Scotland. *Water Science and Technology*, 39(12), 123–131.

Jefferies, C., Duffy, A., Zuurman, A., and Tingle, S. (2008). Disconnection of surface water drainage – A local authority perspective. In *11th International Conference on Urban Drainage*. Edinburgh, Scotland, UK.

Jones, P., and Macdonald, N. (2007). Making space for unruly water: Sustainable drainage systems and the disciplining of surface runoff. *Geoforum*, 38, 534–544.

Jonkman, S. N. (2005). Global perspectives on loss of human life caused by floods. *Natural Hazards*, 34, 151–175.

Kellagher, R. B. B., and Maneiro Franco, E. (2005). Rainfall Collection and Use in Developments. Briefing Report for WaND Project. HR Wallingford Report No. SR677.

Keller, E. A., and Hoffman, E. K. (1977). Urban streams: Sensual blight or amenity? *Journal of Soil and Water Conservation*, 32(5), 237–240.

Kirby, A. (2004). SUDS – Innovation or a tried and tested practice? *Municipal Engineer*, 158(ME2), 115–122.

Kloss, C., and Calarusse, C. (2006). *Rooftops to rivers: Green strategies for controlling stormwater and combined sewer overflows*. New York: Natural Resources Defense Council. Available at <http://www.nrdc.org/water/pollution/rooftops/rooftops.pdf> [Accessed April 2010].

Lambeth Council (no date). Available at <http://www.lambeth.gov.uk/Services/HousingPlanning/Planning/EnvironmentalIssues_EXTRA.htm> [Accessed April 2010].

Leopold, L. B. (1968). *Hydrology for urban planning – A Guidebook on the Hydrologic Effects of Urban Land Use*. Geological Survey Circular 554. United States Department of the Interior, Geological Survey, pp. 119–125.

London Assembly (2005). Crazy Paving. The Environmental Importance of London's Front Gardens. London: Greater London Authority. 25 pp.

Loster, T. (1999). Flood Trends and Global Change. Available at <http://www.iiasa.ac.at/Research/RMS/june99/papers/loster.pdf> [Accessed April 2010].

MacDonald, K., and Jefferies, C. (2003). Performance and design details of SUDS. *National Hydrology Seminar*, pp. 93–102.

MacMullan, E., and Reich, S. (2007). The Economics of Low-Impact Development: A Literature Review. Available at <http://pepi.ucdavis.edu/mapinfo/pdf/Economics-Literature-Review.pdf> [Accessed April 2010].

Malcom, M., Woods-Ballard, B., Weisgerber, A., Biggs, J., and Apostolaki, S. (2003). The hydraulic and water quality performance of sustainable drainage systems, and the application for new developments and urban river rehabilitation. *National Hydrology Seminar*, pp. 83–92.

Marsalek, J., Watt, W. E., Zeman, E., and Sieker, F. (Eds.) (2000). *Flood Issues in Contemporary Water Management*. Dordrecht, the Netherlands: Kluwer Academic Publishers.

McCarthy, J. J., Canziani, O. F., Leary, N. A., Dokken, D. J., and White, K. S. (2001). *Climate Change 2001: Impacts, Adaptation, and Vulnerability. Intergovernmental Panel on Climate Change*. Cambridge: Cambridge University Press.

Mentens, J., Raes, D., and Hermy, M. (2006). Green roofs as a tool for solving the rainwater runoff problem in the urbanised 21st century? *Landscape and Urban Planning*, 77, 217–226.

Met Office (n.d.). Floods 2008. Available at <http://www.metoffice.gov.uk/corporate/verification/case_studies_08.html> [Accessed April 2010].

Mitchell, G. (2005). Mapping hazard from urban non-point pollution: A screening model to support sustainable urban drainage planning. *Journal of Environmental Management*, 74, 1–9

Myatt-Bell, L. B., Scrimshaw, M. D., Lester, J. N., and Potts, J. P. (2002). Public perceptions of managed retreat: Brancaster West Marsh, North Norfolk, UK. *Marine Policy*, 26(1), 45–57.

Ogba, C. O., and Utang, P. (2008). Integrated Approach to Urban Flood Adaptation in the Niger Delta Coast of Nigeria. Available at <http://www.Figurenet/pub/fig2008/papers/ts08e/ts08e_02_ogba_utang_2736.pdf> [Accessed April 2010].

Parkinson, J., Tayler, K., and Mark, O. (2007). Planning and design of urban drainage systems in informal settlements in developing countries. *Urban Water Journal*, 4(3), 137–149

Perry, K. R. (2009). *Transforming Gray Space into "Green Space": Integrating Stormwater with Urban Design*. Available at <http://www.epa.gov/npdes/pubs/gi_conn_retrofit.pdf> [Accessed April 2010].

Perry, T., and Nawaz, R. (2008). An investigation into the extent and impacts of hard surfacing of domestic gardens in an area of Leeds, United Kingdom. *Landscape and Urban Planning*, 86(1), 1–13.

Pitt, M. (2007). *Learning lessons from the 2007 floods*. The Cabinet Office. 165 pp.

POST (Parliamentary Office of Science and Technology) (2007). Postnote 289: Urban Flooding . Available at <http://www.parliament.uk/documents/upload/postpn289.pdf> [Accessed April 2010].

RCEP (Royal Commission on Environmental Pollution) (2007). *The Urban Environment*. Available at <http://www.rcep.org.uk/reports/26-urban/documents/urban-environment.pdf> [Accessed April 2010].

Reed, S. C., Crites, R. W., and Middlebrooks, J. (1998). *Natural systems for Waste Management and Treatment* (2nd edition). New York: McGraw-Hill.

Rose, L. S., Akbari, H., and Taha, H. (1999). Characterizing the Fabric of an Urban Environment: A Case Study of Houston, Texas. LBNL-51448. Lawrence Berkley National Laboratory, Berkley, CCA. 24 pp.

RHS (Royal Horticultural Society) (2005). Gardening Matters: Front Gardens. Are We Parking on Our Gardens? Do Driveways Matter? Royal Horticultural Society. 14 pp.

RRC (River Restoration Centre) (1999, 2002 Update). *River restoration: Manual of techniques*.

Scholz, M., and Grabowiecki, P. (2007). Review of permeable paving systems. *Building and Environment*, 42, 3830–3836.

Schueler, T., Hirschman, D., Novotney, M., and Zielinski, J. (2007). *Urban stormwater retrofit practices*. Ellicott City, MD: Center for Watershed Protection. Available at <http://www.cwp.org/Store/usrm.htm#3> [Accessed April 2010].

Semadeni-Davies, A., Hernebring, C., Svensson, G., and Gustafsson, L.-G. (2008). The impacts of climate change and urbanization on drainage in Helsingborg, Sweden: Suburban stormwater. *Journal of Hydrology*, 350, 114–125.

SEPA (Scottish Environment Protection Agency) (2005). A DOs and DON'Ts Guide for Planning and Designing Sustainable Urban Drainage Systems (SUDS). Available at <http://www.sepa.org.uk/customer_information/idoc.ashx?docid=871d1682-f4de-4b49-acd0-013e8b7ce4ee&version=-1> [April 2010].

SEPA (Scottish Environment Protection Agency) (n.d.). Sustainable Urban Drainage Systems: Setting the Scene in Scotland.

Shaver, E., Horner, R., Skupien, J., May, C., and Ridley, G. (2007). *Fundamentals of Urban Runoff Management: Technical and Institutional Issues* (second edition). Madison, WI: North American Lake Management Society. Available at <http://www.nalms.org/nalmsnew/Nalms_Publication.aspx?Sid=3> [Accessed April 2010].

Shaw, R., Colley, M., and Connell, R. (2007). *Climate Change Adaptation by Design: A Guide for Sustainable Communities*. London: TCPA.

Shutes, B. (2008). SWITCH Deliverable 2.2.1B: The Potential of BMPs to Integrate with Existing Infrastructure (i.e. Retro-Fit/Hybrid Systems) and to Contribute to Other Sectors of the Urban Water Cycle. Available at <http://www.switchurbanwater.eu/ outputs/pdfs/W2-1_DEL_Part_B_Integration_and_contribution.pdf> [Accessed April 2010].

Sidek, L. M., Takara, K., Ab Ghani, A., Zakaria, N. A., and Abdullah, R. (2002). Bio-Ecological Drainage Systems (BIOECODS): An Integrated Approach for Urban Water Environmental Planning. Available at <http://redac.eng.usm.my/html/publish/2002_11. pdf> [Accessed April 2010].

Sieker, H., Bandermann, S., Becker, M., and Raasch, U. (2006). Urban Stormwater Management Demonstration Projects in the Emscher Region. Available at <http://switchurbanwater. lboro.ac.uk/outputs/pdfs/CEMS_PAP_Urban_stormwater_management_demo_projects_Emscher.pdf> [Accessed April 2010].

Singh, R., Jefferies, C., Stovin, V., Morrison, G., and Gillon, S. (2005). Developing a planning and design framework for retrofit SUDS. In *10th International Conference on Urban Drainage*, Copenhagen, Denmark, 21–26 August.

Smith, K. (2004). *Environmental Hazards: Assessing Risk and Reducing Disaster* (4th ed.). London: Routledge.

SNIFFER (Scotland and Northern Ireland Forum for Environmental Research) (2004). *SUDS in Scotland — The Monitoring Programme. Project SR (02)51.* Available at <http://sudsnet. abertay.ac.uk/documents/SNIFFERSR_02_51MainReport.pdf> [Accessed April 2010].

SNIFFER (Scotland and Northern Ireland Forum for Environmental Research) (2006).*Retrofitting Sustainable Urban Water Solutions. Project UE3(05)UW5.* Available at <http://retrofitsuds.group.shef.ac.uk/downloads/UE3(05)UW5%5B1%5D.pdf> [Accessed April 2010].

SNIFFER (Scotland and Northern Ireland Forum for Environmental Research) (2007). Retrofit SUDS: Workshop Report. Available at <http://sudsnet.abertay.ac.uk/documents/ RetrofitworkshopFINAL120717.pdf> [Accessed April 2010].

Stahre, P. (2008). *Blue-green fingerprints in the city of Malmö: Malmö's way to a sustainable urban drainage* . Malmö, Sweden: VA SYD. Available at <http://www.vasyd.se/ SiteCollectionDocuments/Broschyrer/Publikationer/BlueGreenFingerprints_Peter. Stahre_webb.pdf> [Accessed April 2010].

Steinberg, F. (2007). Jakarta: Environmental problems and sustainability. *Habitat International,* 31, 354–365.

Stovin, V., Swan, A., and Moore, S. (2007). *Retrofit SUDS for urban water quality enhancement.* Available at <http://retrofit-suds.group.shef.ac.uk/downloads/EA&BOCF%20 Retrofit%20SUDS%20Final%20Report.pdf> [Accessed April 2010].

Tucci, C. E. M. (2002). *Improving flood management practices in South America: Workshop for decision makers.* Available at <http://www.wmo.int/pages/prog/hwrp/documents/ FLOODS_IN_SA.pdf> [Accessed April 2010].

Urban Water Project Partnership (2008). *Urban water—Living cities: Spotlights and experiences from the Urban Water Project 2003–2008.* Essen, Germany: Emschergenossenschaft. Available at <http://www.urban-water.org/cms/fileadmin/user_upload/008_project_ results/uw_Report.pdf> [Accessed April 2010].

US EPA (United States Environmental Protection Agency) (n.d.) National Menu of Stormwater Best Management Practices. Available at <http://www.epa.gov/npdes/stormwater/menu ofbmps> [Accessed April 2010].

Villarreal, E. L., Semadeni-Davies, A., and Bengtsson, L. (2004). Inner city stormwater control using a combination of best management practices. *Ecological Engineering, 22,* 279–298.

Water UK (2008). Lessons learnt from Summer Floods 2007: Phase 2 report—Long-Term Issues. Water UK's Review Group on Flooding.

Weinstein, N. (2002). Low impact development retrofit approach for urban areas. In *Proceedings of the Seventh Biennial Stormwater Research and Watershed Management Conference*. Brooksville, FL: Southwest Florida Water Management District. May 22–23, pp. 98–104. Available at <http://www.swfwmd.state.fl.us/documents/publications/7thconference_proceedings.pdf> [Accessed April 2010].

Wheater, H., and Evans, E. (2009). Land use, water management and future flood risk. *Land Use Policy*, 26S, S251–S264.

White, I., and Howe, J. (2004). The mismanagement of surface water. *Applied Geography*, 24, 261–280.

Whitford, V., Ennos, A. R., and Handley, J. F. (2001). City form and natural process —Indicators for the ecological performance of urban areas and their application to Merseyside, UK. *Landscape and Urban Planning*, 57, 91–103.

Wilby, R. L. (2007). A review of climate change impacts on the built environment. *Built Environment*, 33(1), 31–45.

Wilson, S., Bray, R., and Cooper, P. (2004). *Sustainable drainage systems: Hydraulic, structural and water quality advice. C609*. London: Construction Industry Research and Information Association.

WMO (World Meteorological Organization) (2008). Urban flood risk management; a tool for integrated flood management. Associated programme on flood management. Available at <http://www.apfm.info/pdf/ifm_tools/Tools_Urban_Flood_Risk_Management.pdf> [Accessed April 2010].

Woods Ballard, B., Kellagher, R., Martin, P., Jefferies, C., Bray, R., and Shaffer, P. (2007). *The SUDS manual (C697)*. London: Construction Industry Research and Information Association.

Zakaria, N. A., Ab Ghani, A., Ayub, K. R., and Ramli, R. (2007). Sustainable Urban Drainage Systems (SUDS). In *Rivers'07: 2nd International Conference on Managing Rivers in the 21st Century*, Sarawak, Malaysia June 6–8. Available at <http://redac.eng.usm.my/html/publish/2007_13.pdf> [Accessed April 2010].

16 Land Use Planning Issues

Bill Finlinson

CONTENTS

16.1 INTRODUCTION: THE IMPORTANCE OF LAND USE IN FLOOD MANAGEMENT

The way we use our land, and plan for its use, are central to the way we manage floods and thereby minimize the risk of flooding to our communities and the infrastructure that supports them. Land has two main functions in relation to flooding: (1) it is the land that generates the floods through runoff of surface water and (2) it is the land that provides areas in which floodwaters can be temporarily conveyed or stored.

Flood management seeks to minimize damage to properties, infrastructure, and other assets through a wide variety of interventions ranging from hard engineering (e.g., built defenses) through "softer" solutions (such as provision of flood storage wetlands) to community responses such as flood warning plans or provision of evacuation routes and shelter refuges.

Land use planning can contribute greatly to minimizing flood risk by regulating the use to which land is put and the type of development allowed and by ensuring that new development is sited away from areas of flood risk wherever possible. Planning regulations can also be used to minimize the amount of runoff generated from an area of land by enforcing the use of the latest sustainable drainage sustainable urban drainage system (SuDS) techniques and making sure that sufficient room is left for floodwaters to flow or be stored.

There are a number of sources of flooding. These include coastal flooding from the sea and estuaries and inland flooding where sources include watercourses (fluvial flooding), heavy rainfall causing overland flow (pluvial flooding), sewer flooding (when the capacity of the piped drainage network is exceeded), and flooding due to high groundwater levels emerging as springs.

The severity of an inland flood is influenced by a number of factors, including the intensity and duration of the storm that causes it, the geology or soils of the land on which the rain falls, how wet the catchment is before the storm begins, and the form of the catchment (e.g., whether it is short and steep-sided or long and shallow). Groundwater flooding (water that emerges from the ground due to an exceptionally high water table) may be caused by an abnormally extended period of rainfall. The amount of surface runoff is also affected by the use to which the land is put—for example, whether it is urban or rural, whether paved or built over, the type of farming, forestry, conservation area, etc., and any flood mitigation measures associated with these uses. In other words, the way we use the land plays an important part in determining the percentage of stormwater that is converted to runoff and hence may contribute to a flood further down the catchment.

Water running off land is usually collected in drainage networks that may consist of natural watercourses, man-made drainage networks, or a combination of both. In early times, people were aware of the tendency of certain areas to flood and it is noticeable that the old parts of towns, and especially churches, often seem to be sited on higher ground. In more modern times with more intense development and mobility of population, people have tended to forget about flooding and the need to avoid floodplain areas. Until about 25 years ago, there was little planning consideration given to river corridors and the floodplains associated with them. As a result, some corridors became overly constricted by flood defenses, and floodplains were built over by new development. It is now realized that land must be set aside for conveying and storing floodwater, whether from coastal flooding or inland flooding and that this must apply to all stages in the runoff process, including water that has not yet reached the drainage network.

In a few urban areas, where existing communities are already at risk of flooding, there is often little opportunity to retrofit such mitigation measures. In these cases, spatial planners may need to work with emergency planners to look for other solutions such as the use of walls and embankments as secondary defenses, creating flood-resilient properties, setting aside space for refuge areas and implementing evacuation plans activated by flood warning schemes.

With current pressures on land use and mounting concern over the increased severity of flood events due to climate change, it is realized that forward planning must account for more severe flood risk and coastal change in the design and future-proofing of our communities and key national assets. This will mean taking a longer-term strategic view and involve tackling some difficult social and political issues. These may include abandoning areas that are no longer economical to defend, and in the long term relocating communities to less vulnerable areas.

16.2 THE PLANNING CHALLENGE

There is a general realization that we need to have more effective integration of our longer-term community planning and our environmental resource planning. New developments must be viable in terms of the resources they need and the waste that they create, and this includes trying to ensure that the level of development can be supported without detriment to the environment, including flood risk.

Traditionally, flood risk management and water (or wastewater) resource planning have been carried out largely in isolation from town and country planning. This situation is now changing through the introduction of a number of studies commissioned by local authorities as part of the evidence base for supporting their local development framework. These include water cycle studies that look at all water-related aspects of the development strategy by directly examining the viability of proposed development plans in terms of available water resources, sufficiency of wastewater infrastructure, the ability of receiving watercourses to receive treated water, and the risk of flooding.

Nevertheless, at present there is a lack of integration between coastal and inland flood risk management planning (as carried out by the Environment Agency [EA] in England and Wales) and other flood operating authorities) and the town and country planning system. Flood risk within local authority spatial planning is largely represented in terms of constraints through development control rather than plans for active intervention to improve flood risk in the existing community as a whole. One challenge is to knit together the current catchment flood management plans (and more detailed surface water management plans currently under development) with the planning authorities' land use plans (e.g., site allocations development plan documents [DPDs]) and reflect their long-term objectives within the council's emergency and spatial (land use) planning. A further challenge is to manage land-use changes that fall outside the remit of the planning process—for example, agricultural use in the wider countryside and minor urban improvements at a household level such as driveway or garden alterations.

The remainder of this chapter looks first at the historical and legislative background that has led to significant new legislation for water and flood management in England, Wales, and Scotland (Section 16.3). Section 16.4 examines the current hierarchy of shoreline and catchment flood management plans that provide the overarching strategies for flood and coastal management. The current procedures for managing flood risk for new development are discussed in Section 16.5.

Finally, Section 16.6 looks at changing responsibilities as a result of the new Floods and Water Management Act (OPSI, 2010) and Flood Risk Regulations (OPSI, 2009) and, in particular, at the requirements of local authorities to develop and lead multi-stakeholder surface water management plans that will provide the basis for more integrated flood management plans at the local scale as required by the Floods directive. This is setting new challenges for local authorities in organizing multi-funded projects and in finding the resources, skills, and funding necessary to fulfill their new role in flood management.

16.3   HISTORICAL AND LEGISLATIVE BACKGROUND

Following extensive river flooding in 1946/1947 and the Lynmouth disaster in 1952, the 1953 East Coast and Thames tidal flooding event was a major catalyst for change. The damages caused by the 1953 event provided the impetus for a wave of defense building along the eastern coastline and eventually led to the creation of the Thames

Tidal Defences that we see today. Between the mid-1950s and late 1990s, flood management in the United Kingdom was largely an engineering matter and focused on securing and protecting land as well as property through the provision of hard flood defenses allied to improved drainage. This period happened to coincide with a number of largely flood-free decades. The formation of the National Rivers Authority (later the Environment Agency) began the process of formalizing development control requirements. In the early 1990s, flood management as known today began to develop in harness with a new concern for the aquatic environment, and this was greatly accelerated by the extreme floods that affected large areas of England and Wales Easter 1998 and during 2000. These floods helped to drive the development of the first national floodplain extent maps, which for the first time provided a nationally consistent spatial measure of flood risk that could be accessed by planners and public alike.

During the past decade, there have been further extreme flood events (e.g., Boscastle, Carlisle, and the summer 2007 floods) that, along with longer-term climate predictions (Evans et al., 2004), have heightened concern that climate change is altering the weather patterns toward more intense and prolonged storm events as well as causing sea levels to rise. As a result, planning controls on development have become more stringent through the introduction of PPG 25 (2001) and later by Planning Policy Statement 25 (PPS25), Development and Flood Risk, (DCLG, 2006; revised 2010). This is supported by its accompanying practice guide (DCLG, 2007). PPS25 sets out requirements for both local authorities and developers with respect to new development. In Wales, similar requirements are set out in Technical Advice Note 15 TAN 15 (Development and Flood Risk) (Planning Policy Wales, 2004). The Scottish government's planning policies have recently been subject to extensive revision and are now set out in the second National Planning Framework (NPF2), which is supported by a revised and consolidated Scottish Planning Policy (SPP) 7 (Scottish Executive, 2010) and planning advice notes (PANs).

The floods of summer 2007 caused national concern and were a catalyst for far-reaching changes that are now being implemented. The novel features of the 2007 floods that caused the most concern were the duration and intensity of storms causing intense flooding in smaller catchments, both urban and rural, including major urban flooding in Hull and other Yorkshire cities, and the threat to key infrastructure, notably water supplies and a power station, closure of the M5, and damage to Uley Reservoir in Yorkshire. A number of major consequential losses, including the evacuation of large populations in Gloucestershire, were only narrowly avoided.

The ensuing independent review (Pitt, 2008) was hard-hitting and set out a number of key areas for improvement including 53 specific recommendations. These included greater coordination between the Met Office and the Environment Agency that culminated in the establishment of a new joint Flood Forecasting Centre. Many of the other recommendations related to new responsibilities for the Environment Agency and Local Authorities and have been incorporated within England and Wales into the Flood & Water Management Act (OPSI, 2010). This is discussed further in Section 16.6.

In parallel with this, European legislation has developed, through the introduction of the Floods Directive (European Commission for the Environment, 2007), a companion to the Water Framework Directive, and driven partly by similar extreme flood events in Europe that occurred over the same period as the UK floods. The Directive sets out requirements for reporting on flood risks and plans to deal with them with a program up until 2015. This has now been translated into law in England and Wales by the Flood Risk Regulations (OPSI, 2009) and in Scotland by the Flood Risk Management (Scotland) Act (Scottish Parliament, 2009). A significant feature regards the new responsibilities placed on the Environment Agency and local authorities.

16.4   SHORELINE AND CATCHMENT FLOOD MANAGEMENT PLANS

Floods are best managed within the context of the natural environment in which they originate rather than the administrative area in which they occur, and plans are therefore based on stretches of coastline (in the case of coastal processes and flooding) or catchments (in the case of inland flooding). Spatial planners are therefore faced with integrating the requirements of flood and coastal management into a planning system that is based on regional and local planning authority boundaries rather than natural boundaries.

A program of shoreline management plans (SMPs) and catchment flood management plans (CFMPs) has been implemented over the past decade. These have been overseen by the Environment Agency and their preparation has involved a wide range of stakeholder consultation, including local authorities, internal drainage boards, water companies, and others (Defra, 2006, 2010; EA, 2010).

In the past 10 years significant progress has been made in understanding and mapping coastal processes through the first generation of SMPs that cover the 6,000 kilometers of coastline in England and Wales. SMPs provide a large-scale assessment of the risks associated with coastal processes and present a long-term policy framework to reduce these risks to people and the developed, historic, and natural environment in a sustainable manner.

The length of shore covered by each plan is primarily dictated by the coastal geology and processes at work, and may cover a number of local authority areas. SMPs have, in general, been led by local authorities who are often responsible for sea defense and coastal repair in their area. SMPs set out the strategy for flood management and coastal erosion for their section of coast—for example, if the coast is currently low-lying and defended, whether the defenses will be maintained in the future ("hold the line") or no longer maintained where it is no longer economical to defend the land ("managed realignment") based on future predictions of sea level rise and coastal erosion.

In a similar way, England and Wales are covered by 77 CFMPs that are based on catchment boundaries and cover the catchment as far down as its boundary with the relevant SMP, usually at the tidal limit of the main watercourse. In some CFMPs where the catchments are quite small, several catchments have been combined into one CFMP. CFMPs set out the preferred policies for flood management in different parts of the catchment. Policies set out preferred flood risk management strategies, and these may be related to land use; examples include watershed or upland management of land in the catchment headwaters or planned use of parts of the floodplain to act as washlands to store floodwaters. Some of these policies have multiple benefits (e.g., water quality improvements) and hence assist in meeting Water Framework Directive requirements as well. There are also likely to be policies to defend urban development and critical infrastructures.

The first generation of CFMPs was completed for England in December 2008 and Wales in March 2009, although 8 plans were still under review at the time of writing. The remaining 59 English and 9 Welsh plans are available online in both summary and complete form (Defra, 2010; EA, 2010). These CFMPs have a number of implications for longer-term land-use planning.

16.5　CONTROLLING DEVELOPMENT

PPS25 and its Companion Guide set out the requirements for flood risk and development control in England. PPS25 places responsibilities on the Environment Agency, local authorities, and developers in relation to new development. At present, the Environment Agency is a statutory consultee for all matters pertaining to flood risk and development control in England and Wales.

At the heart of the PPS25 guidance is a sequential approach to land-use planning in relation to flood risk that is to be applied at all stages—from national and regional planning down to the siting of buildings and other infrastructure within site boundaries. This approach seeks, wherever possible, to site new development away from areas of higher flood risk, thereby working with water and nature to allow a more "natural" management of surface water with the opportunity to offer biodiversity and public space enhancements as well. This is managed through the application of the Sequential and Exception Tests, which new development must "pass" before it is permitted (DCLG, 2006).

As part of this process, local planning authorities (LPAs) are required to undertake strategic flood risk assessments (SFRAs), which assess flood risk within their boundaries, and to develop sustainable approaches to flood risk management. The SFRA is one of a suite of documents that collectively form the evidence base for the LPA's Core Strategy/Local Development Framework. Included within these spatial plans is a site allocation DPD that uses, among other environmental and social constraints, flood risk to assign land use types to identified plots of land. Examples of other documents within the evidence base that relate to flooding include green infrastructure studies and landscape assessments.

The SFRA program began around 2003 and there are now SFRAs covering virtually the whole of the country although their scope and consistency vary considerably, depending on the date of commission and other factors such as whether they predate PPS25 and the quality and availability of flood risk data—particularly data on surface water flooding. Many authorities have carried out more detailed SFRAs, and others are combining these with other studies such as water cycle studies (see Section 16.2) and surface water management plans (see Section 16.1).

In Wales, a similar function is performed by TAN15, which specifies requirements for flood consequence assessments (FCAs). In Wales, relatively few strategic flood consequence assessments (SFCAs) have been carried out to date and, where these are in place, they tend to center around specific development areas. The Scottish government's planning policies have recently been subject to extensive revision and are now set out in the second National Planning Framework (NPF2), which is supported by a revised and consolidated Scottish Planning Policy (SPP) and PANs. The Scottish Environment Protection Agency (SEPA) also recommends that development plans should be underpinned by an SFRA, although this is not a statutory requirement.

Within England, SFRAs provide the starting point for any new planning application within the local authority boundary as the SFRA sets out which flood zone the proposed new development will be situated within and hence which land uses are appropriate for that level of flood risk, depending on their vulnerability as set out in Tables D1 to D3 of PPS 25.

New developments proposed within higher risk flood zones or larger than one hectare in extent are required to undergo a flood risk assessment (FRA), and preparation of this FRA is the responsibility of the developer. FRAs must set out clearly the justification for the development in accordance with PPS25 and take into account the advice given within the SFRA.

SFRAs should take into account the flood management policies within the relevant SMP and CFMP where applicable and where these have spatial planning implications. Unfortunately, many of the first-generation CFMPs were not available when the SFRAs were being drawn up as the processes were being carried out over approximately the same period. Where CFMPs were available, the policies were not always in a form where they could readily be taken on board within the SFRA.

16.6   CHANGING RESPONSIBILITIES

16.6.1   SURFACE WATER MANAGEMENT PLANS

One of the drawbacks with the first-generation CFMPs and many of the earlier SFRAs was the lack of information on sources of flooding other than fluvial and coastal flooding and, in particular, overland (pluvial) flooding and sewer flooding. One of the main conclusions of the Pitt Review (Pitt, 2008) was that there was insufficient information on these major sources, particularly in urban areas. Since 2007 there have been major advances in modeling and mapping capability, including the production of high-level national pluvial flooding maps and, while techniques are still under development, these have facilitated the production of "total" flood maps that are able to provide an overview of all flood risk in a local authority area.

It was furthermore recognized that urban flooding (in particular) often arises from a combination of flooding sources and that responsibilities for the flooding were unclear. While sewer flooding is the concern of the water company, main river flooding is the concern of the Environment Agency, ordinary watercourses of the local authority or (IDB Internal Drainage Board), highways drainage of the highways authority, and so on, it is often difficult to attribute flood damage to any one cause and hence responsibilities are unclear.

To remedy this, central government funding has been made available since 2007 to develop an integrated approach to urban flooding using new techniques to combine fluvial, overland, and surface water flooding (and groundwater where relevant). Local authorities are now beginning to develop surface water management plans (SWMPs), where they take the lead role in coordinating different stakeholders and funding streams. The plans involve more detailed mapping of flood risk through a combination of fluvial, pluvial, and sewer modeling to produce integrated flood maps across an Authority's entire area, identifying existing areas of flood risk as well as the effects of new development. These then form the basis for formulating alleviation proposals that may involve a number of different stakeholders and sources of funding.

While the Environment Agency retains overall responsibility for flood risk within England, the Floods and Water Management Act (OPSI, 2010) gives local authorities the role of coordinating these plans and this will require many authorities to rebuild capacity and skills in this area. Many of the authorities have lost most of their drainage operational skills over the past few years and will need to reequip themselves with flood management skills to the extent they can run the new modeling platforms and interpret the results for their spatial and emergency planning colleagues.

16.6.2 FLOOD RISK REGULATIONS

In parallel with their new responsibilities under the Flood and Water Management Act, the Flood Risk Regulations (Defra, 2009) require lead local flood authorities (LLFAs)* to identify significant flood risk areas that require flood hazard and flood risk mapping and management through flood risk management plans. The process involves a preliminary assessment of all local sources of flooding (exceptions are main rivers, the sea, and reservoirs) and identification of flood risk areas to be carried out by June 2011, hazard and flood risk maps to be prepared for flood risk areas by June 2013, and flood risk management plans by June 2015. This is a requirement of the Floods Directive. The preliminary flood risk assessment (PFRA) marks the first stage in the process and will require the gathering of mapping of all sources of flood risk on a catchment basis and summarizing this together with statistics on the numbers of properties and people at risk together with level of hazard and vulnerability. This information will be combined with coastal and fluvial information prepared by the Environment Agency for onward submission to the European Commission in Brussels.

This is the first stage of production of flood risk management plans for recognized flood risk areas that should bring together Environment Agency strategies for main rivers and recommendations from SWMPs. The plans will set out objectives and measures to prevent flooding and protect communities against the consequences of flooding at a local level. These plans should provide a sounder basis to inform planning and land-use decisions taken by emergency and spatial planners moving forward.

16.7 CONCLUSION

This chapter examined the significance of land-use planning in modern flood risk management through its historical and legislative development to the current hierarchy of coastal and catchment flood management plans and their relation to the

---

* A lead local flood authority is defined as the unitary authority or county council for an area in England and as the county or county borough council for an area in Wales.

planning system. The chapter also examined the changing responsibilities as a result of the new Floods and Water Management Act and EC Floods Directive. In particular, local authorities face significant new challenges in their new role as lead local flood authorities to ensure that the requirements of flood management are reflected within wider spatial planning and to lead local flood action plans. However, these plans should in the longer term provide a sounder basis for a more coordinated approach to flood management.

REFERENCES

DCLG (Department of Communities and Local Government) (2006). Planning Policy Statement (PPS)25: Development and Flood Risk. London: The Stationery Office.

DCLG (Department of Communities and Local Government) (2007). Development and Flood Risk: A Practice Companion to PPS25 'Living Draft.' London: Department for Communities and Local Government.

Defra (Department of the Environment Food and Rural Affairs) (2006). Shoreline Management Plans Guidance. Volume 1: Aims and Requirements. London: Defra. Available at <http://www.defra.gov.uk/environment/flooding/policy/guidance/smp.htm>.

Defra (Department of the Environment Food and Rural Affairs) (2009). Land use planning – Assessing the Quality and Influence of Strategic Flood Risk Assessment (SFRAs). R&D Technical Report FD2610/TR. Report produced by Entec UK, JBA Consulting, and Flood Management Support Services Ltd.

Defra (Department of the Environment Food and Rural Affairs) (2010). Policy Evaluation of the First Generation Catchment Flood Management Plans(CFMPs). Technical Report RMP 5357/TR. London: Defra. Available at <http://www.defra.gov.uk/environment/flooding/documents/interim2/cfmp-policy-eval-tcchreport.pdf>.

EA (Environment Agency) (2010). Working with Natural Processes to Manage Flood and Coastal Erosion Risk. A Guidance Document. Environment Agency, March 2010. Available at <http://www.environment-agency.gov.uk/research/planning/116707.aspx>.

European Commission for the Environment (2007). Directive 2007/60/Ec of the European Parliament and of the Council of 23 October 2007 on the Assessment and Management of Flood Risks. Brussels: EU. Available at <http://eur-lex.europa.eu/LexUriServ/LexUriServ.do?uri=OJ:L:2007:288:0027:0034:EN:PDF>.

Evans, E., Ashley, R., Hall, J., Penning-Rowsell, E., Saul, A., Sayers, P., Thorne, C. and Watkinson, A. (2004). Foresight future flooding, scientific summary: Volume 1, Future risks and their drivers. *Foresight Future Flooding*. London: Office of Science and Technology.

OPSI (Office of Public Sector Information) (2010). The Flood and Water Management Act 2010.

OPSI (Office of Public Sector Information) (2009). The Flood Risk Regulations 2009. Statutory Instruments 2009. No. 3042. Environmental Protection.

Pitt, M. (2008). The Pitt Review: Learning Lessons from the 2007 Floods. London: Cabinet Office.

Planning Policy Wales (2004). Technical Advice Note 15: Development and Flood Risk. Welsh Assembly Government. Available at <http://wales.gov.uk/docs/desh/publications/040701tan15en.pdf>.

Scottish Executive (2010). Scottish Planning Policy SPP 7 — Planning and Flooding. Available at http://www.scotland.gov.uk/consultations/planning/spp7-04.asp>.

Scottish Parliament (2009). Flood Risk Management (Scotland), Bill (SP Bill 15). Available at <http://www.scottish.parliament.uk/s3/bills/15-FloodRisk/index.htm>.

17 Flood Resilience for Critical Infrastructure and Services

Ben Kidd

CONTENTS

17.1 INTRODUCTION

The loss of essential services such as power, water, transport, and telecommunications during large-scale flood events has direct effects on both society and the economy. In summer 2007, the flooding of Mythe water treatment works in Tewkesbury created one of the United Kingdom's worst post-World War II emergencies, leading to the loss of piped water supply to 350,000 customers in over 138,000 properties in the Gloucester area from July 22, 2007, for over two weeks (Pitt, 2008).

The experience of this and other large-scale extreme weather events, including the significant disruption caused by a large snowfall in December 2009 and volcanic ash in May 2010, have raised the profile of the need for greater focus on the resilience of the United Kingdom's critical infrastructure. This chapter provides an overview of the legislative framework and current approaches to civil contingency planning, business continuity planning, and flood risk management for infrastructure.

17.2 KEY CHALLENGES

There are many challenges facing UK infrastructure in the next century and in this section the three main ones considered are (1) legacy infrastructure and regulation, (2) the changing climate, and (3) socioeconomic challenges.

The United Kingdom has a large stock of legacy infrastructure, much of which dates back to Victorian times when engineering designs were good, but which have since fallen into disrepair through lack of maintenance or insufficient funding. Devolution and privatization of the utilities sector confounded an already-complex coordination landscape, increasing the number of stakeholders and geographical or legislative boundaries to overcome. The difficulty in coordinating the prioritization of investment, with each infrastructure sector having different funding sources and regulatory regimes, increases the scale of the challenge.

The latest climate projections for the United Kingdom (UKCIP, 2010) provide insight into the potential impacts that the nation's infrastructure will have to deal with in the future. The UK Climate Projections (UKCP09) are provided as a probabilistic tool with many variables and differing impacts in different regions of the United Kingdom and hence it is difficult to give headline projections of what we might expect. However, if the median of all of these variables is taken and assuming a medium emissions scenario, the 50% probability level UKCP09 Projections (or *central estimate* as they are referred to) are that in the 2080s, the United Kingdom will see

- An increase in mean daily maximum temperatures across the United Kingdom, with an increase in the summer average of up to 5.4°C in parts of southern England and 2.8°C in parts of the north of the United Kingdom.
- Changes in winter precipitation, with the biggest changes predicted along the western side of the United Kingdom with increases of up to 33%.
- Summer precipitation reduction in southern England of as much as 40%—a significant challenge with expected population growth in the southeast creating increasing demand.
- Absolute sea level rise of up to 39 centimeters—sea level rise is expected to be greater in the south of the United Kingdom than in the north.

The projection that is common under all assumptions is that severe, extreme weather events such as heat waves, droughts, storms, and flooding incidents are expected to become more frequent and to be more intense. Management of these extreme weather events will be the real challenge and is the one that has been given the most emphasis in both the "Stern Review of the Economics of Climate Change" (Stern, 2005), and through various government strategy reports such as "Foresight Future Flooding," released by the Department of Trade and Industry in 2004 (DTI, 2004).

The effect of more frequent and more intense flooding events on critical infrastructure could be severe. Existing legacy infrastructure was typically designed and built with the climate at the time of construction in mind. Hence, it is seldom equipped to cope with the impacts of climate change. It is therefore now recognized as vital that critical infrastructure be adapted, through refurbishment and retrofit, to cope with a future changing climate.

In addition to climatic changes, there will be other social and economic challenges. The rate of replacement and renewal of infrastructure is restricted by societal barriers. For example, rail companies are pressed by the media and regulators alike to reduce the amount of time they take "possession" of a line of track for maintenance and renewal works. This constrains the options available for refurbishment and hence the capacity for adaptation.

When considering future scenarios, one must also consider changing demographics and potential changes to how our infrastructure is used in the future. For example, an increased number of vehicles on roads and railways, each carrying heavier loads, will increase the burden on the geotechnical and structural assets underpinning the function of the infrastructure asset. Failure of just one of these assets can have a direct impact on the operation of other assets. An example of this is the structural failure of a drainage culvert due to increased loading on the barrel section from above, ultimately leading to blockage of the drainage pathway and flooding of the infrastructure it serves.

17.3   DEFINITION OF CRITICAL INFRASTRUCTURE

The Centre for the Protection of National Infrastructure (CPNI, 2010), the government authority for protective security advice for national infrastructure, defines "national" infrastructure as

> those facilities, systems, sites and networks necessary for the functioning of the country and the delivery of the essential services upon which daily life in the UK depends.

The sectors considered to deliver essential services are communications, energy, finance, food, government, emergency services, health, transport, and water. The CPNI goes on to define *critical* elements of national infrastructure:

> the loss or compromise of which would have a major impact on the availability or integrity of essential services leading to severe economic or social consequences or to loss of life in the UK. These critical elements make up the critical national infrastructure.

The Cabinet Office, coordinators of policy and strategy across government departments, uses a criticality scale that assigns categories for different degrees of severity of impact. The scale runs from CAT 5, the loss of infrastructure that would have a catastrophic effect on the United Kingdom, to CAT 0, the loss of infrastructure that would be minor on a national scale. Three factors provide the means to distinguish between different degrees of severity on the Cabinet Office criticality scale:

1. The degree of disruption to an essential service.
2. The extent of the disruption, in terms of population affected or geographical spread.
3. The length of time the disruption persists.

Sponsor government departments lead on identifying what infrastructure in their sector may be considered critical, in conjunction with sector experts at the CPNI and the infrastructure asset owners themselves.

17.4 REGULATORY FRAMEWORK

Those elements of infrastructure that have been identified as critical are thus subject to control and regulation to ensure the safety of populations. Key regulations are summarized below.

17.4.1 CIVIL CONTINGENCIES ACT 2004

The legislative framework for civil protection across the United Kingdom is defined in the Civil Contingencies Act (CCA) 2004. The CCA defines roles and responsibilities for emergency planning and divides local responders into two categories:

- Category 1 responders who are at the core of the emergency response and include the emergency services, local authorities, National Health Service (NHS) authorities, and environmental regulators.
- Category 2 responders include the majority of infrastructure owners and operators who have a lesser set of duties under the CCA but are required to assist in planning and incident management through sharing of relevant information with other Category 1 and 2 responders.

Coordination and sharing of information between Category 1 and 2 responders is undertaken through local resilience forums (LRFs). LRFs are typically based on police areas. The LRFs in England are all linked by the Civil Contingencies Secretariat (CCS), which sits in the Cabinet Office and has overall responsibility for civil contingency planning. The CCS works in partnership with government departments and the devolved administrations; that is, the Scottish government, the Welsh Assembly Government, and the Northern Ireland Office, each of which has its own national resilience forums.

The CCS operates a National Resilience Extranet, a secure web-based browser tool that enables responders to have access to key information up to and including restricted level, for multi-agency working and communication (Cabinet Office, 2010a). LRFs regularly use the National Resilience Extranet as a means to keep up to date with both national and local information pertaining to emergency planning.

Increasing awareness of interdependencies between infrastructure assets and the need for coordinated efforts to build levels of resilience led, in 2009, to the formation of the Natural Hazards Team (NHT) within the Cabinet Office; the NHT sits as part of the existing CCS. Given the coordination challenges described above, the NHT is well placed to effect change right across the United Kingdom because the CCA provides a common framework for all stakeholders and devolved powers.

Key national sources of information on emergency planning include the following:

- UK Resilience: <http://www.cabinetoffice.gov.uk/ukresilience.aspx>
- Cabinet Office Natural Hazards Team: <http://www.cabinetoffice.gov.uk/ukresilience/infrastructureresilience.aspx>
- Wales Prepared: <www.walesprepared.org>
- Ready Scotland: <http://www.scotland.gov.uk/Topics/Justice/public-safety/ready-scotland>
- Northern Ireland Civil Contingencies Policy Branch: <http://www.ofmdf-mni.gov.uk/emergencies>

The NHT has published a "Strategic Framework and Policy Statement" (Cabinet Office, 2009) that gives details of the process, timetable, and expectations for its Critical Infrastructure Resilience Programme (CIRP). During 2009, the government departments responsible for each of the nine sectors of national infrastructure assessed the current vulnerability of their sector to flooding. A summary of the findings, produced as Sector Resilience Plans (Cabinet Office, 2010b), was published in March 2010, two years after being recommended in the "Pitt Review — Learning Lessons from the 2007 Floods" (Pitt, 2008). The Cabinet Office has also published "Interim Guidance for the Economic Regulated Sectors" (Cabinet Office, 2010c), identifying considerations for industry on how they may be able to support resilience building.

17.4.2 CLIMATE CHANGE ACT 2008

There are a number of other cross-government initiatives that relate to the infrastructure resilience agenda, none more so than the Climate Change Adaptation program. This program is bringing about a step-change in adaptation, and thus resilience, to climate change through the requirement for formal reporting under the Climate Change Act 2008. As a result of the Climate Change Act (2008), the United Kingdom is the first country in the world to have a legally binding, long-term framework to cut carbon emissions and adapt to climate change.

The Committee on Climate Change (CCC) is an independent body established under the Climate Change Act to advise the government on emissions targets and to report to Parliament on progress made in reducing greenhouse gas emissions (Committee on Climate Change, 2010). The Adaptation Sub-Committee (ASC) is a subcommittee of the CCC and provides expert advice and scrutiny through the CCC to ensure that the government's program for adaptation enables the United Kingdom to prepare effectively for the impacts of climate change.

Part of the Climate Change Act requires the Secretary of State to present to Parliament assessments of the risks posed to the United Kingdom by climate change.

This is being undertaken within the Adapting to Climate Change (ACC) cross-government program (Defra, 2010a). The ACC is led by Defra and includes the first UK Climate Change Risk Assessment (CCRA) to be presented to Parliament in January 2012. The current work of the CCRA is developing a robust framework for consideration of both physical impacts and the consequences of these impacts. This framework will then be used in subsequent risk assessments undertaken in five-year cycles.

The CCRA has been commissioned by Defra and the devolved Administrations—the "co-funders"—and is being led by HR Wallingford. The project is being undertaken in several phases; the first phase started in September 2009 and focused on developing the assessment method. For the second phase of the project (which started in January 2010), the project team is undertaking sector-level analyses for the United Kingdom across eleven sectors, including the built environment and infrastructure.

A review undertaken on the anticipated climate change impacts on four infrastructure sectors (energy, telecommunications, water, and transport) has been published by Defra (Rogers et al., 2010). This review, together with the CCRA, has identified the following key climate-related risks to infrastructure:

- Overheating
- Water availability and management
- Flooding
- Ground stability
- Windstorms
- Rainwater penetration
- Mold and pests
- Freeze/thaw
- Fire
- Snow

17.4.3 Flood Risk Management Legislation

The overarching legislation for flood risk management applied to infrastructure in England and Wales is the Flood and Water Management Act 2010. This sets out roles and responsibilities across the Environment Agency, Local Authorities, and other parties, and puts a duty on water companies to share information on flood risk with Local Authorities.

Flood risk management legislation across the United Kingdom includes

- England:
  - The Flood and Water Management Act (2010)
  - Planning Policy Statement 25 (PPS25) "Development and Flood Risk" (CLG, 2006)
- Wales:
  - The Flood and Water Management Act (2010)
  - Welsh Technical Advice Note 15 (TAN15) "Development and Flood Risk" (Welsh Assembly Government, 2004)

- Scotland:
  - Flood Risk Management (Scotland) Act, 2009
  - Scottish Planning Policy 7 (SPP7) "Planning and Flooding"
  - Planning Advice Note (PAN) 69 "Planning and Building Standards Advice on Flooding" (The Scottish Government, 2004a,b)
- Northern Ireland:
  - Water Environment (Floods Directive) Regulations Northern Ireland 2009
  - Planning Policy Statement 15 (PPS15) "Planning and Flood Risk" (Northern Ireland Planning Service, 2006)

17.4.4 REGULATION OF PRIVATE UTILITY COMPANIES

In the United Kingdom, private utilities operate under licenses issued, and economically regulated, by independent agencies. These agencies control monopoly power, protect consumers, and promote competition. The principal utility regulators are Ofwat (water), Ofgem (energy), and Ofcom (telecommunications). In England and Wales, Network Rail is regulated by the Office of Rail Regulation (ORR).

All essential service providers are required to provide continuity of service. For example, the statutory guaranteed standards scheme (GSS) establishes minimum standards of service that each water company must provide to consumers. When disruption is experienced on the rail network, Network Rail is liable for the time railway lines are not available to train operating companies.

17.4.5 REGULATION OF PUBLICLY OWNED SERVICE PROVIDERS

Key publicly owned service providers in the United Kingdom include the highways network and, in Scotland, the rail network. Water services are also publicly owned in Scotland and Northern Ireland.

In England, the Highways Agency operates as an executive agency of the Department for Transport (DfT) and is responsible for operating, maintaining, and improving the strategic road network. In Wales this work is undertaken by Transport Wales acting for the Welsh Assembly Government, and in Northern Ireland by the Roads Service within the Northern Ireland Executive. Transport Scotland is the national transport agency for Scotland, working directly for the Scottish government.

In Scotland and Northern Ireland, the water industry is still publicly owned. The Water Industry Commission for Scotland (WICS) is a nondepartmental public body with statutory responsibility for the regulation of water and sewerage services. Scottish Water is responsible for providing water and wastewater services, delivering the investment priorities of Scottish ministers within the funding allowed by the Water Industry Commission for Scotland. Northern Ireland Water provides similar services in Northern Ireland, economically regulated by the utilities regulator.

17.4.6 SPATIAL PLANNING

Flood risk management regulation for infrastructure is largely brought about through the spatial planning process. In England, Planning Policy Statement 25

(PPS25) "Development and Flood Risk" provides a comprehensive framework for the national, regional, and local consideration of flood risk (CLG, 2006). Its practice guide (CLG, 2009) provides useful guidance on the hierarchy of flood risk management measures:

- Assess
- Avoid
- Substitute
- Control
- Mitigate

Through the implementation of this best practice, we can be confident that the next generation of infrastructure projects will incorporate appropriate levels of flood resilience.

17.5   FLOOD RISK ASSESSMENT FOR CRITICAL INFRASTRUCTURE

As part of routine asset management, emergency planning, and business continuity management, an infrastructure owner or operator will regularly undertake, review, and update an analysis of its vulnerability to particular natural hazards. An organization's business continuity plan should list all possible hazards that are likely to cause significant disruption to its service. Consideration should be given to wider organizational and system resilience; for example, whether the flood hazard has the potential to cause wider impacts on an organization's operations by affecting movement of staff and resources. Guidelines on business continuity management are set out in the relevant British Standard, BS 25:999 (BSI, 2006).

Flood risk assessments will also be undertaken for all new developments and, where alterations or remedial work is undertaken, on existing infrastructure. The Environment Agency in England and Wales (SEPA in Scotland and Rivers Agency in Northern Ireland) works closely with infrastructure owners and operators to ensure that robust vulnerability assessments are undertaken and to provide all the latest information on flood hazards. For example, information on surface water flood risk, identified as a key risk of flooding for a number of infrastructure assets, is often made available for the development of surface water management plans by local authorities and ultimately for use in multi-agency flood plans. Tables D1, D2, and D3 of PPS25 give flood risk vulnerability classifications for different types of infrastructure (CLG, 2006).

The National Flood Risk Assessment (NaFRA) database was prepared by Defra and the Environment Agency and identifies the indicative number and type of existing infrastructure assets in the floodplain areas. These data, like the national flood maps for Scotland and Northern Ireland, cover flooding from rivers and the sea.

New techniques to establish susceptibility to surface water flooding have been used to map England, Wales, and Scotland. These techniques use digital terrain models, often based on aerial light detection and ranging (LiDAR) survey data, in conjunction with two-dimensional hydraulic models to provide an indication of the flood paths likely to be taken by rainfall runoff.

There is a wealth of data and guidance available on the consequences of flooding on residential and commercial property, but specific guidance on the social, economic, and environmental impacts of disruption to essential services is limited. Ofwat's framework (Ofwat, 2009) does list the following as potential consequences of flooding-related disruption of water infrastructure:

- Loss of state revenues due to nonfunctioning of the private sector
- Costs associated with state support for provision of emergency supplies if interruption is substantial
- Inconvenience of interruptions due to service loss
- Health risk due to contamination of water supply and the environment
- Extra clean-up costs due to wastewater mixing with flood water and entering property
- Environmental pollution due to wastewater mixing with floodwater.

Information on the current assessment methodologies across different infrastructure sectors is contained in CIRIA C688 "Flood Resilience and Resistance for Critical Infrastructure" (McBain et al., 2010). An example of the assessment process taken in the energy sector follows:

> In 2005, partly in response to the Carlisle floods, the government ordered a study to identify the top 1,000 electricity substations most at risk from flooding. The Energy Networks Association (ENA) undertook this work and subsequently developed guidance that formalizes the approach to risk assessment and flood defense prioritization (ENA, 2008). The ENA review identified that much of the vulnerable high-risk equipment in electricity substations, such as the primary circuits, are at a safe height but many control circuits and secondary wiring are at a lower level that can still cause the circuit to trip. A major design issue for many of these sites is the location of the switchboard. Although much of the equipment is remotely operated, if staff cannot access the switchboard due to flooding, then there is still a possibility of failure.

17.6 INTERDEPENDENCIES ASSESSMENT

Infrastructure systems are highly interconnected and the failure of one asset system can have both direct and indirect cascade effects on other essential services. For example, where water supply is pumped, the loss of power may result in a loss of water supply. Infrastructure providers will assess their critical dependencies through the business continuity planning process and, where necessary, these will be shared through the CCA framework.

17.6.1 DEPENDENCY ON INFORMATION AND COMMUNICATION TECHNOLOGY

A key issue in flood emergency management and business continuity planning is the absolute dependency of all infrastructure sectors on information and communication technology (ICT). The web of communications technologies and devices on which modern business and the operation of society rely is vast. These include everything from traffic lights to closed-circuit television (CCTV) and automated control systems that switch pumps on and off at remote sites. This web of data transmission–reliant automated control systems that "run" the UK infrastructure is extensive and largely hidden from view.

The operation of civil infrastructure also increasingly relies on the power and ICT transmission systems to provide energy and connectivity for control and management. This is particularly evident as greater efficiencies in maintenance regimes are demanded, requiring reduced man-hours for site inspection. Remote monitoring is regularly used by the Environment Agency, Highways Agency, and others to monitor flood levels at key flood risk management assets such as culverts. Mobile-operated cameras are used to give real-time information on flood levels and condition of the asset. This dependency on power and ICT will need to be reviewed against emergency planning scenarios and incorporated into business continuity plans.

17.6.2 COLLABORATION BETWEEN PROVIDERS

In recognition of interdependencies and the need for greater collaboration between infrastructure providers, the Environment Agency set up a Memorandum of Understanding (MoU) between itself and key infrastructure providers such as the Highways Agency and Network Rail. These MoUs enable the transfer of information in confidence and overcome sensitivities to data protection encountered through Category 2 responder engagement with LRFs in the past.

Other partnerships are being formed to aid strategic collaboration, including the Natural Hazards Partnership (NHP), which includes collaboration between the Environment Agency, the Met Office, and the Cabinet Office's NHT. The NHP is looking to harness a greater understanding of natural hazards through partnership with the British Geological Survey and others, in order to better predict natural hazard events in the future.

17.7 FLOOD RESILIENCE AND RESISTANCE FOR CRITICAL INFRASTRUCTURE

Providers of essential services who are exposed to flood hazards will need to adopt a range of measures (both structural and nonstructural) to manage the risks, utilizing the flood risk management hierarchy mentioned in Section 17.4.6.

17.7.1 NONSTRUCTURAL MEASURES

The first nonstructural mitigation measure to consider, as outlined in the flood risk management hierarchy, is *avoidance*. This approach will be crucial for the development of new infrastructure and is fundamental in the spatial planning process. However, given that large swathes of *existing* critical infrastructure are at risk of flooding, we focus here on other nonstructural measures available in an infrastructure asset manager's toolkit—namely, redundancy and reserve capacity, flood forecasting and warning, incident management, and emergency exercises.

Reserve capacity is the operational capacity available within a system that is not strictly required to meet demand. For example, the national grid will have extra "redundant" capacity such that, if an unexpected failure occurs on one element of the system (i.e., one substation becomes inundated), there will be sufficient supply available through the redirecting of supply to meet customer demand. Provision of such "buffer" storage is also a technique used in the water industry where sufficient potable water supply is kept in reserve for the eventuality that a water treatment plant fails, thereby reducing the risk of supply disruption.

The Environment Agency has a dedicated flood warning service for infrastructure operators that covers flooding from rivers and the sea. As mentioned previously, infrastructure owners may also utilize their own network of remotely operated sensors to monitor flood levels at critical sites. Further information on flood warnings is given in Chapter 3.

An effective communication strategy is required for incident management that takes into consideration the implications of flooding of owned assets as well as flooding of those assets upon which they depend. This strategy will be designed to ensure that all parties, including the general public, are aware of what actions are appropriate at any one time. A National Flood Emergency Framework has been developed for England (Defra, 2010b). This sets out the roles and responsibilities for emergency planners at a national and local level, and gives guidance on key considerations and detailed planning arrangements during a flood.

There is no better way of learning how best to manage the effects of flooding on infrastructure than through actual experience. However, it is important that regular exercises are undertaken to ensure that all parties and individuals involved in the coordination of the incident response are trained and prepared for a real-life scenario. Regular local exercises are undertaken, coordinated by LRFs and involving both Category 1 and 2 responders. At a national level, a large exercise is currently being planned entitled "Exercise Watermark," which is due to take place in March 2011 and will involve wide engagement of both Category 1 and 2 responders. Exercise Watermark is seen by many as the first opportunity for an in-depth analysis of interdependencies between infrastructure operators and to test plans that have been put in place since the major flood events of 2007.

17.7.2 STRUCTURAL MEASURES

Fixed defenses can include floodwalls or embankments, and these are widely used across the United Kingdom by flood defense operators. Design and construction of flood defenses is a specialized field requiring input from hydrologists, and geotechnical and civil or structural engineers. The Concrete Centre has guidance on the use of flood-resilient concrete materials for permanent flood protection measures (The Concrete Centre, 2009). Examples of permanent flood defense schemes installed by infrastructure asset owners can be found in CIRIA C688 (McBain, Wilkes, and Retter, 2010).

The Environment Agency (EA) has a policy statement on the appropriate use, deployment, and management of temporary and demountable flood defenses (EA, 2010). Important issues relate to the need to understand and minimize the obstacles to their successful deployment and also recognizing that the whole-life costs (including operational and maintenance costs) are higher for these systems than they are for fixed structures. However, they have been used successfully in large-scale flood incidents and were crucial in protecting the Walham electricity substation from flooding in 2007. Details on the use of temporary flood defenses at the Walham substation, as well as other examples, are included in CIRIA C688 (McBain, Wilkes, and Retter, 2010).

17.8 SUMMARY

The protection of critical infrastructure in emergency situations has received much attention in recent years following multiple high-profile events that have resulted in severe disruption. There have also been several "near misses" where swift action, careful planning, and lucky circumstances have prevented more serious conse- quences. Much of the emergency planning legislation discussed has been built on experience derived from flood events, and flooding remains one of the major threats to critical infrastructure in the United Kingdom.

The fact that a large part of what is regarded as critical infrastructure is under the control of private companies has meant that protection is complicated and requires cross-communication between agencies, government organizations, and private companies on an unprecedented scale. The number of emergency responder organi- zations is vast. This complexity also makes information about the protection level of our infrastructure difficult to obtain or verify.

This chapter summarized the state of protection of critical infrastructure against flooding and suggested the way forward to increased resilience in the future; however, further information about critical infrastructure protection can be gained from the following websites:

- Cabinet Office: http://interim.cabinetoffice.gov.uk/ukresilience/infrastructure resilience.aspx
- UKCIP: http://www.ukcip.org.uk/
- Infrastructure UK: http://www.hm-treasury.gov.uk/ppp_infrastructureuk.htm
- Defra Adapting to Climate Change programme: <http://ww2.defra.gov.uk/ environment/climate/government/>
- CIRIA C688 Flood resilience and resistance for critical infrastructure http://www.ciria.org/service/c688
- Climate UK: <http://www.climateuk.net/
- Engineering the Future: http://www.theiet.org/publicaffairs/ete/index.cfm
- Improvement & Development Agency (for local authorities): http://www. idea.gov.uk/idk/core/page.do?pageId=22556485

REFERENCES

BSI (British Standards Institution) (2006). BS 25:999 Code of Practice for Business Continuity Management. British Standards Institution.

Cabinet Office (2009). Strategic Framework and Policy Statement (Natural Hazards Team). London: Cabinet Office. Available at <http://www.cabinetoffice.gov.uk/ukresilience/infrastructureresilience.apsx>.

Cabinet Office (2010a). National Resilience Extranet. Available at <http://www.cabinetoffice.gov.uk/ukresilience/preparedness/resilient_telecommunications/nre.aspx>.

Cabinet Office (2010b). Sector Resilience Plans. London: Cabinet Office. Available at <http://www.cabinetoffice.gov.uk/ukresilience/infrastructureresilience.apsx>.

Cabinet Office (2010c). Interim Guidance for the Economic Regulated Sectors. London: Cabinet Office. Available at <http://www.cabinetoffice.gov.uk/ukresilience/infrastructureresilience.apsx>.

CLG (Communities and Local Government) (2006). Planning Policy Statement 25: Development and Flood Risk, London: Communities and Local Government.

CLG (Communities and Local Government) (2009). Planning Policy Statement 25: Development and Flood Risk — Practice Guide. London: Communities and Local Government.

Committee on Climate Change (2010). Independent Advice to Government on Building a Low-Carbon Economy [online]. Available at <http://www.theccc.org.uk/>.

CPNI (2010). Glossary of Terms. London: Centre for the Protection of National Infrastructure. Available at <http://www.cpni.gov.uk/glossary.aspx>.

Defra (Department for Environment, Food and Rural Affairs) (2010a). Adapting to Climate Change Programme [online]. Available at <http://www.defra.gov.uk/environment/climate/programme/index.htm>.

Defra (Department for Environment, Food and Rural Affairs) (2010b). National Flood Emergency Framework for England. Defra.

DTI (Department for Trade and Industry) (2004). The Foresight Future Flooding Report. Available at <http://www.environment-agency.gov.uk/research/library/publications/33923.aspx>.

EA (Environment Agency) (2010). Kitemark Scheme for Flood Protection Products—Environment Agency Advice. Available at <http://www.environment-agency.gov.uk/homeandleisure/floods/113219.aspx>.

ENA (Energy Networks Association) (2008). Resilience to Flooding of Grid and Primary Substations. Issue 1, ETR 138, Final Engineering Technical Report to the Energy Emergencies Executive. London: Energy Network Association.

McBain, W., Wilkes, D., and Retter, M. (2010). Flood Resilience and Resistance for Critical Infrastructure. London: Construction Industry Research and Information Association, C688. www.ciria.org/service/C688

Northern Ireland Planning Service (2006). Planning Policy Statement (PPS 15): Planning and Flood Risk. Belfast, Northern Ireland: Department of Environment. Available at <http://www.planningni.gov.uk/index/policy/policy_publications/planning_statements/pps15.htm>.

Ofwat (Water Services Regulation Authority) (2009). Asset Resilience to Flood Hazards: Development of an Analytical Framework. Birmingham: Ofwat. Available at <http://http://www.ofwat.gov.uk/pricereview/pr09phase2/ltr_pr0912_resilfloodhaz>.

Pitt, M. (2008). The Pitt Review—Learning Lessons from the 2007 Floods. London: Cabinet Office.

Rogers, W., et al. (2010). Adapting Energy, Transport and Water Infrastructure to the Long-Term Impacts of Climate Change. Ref. No RMP/5456. Defra. Available at <http://www.defra.gov.uk/environment/climate/programme/infrastructure.htm>.

Stern, N. (2005). Stern Review on the Economics of Climate Change. London: Cabinet Office. Available at <http://www.direct.gov.uk/en/Nl1/Newsroom/DG_064854>.

The Concrete Centre (2009). Concrete and Flooding, Ref: TCC/05/019. Blackwater, UK: The Concrete Centre.

The Scottish Government (2004a). Scottish Planning Policy 7 (SPP7) Planning and Flooding. Edinburgh: The Scottish Government. Available at <http://www.scotland.gov.uk/Home>.

The Scottish Government (2004b). Scottish Planning Advice Note PAN 69: Planning and Building Standards Advice on Flooding. Edinburgh: The Scottish Government. Available at <http://www.scotland.gov.uk/Publications/2004/08/19805/41594>.

UKCIP (United Kingdom Climate Impacts Programme) (2010). Observed UK Climate Trends [online]. Available at <http://www.ukcip.org.uk/index.php?option=com_content&task= view&id=243&Itemid=337>.

Welsh Assembly Government (2004). Technical Advice Note (TAN) 15: Development and Flood Risk. Cardiff: Welsh Assembly Government. Available at <http://wales.gov.uk/ topics/planning/policy/tans/tan15?lang=en>.

18 Protection and Performance of Flooded Buildings

Mike Johnson

CONTENTS

18.1 INTRODUCTION

Keeping the weather out is an essential function of the building envelope in almost any climate. In temperate zones the most important aspect of climate management through the building envelope is resistance to moisture; this is currently dealt with in Part C of the Building Regulations for England and Wales. The role of precipitation in all its forms was given greater emphasis in the 2004 revision of Part C (Building [Amendment] Regulations, 2004). Part C does not currently include flood performance as flooding may be from other sources such as blocked drains or tidal inundation. Furthermore, flooding can be the manifestation of an accumulation of precipitation rather than an immediate effect. However, it has been determined that flooding could be brought into the future scope of Part C; this will be considered in the next review of Part C.

The fact that flood performance does not form part of national building codes does not necessarily mean that there is a lack of official technical guidance. Guidance has been available for new construction (Scottish Building Standards Agency, 1996) and flood repair for more than 10 years (BRE, 1997). Following the floods of autumn 2000, the UK Parliamentary Select Committee for the Environment commented that this advice was not as accessible to the public as it might be. This led to the production of an interim guide for property holders entitled "Preparing for Floods" (DTLR, 2002). Since then, the Environment Agency has sponsored the production of several flood guides and information leaflets (www.environment-agency.gov.uk).

Some resistance to flooding may be provided by building materials in their constructed form in their role of excluding or offering resistance to moisture. The mitigation of the effects of moisture may be by varying degrees:

- *Avoidance:* moving the exposed element from the source of moisture or by sheltering it.
- *Resistance:* applying a covering that repels or excludes moisture in masonry or other mass construction (known as the *raincoat effect*; moisture runs off the surface.

Most building materials use the "overcoat effect" by absorbing moisture gradually, whereby the degree of protection is a function of the rate of moisture transmission and the thickness of the element.

Similar forms of mitigation apply to flood performance. The most immediate means of flood avoidance is to build elsewhere. In some locations this may not be practicable for reasons of land supply, topography, or heritage. A further method is to raise the building above estimated flood levels but this carries a number of disadvantages. If solid fill is used to raise flood levels, it may reduce floodwater storage in the catchment, new buildings may obstruct the natural overland flow of floodwater and cause the flood to spread elsewhere, or the buildings could be stranded in rising floodwater and place burdens on already stretched emergency services. Provided occupation can be sustained during flooding, alternative building layout may have a role to play, such as placing living rooms and kitchens on the upper floor and bedrooms on the lower floor. However, sleeping accommodations should never be located in basements or places subject to very sudden inundation. As a safety measure, consideration can be given to using an automated flood warning system.

Flood resistance is also known as *dry-proofing* (DTLR, 2002), where water is kept out of the building by either barriers or materials that have a high level of water resistance. This method must be used with caution as there are physical limitations on how great a depth of water can be kept out in terms of the structural stability in most common forms of construction. There is also a risk of sudden inundation through apertures such as windows.

Forms of construction that control the rate at which water enters the buildings are providing flood resilience; this is also known as *wet-proofing*. Such forms of construction should have the ability to recover from flooding with minimum intervention or replacement.

18.2   PLANNING AND FORESIGHT

Whatever form of flood performance is chosen for a building, due consideration should be given to the planning regime (England PPS25, Scotland NPPG7, and Wales TAN15). Building should only take place in locations where flooding is less likely and not severe. Any departures from this premise should be subject to an exception process that sets out the reasons for building in a higher risk location and how the risk will be accommodated. Development in areas subject to flooding are exceptional; flood risk assessments are required to show that the development is necessary and that there are no other viable options.

To some extent, flood risk is comparable to other natural and man-made hazards. This can be demonstrated by the concept of linkages between the vulnerable item and the hazard. The usual model is source pathway and receptor. The need to carry out any risk assessment can be determined by cogent facts. In crude terms, this involves asking simple questions: Could there be a problem? How likely is it to become a hazard? Is there potential for harm? How severe could that harm be? Can the effects be mitigated? What is the best option for mitigating harm? How can this process of mitigation be realized? This process can be illustrated by flowcharts. Some of the most useful are in guidance for dealing with pollutants (EA, 2004). The sequential approach in Planning and Policy Statement 25 (PPS25) has some similarities.

Where the depth of water is sufficient to large mobile debris, such as tree trunks, great care should be taken when considering building near watercourses that may become very deep or fast flowing or in places where water levels could rise very quickly. This should be considered in the "Flood Risk Assessment"; and if the hazard cannot be accommodated, development should not take place. An extreme example of potential damage is that during the floods in central Europe in 2002, several buildings were demolished by large debris that included objects as large as freight containers. Some degree of protection from debris in floodwater can be provided by robust hedges or fenders, such as gabions (Norwich Union, 2008). These systems have the benefit of being effective barriers while still allowing water to flow by.

18.3   FLOOD PROTECTION AND PERFORMANCE

Flood depth will usually tend to be a deciding point for the method of flood protection. Hydrostatic head will place stresses on walls and any other vertical elements; it will also be the driving force that effectively pushes the floodwater through the wall. If the flood depth is predicted to be greater than 900 millimeters, flood-proofing is unlikely to be feasible unless special construction, such as reinforced masonry or solid concrete construction, is used. It is improbable that such forms of building will be used in volume housing or small nondomestic buildings. In many cases, 600 millimeters will be the cut of depth for flood-resistant construction, particularly if there is no buttressing of outside walls from crosswalls or partitions. The body of opinion is that the cut-off depth for wet-proof construction should be 600 millimeters (USACE, 1988).

Water depth greater than 600 millimeters also presents difficulties in evacuation. For instance, in shallow water, rescuers can walk knee or thigh deep and can carry or lead their charges, whereas greater depths tend to involve the use of boats, where the actual rescuers have to wade in the water both to get people into the boats and to keep their boats in place while effecting the rescue. This is demanding in terms of manpower and has a high potential for accidents.

Where the predicted depth is greater than 600 millimeters, a water entry strategy should be adopted and resilient construction used for the building. Between 300 and 600 millimeters, flood-proofing may be possible but the interior finishes may need to be resilient; if there is any doubt about the ability of the exterior elements' ability to exclude water, the whole construction should be resilient. At depths up to 300 millimeters, flood-proof construction is feasible providing that the right methods are chosen and the interface between the walls and floors is correctly detailed. Any floodwater that enters a building has the potential to cause damage; it can damage or destroy floor coverings (e.g., carpets and furniture) and fittings (e.g., kitchen units). Damage to kitchen units can be mitigated by the use of casing on stubby plastic legs that raises the units by approximately 150 millimeters. This may be sufficient for very shallow floods or events of short duration where the water runs in and out again. Replacement may be limited to plinth facings and end panels.

Depth is not the only criteria in the choice of flood-proofing. Except in the case of jointless mass construction, water will gradually seep through most building materials. The contact or immersion time must be considered when choosing materials and methods. Absorption rates may vary, and this is often time related. Generally, tidal inundation or sewer flooding does not last more than a few hours and rarely exceeds 48 hours. There may be local variations where buildings are located in depressions and these may be flooded for several days. In 2007, some homes in Doncaster were under water for several weeks. The flood depth is usually linked to the source of flood (Figure 18.1).

Flood duration will increase the time of exposure to water and in due course the amount of water absorbed. This can be critical in the choice of materials. For example, dense concrete blocks tend to have an open-grain structure that tends, initially, to let water seep through but the rate slows over time. Aerated concrete tends to be slow to absorb water but, once saturated, is slower to dry out. These detailed differences are crucial in the choice of material.

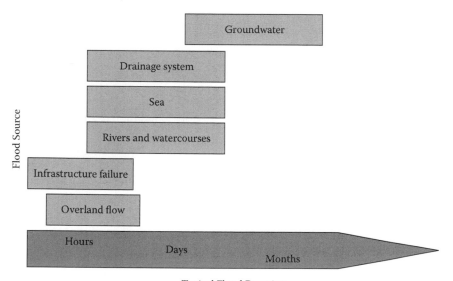

FIGURE 18.1 Typical flood durations. (Adapted from CLG Department for Communities and Local Government) (2007). Improving the Flood Performances of new Buildings—Flood Resilient Construction. London: Department for Communities and Local Government.)

The ability of different materials to limit water entry should not be assessed empirically. It might be very tempting to assume that engineering bricks or other masonry units that have low permeability used with closed-cell insulation would provide fairly resistant walls, but the key to success lies in the detailing and interfaces between materials and building elements. Most tests for masonry are either for vertical water movement to assess the risk of water rising by capillary action or absorption, or water retention that may increase the risk of frost damage. The safest way to test for flood resistance is to subject sample units to water pressure over time and then to incorporate them in sample constructions. Physical testing (CIRIA, 2006) was one of the key elements in developing guidance for flood-resilient construction (CLG, 2007).

The duration of the flood period is also a consideration to what extent flood resistance should be provided. A high level of flood protection may give a false sense of security to occupants who may decide to remain in their homes. If mains services are disrupted or food runs out, sustaining or evacuating stranded residents will become a burden on emergency services.

In terms of the building envelope, the materials used for flood-resistant and flood-resilient construction are similar. Differences may occur with internal finishes. Gypsum-based products are not flood resistant and thus may not be appropriate for rendered or wet-applied finishes. Cement or lime-based products are more appropriate but both have limitations. Cement-based products seal surfaces and may retain water in saturated masonry backgrounds, thus delaying the drying out.

Lime plasters are vapor permeable so will let underlying layers dry out. However, lime does not set by immediate reaction, as is the case with cement. Lime hardens and cures by the reabsorption of carbon dioxide from the surrounding air. For this reason, it is not possible to test its performance due to very long real-time constraints. Also, its performance in situ depends on the building staying dry for several months. If lime plaster is saturated for less than a month since application, it will debond from the walls so, in some cases, dry lining is the best option because it is easy to remove, to enable drying out, and cheap to replace. The extent of replacement can be reduced if panels are placed horizontally so that only half the room height requires replacement (Scottish Building Standards Agency, <www.sbsa.gov.uk>).

Interfaces and details can be very important. Simple factors such as repointing on completion of work can be important. Scaffolding to masonry is often supported in joints between the bricks (putlogs); these often reach half-way through the masonry and may be set in completely open joints. Therefore, full-depth filling and tooling-off is needed to ensure an effective seal. The base of walls and the interface with the ground floor is important if water is to be excluded from this joint. Rather than using compacted fill, it is more advisable to use concrete at this interface to prevent lateral flow between floor and wall or to prevent water from rising through the floor. If high hydrostatic pressures are anticipated, waterproof construction of the type used for basements may be advisable.

Completely waterproof construction may be advisable where there is a risk of groundwater intrusion. This might involve water under pressure that can drive water through floors or walls in contract with the ground. Additionally, groundwater flooding may last several months as overcharged aquifers continue to flow, with water being carried along fissures that have been opened up by the first flush of floodwater.

Depth and duration are the determining forces for water to find its way into a building. There are many ways water can enter a building; they range from the bulk elements themselves to small features such as joints in the building. When buildings have a suspended floor, it will usually be ventilated and the airbricks provide an entry route (vent covers and seals are now available). Building services also provide entry routes, including backing-up of drains. Typical water entry routes are shown in Figure 18.2.

Walls are often seen to be the first line of defense in keeping floodwater out of buildings. The walls of all buildings have apertures, with doors being the most common. Door boards can be used to protect entrances, and most are very effective and are often tested to rigorous standards (BSI PAS 1188 Flood Protection Products). The limitation on these products is that they must be put in place. This could be problematic when the occupants are absent. There are a few door protection systems that self-deploy but most are bulky and may not be suited to domestic applications. Doors themselves could be made floodproof but none have been tested; indeed, door suppliers are reticent to say whether their products might be flood resistant (author's personal communications with suppliers), although some have interlocking seals and multipoint locking systems. Water entry beneath doors may present difficulties as there is increasing demand for level access to buildings for people with mobility problems.

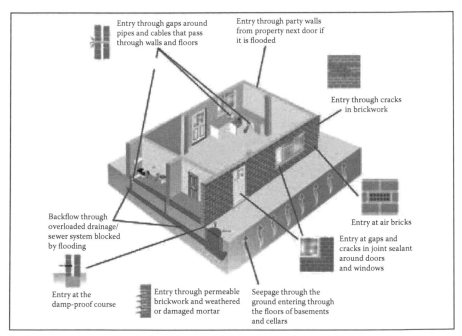

FIGURE 18.2 Potential routes for entry of floodwater into a dwelling. (Adapted from Department for Communities and Local Government (2007). Improving the Flood Performances of New Buildings—Flood Resilient Construction. London: Department for Communities and Local Government.)

Floors are influenced by two different conditions: (1) water entering from below or (2) the effects of standing water. Water entering from below may be subject to hydrostatic pressures and thus may have to be strong enough to resist it. This may involve designed concrete mixes and reinforced, thicker slabs. Where the ground is saturated, and stays saturated for a long time, there is the possibility of mobilizing or activating contaminants in soil or fill beneath the slab. Fill should be selected carefully and the soil checked for high natural content of sulfates or phosphates. Careful consideration should be given to where to place insulation to ground-floor slabs. Rising water may cause insulation to become buoyant; the uplift may be sufficient to cause uplift that may damage the floor. The mass of the overlay should be sufficient to resist such uplift. Standing water that has entered the building through walls may saturate the floor construction. The soakage period can be reduced by sweeping off or pumping out the water as soon as the flood subsides.

18.4 BALANCING THE COSTS

Flood protection is more than just a technical issue. Cost effectiveness is an important consideration. The first stage in mitigating cost is to become aware of the risk and building in protection at the appropriate stage—either initial construction or during major renovation. Retrofitting by itself is much more expensive than providing the measures as part of other works. This may be part of the reason why less than 30% of flooded property owners consider enhanced flood performance at the time of reinstatement (Greenstreet Bergman evidence base for flood protection, 2008). Furthermore, return periods should be considered when evaluating which measures to take (Defra, 2008). Return periods vary between twenty and fifty years but it should be noted that properties at moderate risk may only be reflooded on a 75-year cycle, whereas the most at-risk may flood again in a period of 10 years or less.

For a typical home, the additional cost for flood measures may be in the region of £10,000 to £20,000, effectively doubling the cost. It should be noted that there are secondary costs, such a temporary accommodation. This reduces the marginal cost of flood-proofing. If the property is flooded again, there should be significant savings on temporary accommodations and the reinstatement work may be limited to cleaning and localized redecoration. In the case of new build, the changes in specification to bring in some degree of flood protection are not costly and could be offset by less flamboyant finishing.

The existence of flood-resilient or flood-resistant construction should be made known to future occupants and repairers. Some repairers may routinely hack off internal finishes to facilitate drying out. If a vapor-permeable finish is used, such drastic measures may not be needed, provided that the surface can be cleaned and that the material will allow any moisture absorbed by underlying layers to disperse.

The way that water moves around the outside of the building can be influenced by landscape features and drainage. In the case of erratic overland flow, particularly from overloaded sewers, protection can be afforded by blocking boards under fences, effectively enhanced gravel board associated with solid gates that have bottom seals (Severn Trent Water, CSO alleviation work from 2006 onward). The carriage of water through the development can be better managed by sustainable drainage systems that manage surface water in a more natural manner (www.ciria.org). Where there is a risk of backflow from drains into buildings, anti-flood valves should be provided; this is recommended in Approved Document H to the Building Regulations (Building (Amendment) Regulations, 2001).

Some of the most expensive parts of flood damage claims relate to fittings. In general, sanitary fittings are robust but it should be noted that most plastic baths have a wood-fiber stiffening panel that will likely be damaged by immersion. Kitchens are also vulnerable, due to the extensive use of wood-fiber products. Units can be made more robust by the use of solid wood carcasses or some other form of construction for the units. This ranges from plastic units to pressed metal or even concrete. Traditional fascias can be provided and lift-off hinges allow doors to be removed before water enters.

The psychological effect of using units that have been contaminated by very dirty water should not be ignored. Occupants may not be convinced about post-flood cleaning and may prefer to have units stripped out. Sometimes the simplest option is to use lowest-cost casing with better-finished doors and facings. If only limited water entry or seepage is expected, kitchen units on stubby legs may be sufficient to avoid contact with floodwater. This may also provide justification for built-in refrigerators, as it is very difficult to source refrigerators that have their motors raised above the base. Placing the unit inside a house will lift it about 150 millimeters and provide some protection from shallow floods.

18.5   CONCLUSIONS

Enhancing the floor performance of buildings cannot be considered a technical fix in isolation from other factors. Planning considerations such as public safety, topography, and flood reaction must be considered as the detailed design progresses. Very high investment in structural protection could make most buildings safe to occupy, but consideration must be given to exceptional events where the building might be overwhelmed or the simple fact that the occupiers may run out of provisions or because someone becomes ill for some other reason.

Buildings that have enhanced flood performance are not manufactured products. They are made up of several materials and components put together in layers. Selection of the right material for the right place is essential, as is compliance during the execution of the work. Most British homes are made up of small masonry units that are joined up with mortar. Joints can be the weakest link to the correct mix, so proper bedding, tooling, and pointing are essential if a water-resistant wall is to be provided.

Ad hoc substitution of materials should not be permitted without due consideration. In the case of inner skin materials for walls, the determining factor is likely the duration of the flood. If the flood duration is expected to be greater than twelve hours, dense blocks may be more advisable than foamed concrete due to the difference in drying qualities.

Improving the flood performance of buildings has a role to play in reducing reinstatement costs and impacts on people during and while recovering from floods. These reasons are driving the consideration for bringing flood performance into national building codes. However, for them to be part of an effective policy, they must be achievable, cost effective, and acceptable to users. Enhanced flood performance at the initial build stage will be less costly and provide more benefits.

REFERENCES

BRE (1997). Repairing Flood Damage. Watford, Building Research Establishment Press.
CIRIA (2006). Improving the Flood Resilience of Buildings through Improved Materials and Methods. Report 5C on laboratory tests. London: Construction Industry Research and Information Association.
CLG (2007). Improving the Flood Performances of New Buildings—Flood Resilient Construction. London: Department for Communities and Local Government, p. 96.
EA (Environment Agency) (2004). Model Procedures for the Management of Contaminated Land. Contaminated Land Report 11. London: Department for Environment, Food and Rural Affairs/Environment Agency.
Defra (Department for Environment, Food and Rural Affairs) (2008). Consultation on Policy Options for Promoting Property Level Flood Protection and Resilience. London: Department for Environment, Food and Rural Affairs.
DTLR (Department for Transport, Local Government and the Regions) (2002). Preparing for Floods. London, UK: Department for Transport, Local Government and the Regions.
Norwich Union (2008). *Flood-Proof Houses for the Future: A Compendium*. Leeds: Royal Institute of British Architects (RIBA).
Scottish Building Standards Agency (1996). Design Guidance on Flood Damage to Dwellings. Edinburgh: HMSO.
USACE (U.S. Army Corps of Engineers) (1988). Tests of Materials and Systems for Flood Proofing Structures. Washington, DC: U.S. Army Corps of Engineers, p. 89.

19 Impacts of Flood Hazards on Small and Medium-Sized Companies
Strategies for Property-Level Protection and Business Continuity

Bingunath Ingirige and Gayan Wedawatta

CONTENTS

19.1 INTRODUCTION

Worldwide, floods have become one of the costliest weather-related hazards, causing large-scale human, economic, and environmental damage during the recent past. Recent years have seen a large number of such flood events around the globe, with Europe and the United Kingdom being no exception. Currently, about one in six properties in England is at risk of flooding (EA, 2009), and the risk is expected to further increase in the future (Evans et al., 2004). Although public spending on community-level flood protection has increased and some properties are protected by such protection schemes, many properties at risk of flooding may still be left without adequate protection. As far as businesses are concerned, this has led to an increased need for implementing strategies for property-level flood protection and business continuity, in order to improve their capacity to survive a flood hazard.

Small and medium-sized enterprises (SMEs) constitute a significant portion of the UK business community. In the United Kingdom, more than 99% of private-sector enterprises fall within the category of SMEs (BERR, 2008). They account for more than half of employment creation (59%) and turnover generation (52%) (BERR, 2008), and are thus considered the backbone of the UK economy. However, they are often affected disproportionately by natural hazards when compared with their larger counterparts (Tierney and Dahlhamer, 1996; Webb, Tierney, and Dahlhamer, 2000; Alesch et al., 2001) due to their increased vulnerability. Previous research reveals that small businesses are not adequately prepared to cope with the risk of natural hazards and to recover following such events (Tierney and Dahlhamer, 1996; Alesch et al., 2001; Yoshida and Deyle, 2005; Crichton, 2006; Dlugolecki, 2008). For instance, 90% of small businesses do not have adequate insurance coverage for their property (AXA Insurance UK, 2008) and only about 30% have a business continuity plan (Woodman, 2008). Not being adequately protected by community-level flood protection measures as well as property- and business-level protection measures threatens the survival of SMEs, especially those located in flood risk areas.

This chapter discusses the potential effects of flood hazards on SMEs and the coping strategies that the SMEs can undertake to ensure the continuity of their business activities amid flood events. It contextualizes this discussion within a survey conducted under the Engineering and Physical Sciences Research Council (EPSRC) funded research project entitled "Community Resilience to Extreme Weather—CREW."

19.2   EFFECTS OF FLOOD HAZARD ON SMEs

Several recent studies (Heliview Research, 2008; Norrington and Underwood, 2008; Woodman, 2008) point out that flooding has affected a significant portion of UK businesses during the past couple of years. Heliview Research (2008) reveals that flooding affected 47% of the UK businesses, subjected to their study, in 2007. According to Norrington and Underwood (2008), flooding or rain affected 34% of SMEs in the southeast of England in 2006–2007. Further, Woodman (2008) reports that extreme weather, including flooding, affected 29% of UK businesses in 2007. Such flood events can create a variety of effects on an SME, ranging from direct physical impacts to indirect effects due to their supply chain partners being affected by a flood event. Damage to premises, equipment, and fittings; loss of stock; reduced customer visits and sales; and disruption to business activities were among the most common effects experienced by UK businesses due to flooding (Norrington and Underwood, 2008; Pitt, 2008; Chatterton et al., 2010). An SME affected directly by a flood event may take months to recover and return to normal trading. The recovery process may also be affected by paperwork that is lost in the flooding, as this may lead to delays in completing insurance claims, tracing orders, filing tax returns, etc. (Pitt, 2008). Another recent study, conducted on behalf of the Chartered Management Institute (Woodman, 2008), shows that the effects of flooding were felt well beyond the businesses whose workplaces or premises were actually affected.

On top of the initial direct impacts of flooding, SMEs may experience other forms of crises, such as loss of market share, loss of key personnel, loss of production efficiency, withdrawal of supplies, withdrawal of licenses, and loss of quality or standard accreditation (Aba-Bulgu and Islam, 2007). Some of the indirect effects may include difficulty in securing financing and obtaining insurance coverage at reasonable cost, as investors and credit suppliers will be reluctant to supply financing (Metcalf and Jenkinson, 2005). Insurance companies may demand a higher premium for coverage if the possibility of damage to a business due to flooding is high (Dlugolecki, 2004).

The impact on an SME due to these effects can range from mere inconvenience to significant costs and ultimate business failure. In fact, Wenk (2004) states that 43% of companies experiencing a disaster never reopen and 29% of the remainder close within two years. More importantly, Alesch et al. (2001) have found that while the weakest small businesses tend to fail immediately following a disaster, many owners continue to struggle to recover until, one by one, their resources, energy, and options become exhausted, leading to more economic and social losses. Thus, flood-hit SMEs that struggle to recover but ultimately fail may have to suffer further losses in addition to those initially experienced. While the failure of a single SME might not be significant enough to destabilize a local economy, failure of several such SMEs might significantly affect a local economy. Only a few studies have attempted to look at these issues from a business perspective, when compared with the extent of studies on the household sector.

19.3 PROPERTY-LEVEL FLOOD PROTECTION AND STRATEGIES FOR BUSINESS CONTINUITY

Although community-based structural flood protection measures aim to reduce the vulnerability of high-flood-risk areas, it is neither technically feasible nor economically affordable to make the entire nation completely floodproof (EA, 2009). Therefore, the aim of the research is to focus on property-level flood protection measures and other strategies for business continuity that informs policy making to reduce vulnerability. Property-level flood protection measures can be either resistant or resilient (Bowker, Escarameia, and Tagg, 2007). Resistant measures attempt to prevent floodwaters from entering the property, whereas resilient measures attempt to minimize the impact of floodwaters on property (Bowker et al., 2007). From a business perspective, property-level flood protection measures assist the process of business continuity and fall under the umbrella of business continuity strategies. They represent the "hard," tangible measures for business continuity and act as an integral component of business continuity, as protection and restoration of business premises is fundamental to the successful operation of a business during and after a flood event. Other generic business continuity strategies such as property insurance, having a business continuity plan, online data backup systems, etc., represent the "soft," intangible measures, which allow a business to minimize the negative impacts of flooding as well as to recover smoothly following a flood event. Flood-hit business premises may put an SME out of business for months, making it very difficult to return to normality (Pitt, 2008). Consequently, one of the recommendations of the Pitt Review (Pitt, 2008) is to promote business continuity by encouraging the uptake of property-level flood resistance and resilience measures by businesses.

A study conducted on flood-affected companies in Saxony, Germany, revealed that relocation of utilities and hazardous substances to upper floors of buildings, flood-proofed tanks and air-conditioning, adapted use of property, water barriers, and adapted building structure were the common property-level flood protection strategies implemented by businesses (Kreibich et al., 2008). Flood insurance, emergency plans, and emergency exercises were other strategies commonly implemented by businesses. Crichton (2006) identified that home or flexible working, commercial insurance, reviewing risks to the premises, obtaining more advice, and considering moving elsewhere as actions that UK small businesses were willing to implement to cope with the risk of flooding. In addition to such generic strategies, some businesses with previous flood experiences have implemented various strategies on their own— for example, restaurants with fryers set on a hydraulic system enabling the fryer to rise up above the water level, fridges made of stainless steel with the motors set at the top, and water-sealed ventilation systems (ODPM, 2003).

It is pertinent that a mix of both hard (e.g., flood barriers, stock or equipment relocation) and soft (e.g., insurance, flood plan) measures are necessary for SMEs to prevent flood damage from affecting business, withstand the effects of a flood event, as well as facilitate quick business recovery and continuity following flooding. Because SMEs, by definition, have limited resources, careful analysis is required to weigh the costs and benefits of the options available in order to select the best options based on the risk of flooding. For example, the Environment Agency (Thurston et al., 2008) reveals that property-level flood resistance measures are economically beneficial for a business if the risk of flooding is greater than 4% (25-year return period), whereas property-level resilience measures are beneficial if the risk of flooding is greater than 10% (10-year return period). Furthermore, if the insurance terms do not accurately reflect the flood risk, resistance and resilience measures were identified by businesses as cost beneficial only if the risk of flooding is greater than 10% (Thurston et al., 2008). Hence, it is important for SMEs to assess the risk of flooding and identify the protection measures available, and their costs and benefits to their particular business, before selecting what mechanisms to implement.

19.4   SURVEY OF SMEs

The review and synthesis of literature on the effects of flood hazards on SMEs reveal a developing body of literature on the implementation of resistance and resilient measures based on the level and intensity of risks. It is pertinent therefore to conduct an empirical investigation to identify the flood risks and the measures taken by SMEs. Hence the research team decided to conduct a survey of SMEs within an area that has a high risk of flooding to contextualize the research issue.

19.4.1 Method

A questionnaire survey was carried out, targeting SMEs operating in Greater London, through the Federation of Small Businesses (FSB). According to the London Climate Change Partnership (LCCP, 2002), London is particularly vulnerable to flood hazard because a significant proportion of London lies within the floodplain of the River Thames and its tributaries. Currently, Greater London has the highest number of people at risk of flooding (EA, 2009). Furthermore, frequent intense winter rainfalls are expected to increase the likelihood of flooding in London by rivers and flash floods in the future (LCCP, 2002). The survey looked at the response of SMEs to flood hazard and other extreme weather events (such as strong rainfall and storms) that may lead to a flood hazard. The survey was specifically aimed at identifying the effects of these events on SMEs and the coping mechanisms implemented by SMEs in general. In broad terms, the survey looked at the resilience of SMEs to flood hazard and other weather extremes. The survey template was piloted first via the FSB and then via several borough council representatives of Greater London before administering it online. A total of 140 responses were received from SMEs representing many of the industry sectors. The sectors such as real estate, renting, and business activities (21%); transport, storage, and communication (21%); construction (15%); and wholesale and retail (14%) were well represented, whereas the remainder of the respondents (29%) were from other industry sectors.

19.4.2 Survey Data

The survey revealed that about 29% of the SMEs subjected to the study had been affected during the previous 5 years either by flooding or weather extremes that could lead to flooding. Although a considerable percentage of SMEs had been affected, the level of significance noted was quite low and was predominantly between 2 and 3 on a five-point Likert scale (1 = not affected, 2 = affected a little, 3 = somewhat affected, 4 = much affected, 5 = very much affected). Most of the businesses considered the hazards as minor ones, whereas only a few (15%) considered them major. Nonattendance of employees (49%), loss of sales or production (46%), damage to property or business premises (32%), and a decrease in turnover or profit (27%) were the most common effects experienced by the SMEs surveyed (see Figure 19.1). A relatively lower percentage of businesses experienced direct physical impacts, such as disruption of access to premises (19%) and damage to stock and equipment (16%). The survey results also demonstrate that the effects of flooding and related extreme weather events (EWEs) such as heavy rainfall and storms extend well beyond the businesses whose workplaces or premises were actually affected by the hazard.

Initially, disruptions to the activities of the supply chain outside their organizations were not identified by the SMEs as a commonly experienced effect due to flooding and related EWEs. However, to a subsequent question, a significant number of SMEs provided examples of the supply chain disruptions experienced, highlighting the strong link that exists between the flooding and supply chain disruptions. Based on the examples given for the effect on the supply chain, more than 50% of the SMEs actually experienced supply chain disruptions. Consequently, it was noted that

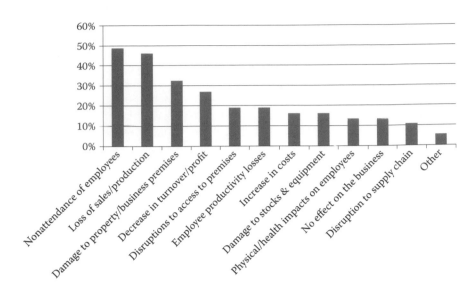

FIGURE 19.1   Effects of flooding: EWEs that could lead to flood hazard on SMEs.

disruptions to the activities of the supply chain outside the organization became the most commonly experienced effect by the SMEs.

When questioned about what actions SMEs that had been affected by floods and related hazards have undertaken to cope with these hazards to date, a significant number (29%) said that they are yet to undertake any action. Commonly implemented actions include improvements to business premises (31%), preparation of business continuity plans (26%), and having a business data backup system (23%). Other main measures implemented include property insurance (17%) and stock or equipment relocation (14%). Only a handful of businesses had implemented specific flood protection measures such as installing flood defenses and developing a flood plan. Stock or equipment relocation within the business premises, developing a business continuity plan, and having a business data backup system were quoted as the actions they will consider implementing in the near future.

In comparison, 53% of the businesses that had not been affected by a flood-related hazard have not taken any action to cope with those hazards. Property insurance (27%), business data backup system (17%), business continuity plan (10%), and business interruption insurance (10%) were the most commonly implemented coping strategies by SMEs. It must be noted that there is a significant difference between the two groups (*with* flood-related hazard experience and *without* such experience) when it comes to the uptake of property-level flood protection strategies. SMEs that have not been previously affected seem to rely on generic business continuity strategies rather than property-level protection strategies. Only a few have opted for property-level protection measures. For example, only about 6% had carried out improvements to their business premises, compared with 31% of flood-affected businesses. None of these SMEs had opted for temporary or permanent flood defenses. It is also noteworthy that some SMEs had thought about their business continuity without any prior experience of flooding or EWEs leading to flood hazard. SMEs with previous flood-related hazard experience were ahead in implementing coping strategies when compared with the SMEs without such experience. However, this was not the case when compared to the uptake of property insurance to cover the risks of flooding and other natural hazards. Only about 17% of SMEs that had been affected by EWEs had opted for property insurance, compared with 27% of those who had not been affected.

Of the SMEs that had not implemented any coping mechanism to date, 41% said that the impacts of the hazards are not significant enough to warrant any action, whereas 39% said that they do not foresee a hazard affecting them in the near future. This suggests that such SMEs do not consider the flood hazard a significant threat to their businesses. Thus, the survey results provide evidence that the potential risk of a flood hazard is not a priority in their business agenda until they are affected by such an event.

19.5 DISCUSSION

The survey results show that the effects of flooding and the EWEs leading to flooding extend well beyond the SMEs whose premises are affected by the hazard itself. For example, the most commonly experienced effect—the inability of employees to report to work—is likely due to disruptions in the transportation network or to their homes or regions being affected. Thus, the results indicate the importance of considering the whole host of such effects when action planning for a flood hazard. The results show an increase in implementing coping mechanisms by SMEs that have experienced a flood hazard, when compared with those that have not experienced any flood hazard. While the SMEs with previous flooding, strong rainfall, or storm experience have looked at property-level protection measures, such as improvements to premises, stock or equipment relocation, and flood defenses, SMEs without such experience have only looked at generic business continuity strategies. This corresponds to the economics of such measures (i.e., costs–benefit), as property-level flood protection measures were found to be economical only when the risk is high (Thurston et al., 2008). However, this also makes them vulnerable to damage in any future flood event.

Initially, the survey respondents failed to note the significance of their supply chain in relation to flooding and other related EWEs. However, in response to a subsequent question, a significant number of SMEs identified that they actually experienced various disruptions in relation to their supply chain. This indicates that although supply chain disruptions do occur due to flooding, SMEs do not necessarily recognize them as important to the well-being of their businesses. Although the importance of supply chain disruptions is often highlighted (Christopher and Peck, 2004; Kleindorfer and Saad, 2005; Huddleston and Eggen, 2007), SMEs tend to place lesser importance on them when making business decisions with regard to flooding and other natural hazards. This may leave SMEs vulnerable to the impacts of flooding, even if they have implemented various coping strategies to protect their business without considering the holistic view including their supply chains.

In general, the survey reported that many SMEs have not implemented coping mechanisms to deal with the flood hazard. On the one hand, this could be partly due to the survey being focused on SMEs located in Greater London rather than high-flood-risk or flood-affected localities within the Greater London area. On the other hand, it could be due to the survey having focused entirely on SMEs. A majority of respondents are micro businesses (84%), which are less likely to take action when compared with medium or large businesses. Several previous studies have revealed that smaller businesses are less likely to take action against natural hazards in comparison to their larger counterparts (Alesch et al., 2001; Crichton, 2006; Woodman, 2008).

A major phenomenon that was encountered in conducting this research was the criticality of the time lag between the disruptive event and the time of the survey. As pointed out by Alesch et al. (2001) and Webb, Tierney, and Dahlhamer (2002), many SMEs that experience a serious impact from a major disruptive event are unlikely to reopen or will fail within a short period of time after reopening. Therefore, it is possible that some of the SMEs that had suffered a major impact may have been automatically omitted in the study's sample set. Thus, the real impact on SMEs could be higher than identified in this study.

19.6   CONCLUSIONS

Flooding in the United Kingdom has become a significant threat to SMEs, and that threat is expected to further increase in the future due to the impacts of climate change. Flooding is capable of creating a variety of effects on SMEs and thereby affects the business continuity, particularly of SMEs that are less likely to be prepared to deal with such a hazard. Thus, the need for strategies in business continuity, including property-level protection measures, is often emphasized. However, uptake of property-level flood protection measures incurs costs to an SME and therefore the economics of such measures must also be considered when implementing them.

Findings of the questionnaire survey revealed very low levels of property-level flood protection measures by SMEs that had not experienced flooding in the past. Such SMEs were relying mostly on general business continuity strategies although the uptake of those strategies was also found to be minimal. This issue is of importance to the policy makers; the Pitt Review (Pitt, 2008), a government-appointed independent review in the aftermath of 2007 flooding in the United Kingdom, recommends that local authorities should promote property-level flood resistance and resilience measures by businesses to ensure business continuity. Given the fact that SMEs, especially the ones that had not experienced a flood-related hazard in the past, do not consider the risk of flooding as a significant threat to their businesses, local authorities will be required to initiate a major shift of SME thinking and decision making in order to achieve this recommendation. Although various initiatives have been started during recent years in populating such schemes (e.g., in a pilot grant scheme by the Department for Environment, Food and Rural Affairs), a concentrated effort is required by the central government, local authorities, environment agency, emergency services, business support organizations, local community forums, etc., to instigate a joined-up strategy and integrated policy for ensuring the business continuity of SMEs amid the ever-increasing risk of flooding.

While the study primarily concentrated on SMEs, the importance of the supply chain emerges as a major factor in determining SME business continuity. Although many SMEs initially did not recognize that they had been affected due to supply chain disruptions, when presented with various sub-effects coming under the umbrella term of *supply chain disruptions*, they acknowledged that they had been affected by them. This indicates that policy makers in this field cannot take a microscopic view of SME business continuity when facing the effects of flooding, but rather must broaden the scope of their actions to include the effects of a wider body of stakeholders that directly and indirectly interacts with SMEs. Such policy making can have significant long-term effects and will be positively aligned with some of the UK government's actions on climate change and community well-being.

REFERENCES

Aba-Bulgu, M., and Islam, S. M. N. (2007). *Corporate Crisis and Risk Management: Modelling, Strategies and SME Application.* Oxford: Elsevier Ltd.

Alesch, D. J., Holly, J. N., Mittler, E., and Nagy, R. (2001). Organizations at Risk: What Happens When Small Businesses and Not-for-Profits Encounter Natural Disasters. Small Organizations Natural Hazards Project, First Year Technical Report, University of Wisconsin–Green Bay. Fairfax, Public Entity Risk Institute.

AXA Insurance UK (2008). Preparing for Climate Change: A Practical Guide for Small Businesses. London: AXA Insurance UK.

BERR (Department for Business Enterprise and Regulatory Reform) (2008). SME Statistics for the UK and Regions 2007. Enterprise Directorate Analytical Unit, Department for Business Enterprise and Regulatory Reform.

Bowker, P., Escarameia, M., and Tagg, A. (2007). Improving the Flood Performance of New Buildings: Flood Resilient Construction. London: Department for Communities and Local Government.

Chatterton, J., Viavattene, C., Morris, J., Penning-Rowsell, E., and Tapsell, S. (2010). Delivering benefits through evidence: The costs of the summer 2007 floods in England. *Flood and Coastal Erosion Risk Management Research and Development Programme.* Bristol: England: Environment Agency.

Christopher, M., and Peck, H. (2004). Building the resilient supply chain. *International Journal of Logistics Management,* 15(2), 1–14.

Crichton, D. (2006). Climate Change and its effects on Small Businesses in the UK. London: AXA Insurance UK.

Dlugolecki, A. (2004). A Changing Climate for Insurance: A Summary Report for Chief Executives and Policymakers. London: Association of British Insurers.

Dlugolecki, A. (2008). Climate change and the insurance sector. *The Geneva Papers,* 3371–3390.

EA (Environment Agency) (2009). Flooding in England: A National Assessment of Flood Risk. Bristol, England: Environment Agency.

Evans, E., Ashley, R., Hall, J., Penning-Rowsell, E., Saul, A., Sayers, P., Thorne, C., and Watkinson, A. (2004). *Foresight. Future Flooding. Scientific Summary: Vol. I. Future Risks and Their Drivers.* London: Office of Science and Technology.

Heliview Research (2008). *Climate Change Effects.* Breda, the Netherlands: Heliview Research.

Huddleston, M., and Eggen, B. (2007). Climate Change Adaptation for UK Businesses: A report for the CBI task group on climate change. Devon: Met Office.

Kleindorfer, P. R., and Saad, G. H. (2005). Managing disruption risks in supply chains. *Production & Operations Management,* 14(1), 53–68.

Kreibich, H., Seifert, I., Thieken, A. H., and Merz, B. (2008). Flood precaution and coping with floods of companies in Germany. In Proverbs, D., Brebbia, C. A., and Penning-Rowsell, E. (Eds.). *Flood Recovery, Innovation and Response,* Southampton: WIT Press.

LCCP (London Climate Change Partnership) (2002). London's warming: The impacts of climate change on London, London: London Climate Change Partnership.

Metcalf, G., and Jenkinson, K. (2005). *A changing climate for business,* Oxford, UK: Climate Impacts Programme.

Norrington, H., and Underwood, K. (2008). Climate Change and Small Businesses: How Directors Are Responding to the Challenges of Climate Change—Research Findings 2008. Guildford, Climate South East.

ODPM (2003). Preparing for floods: Interim Guidance for Improving the Flood Resistance of Domestic and Small Business Properties. London: Office of the Deputy Prime Minister.

Pitt, M. (2008). The Pitt Review—Learning Lessons from the 2007 floods. London: Cabinet Office.

Thurston, N., Finlinson, B., Breakspear, R., Williams, N., Shaw, J., and Chatterton, J. (2008). Developing the Evidence Base for Flood Resistance and Resilience: R&D Summary Report FD2607/TR1. London: Department for Environment, Food and Rural Affairs (Defra).

Tierney, K. J., and Dahlhamer, J. M. (1996). Business Disruption, Preparedness and Recovery: Lessons From The Northridge Earthquake. DRC Preliminary Papers, Disaster Research Center, University of Delaware.

Webb, G. R., Tierney, K. J., and Dahlhamer, J. M. (2000). Business and disasters: empirical patterns and unanswered questions. *Natural Hazards Review,* 1(2), 83–90.

Webb, G. R., Tierney, K. J., and Dahlhamer, J. M. (2002). Predicting long-term business recovery from disaster: A comparison of the Loma Prieta earthquake and Hurricane Andrew. *Global Environmental Change Part B: Environmental Hazards,* 4(2-3), 45–58.

Wenk, D. (2004). Is 'good enough' storage good enough for compliance? *Disaster Recovery Journal,* 17(1), 1–3.

Woodman, P. (2008). Business Continuity Management 2008. London: Chartered Management Institute.

Yoshida, K., and Deyle, R. E. (2005). Determinants of small business hazard mitigation. *Natural Hazards Review,* 6(1), 1–12.

Section IV

The Community Perspective

20 Living with Flood
Understanding Residents' Experiences of Recovery

Rebecca Whittle and Will Medd

CONTENTS

20.1 INTRODUCTION

> Lucy: "The house seemed worse after they came in and gutted it. It didn't seem to be so bad when it was flooded, I know it had to be done."
> Len: "That was the heartbreaking part of it. When they walked down the drive with crowbars in their hands, I thought, 'They aren't going to be nice about this.' ..."
> —Len and Lucy, interview

Len and Lucy's* comments encapsulate the central theme of this chapter. For them, and many other people who have experienced flooding, it is not the floods themselves but rather what comes afterward that is so difficult to deal with. Using examples drawn from a 2-year study of flood recovery in Kingston-upon-Hull (United Kingdom), we explain why this is so and highlight ways forward for research and practice.

We begin with a brief review of the literature before describing the methodology on which the Hull study was based. We then explore key dimensions of recovery from the perspectives of the people taking part in our study before concluding with some reflections on what we can learn about the nature of recovery.

---

* All names used here are pseudonyms.

20.2 FLOOD AND FLOOD RECOVERY

In recent years we have seen a shift in flood risk management policy from an approach based almost exclusively on the provision of large-scale structural defenses in a bid to prevent flooding, to a recognition that floods cannot be prevented entirely. As a result, current policies are geared toward learning to live with floods through an approach based on resilience and adaptation (see, for example, "Making Space for Water" [Defra, 2005]; the Pitt Review [Cabinet Office, 2008]; and the Flood and Water Management Bill [Defra, 2010]).

Floods are expected to become increasingly unpredictable and multisourced as a result of climate change (Hulme et al., 2002; Cabinet Office/HM Treasury, 2006), and building resilience to these events is therefore a vital goal. Fundamental to this aim is the need to understand more about the long-term recovery process and how people can be supported more effectively during this time.

Previous research has worked to document the social, economic, and health impacts of flooding and the relationship between the social and physical parameters of community resilience and preparedness (Kirschenbaum, 2002; Gordon, 2004; Tapsell et al., 2005; Thrush, Burningham, and Fielding, 2005; Twigger-Ross, 2006). However, until recently there has been little research into how people recover from floods in the long term or what effect institutional support and investment in the built environment can have on that recovery. The "post-flood" studies that have been undertaken (e.g., Tapsell and Tunstall, 2001; Werritty et al., 2007) have generally been restricted to retrospective work or repeat visits that only capture single points in time. There was therefore little knowledge of flood recovery as it was lived on a daily basis, in terms of its social, economic, and health impacts and how these might relate to the interaction between different individual, community, institutional, and engineering responses.

Our research in Hull was a direct response to this gap in our understanding, with the main aim being to identify and document key dimensions of the longer-term experience of flood impact and flood recovery. The following section gives some background to the Hull flood and the research that we conducted.

20.3 RESEARCHING RECOVERY IN HULL

The floods of summer 2007 were the worst inland floods to affect England and Wales since 1947. Approximately 48,000 households and 7,300 businesses were flooded, while 13 people lost their lives (Cabinet Office, 2008). In the city of Kingston-upon-Hull, where our study was based, one person was killed and more than 8,600 households and 91 of the city's 99 schools were affected (Coulthard et al., 2007). The biggest event in Hull occurred on June 25th when more than 110 millimeters of rain fell, overwhelming the city's drainage system and resulting in widespread pluvial flooding. However, a smaller number of houses were affected in more localized incidents ten days previously.

The project design was based on the use of weekly diaries, combined with interviews, group discussions, and stakeholder engagement activities. It was adapted from a longitudinal diary-based study into recovery from the foot-and-mouth disease (FMD) disaster that severely affected Cumbria in 2001 (Mort et al., 2004).

Following a process of consultation with local stakeholders and experts in flood vulnerability, we identified a list of key characteristics that we used as a guide when recruiting participants: age (particularly elderly people), gender, type of disruption or displacement experienced (e.g., living upstairs, with relatives, in a caravan, etc.), tenure type, disabilities, uninsured, single parents, and families with young children. With this "profile" in mind, we then recruited a total of 42 residents for the study; the study lasted 18 months. The following statistics give a breakdown of the participants by age, tenure type, and additional considerations.

Age						
20–29	**30–39**	**40–49**	**50–59**	**60–69**	**70–79**	**80–89**
4	9	10	7	5	5	2

Tenure			
Owner Occupied	**Council Rented**	**Private Rented**	**Housing Association**
33	5	3	1

Type of Displacement from Property			
Rental Property	**Caravan**	**Lived in Flood-Damaged Property**	**Lived with Friends or Relatives**
20	7	12	3

Additional Considerations			
Disability or Serious Illness in the Family	**Uninsured**	**Single Parents**	**Families with Children under Five**
9	7	2	12

Upon recruitment, which took place between November 2007 and January 2008, the participants gave an initial semistructured interview that enabled them to tell their story thus far. At this point they were introduced to the weekly diary booklets that we encouraged them to keep for the 18-month duration of the fieldwork. The diaries started with a few simple "warm-up" questions where we asked participants to rate their quality of life, relationships with family and friends, and health using a simple scale ranging from "Very Poor" to "Very Good." There was also a section where they could enter details of what they had done on particular days during the week. The main purpose of these sections was to familiarize participants with writing in readiness for the main, "free-text" part of the diary, where they were encouraged to write whatever they wanted about their lives that week.

Diarists also took part in group discussions at quarterly intervals during the project where they could discuss the issues they were facing with their fellow participants.

In addition to this work with residents, we also undertook extra interviews with eight people who had been involved with the floods through their professional roles.

20.4   KEY ISSUES IN UNDERSTANDING RECOVERY

20.4.1   THE STRIP-OUT

As Len and Lucy's quote at the start of the chapter reveals, the first hurdle to overcome was the "strip-out." This process takes place at the beginning of the recovery period where anything touched by the water (e.g., furniture, carpets, crockery [dinnerware], photographs, plaster) is ripped out and thrown into a trash bin to help the property dry out. Particularly for older people who had been in their houses for decades, the experience of watching their homes and belongings vanish into a skip was deeply traumatic, as Chris, a pensioner in his 70s, described:

> "It was breaking our hearts, the biggest thing is when you see all your work going in a skip, it absolutely, you can't describe it ... I bet we've lived in eight places, flats and bungalow and a caravan and things, this is second time round, we've been together twenty-six years. This was it, this was the dream, the bungalow and it's gone, it's gone. And it broke our hearts ... Betty can't bear to go back and look at the bungalow and see somebody else in it." —Chris, interview

Living in a bungalow that they rented from a housing association meant that there was no upstairs space in which to save things, so they lost everything apart from their television set and its stand. The experience of being flooded was so upsetting that they decided to leave their bungalow and apply for a transfer to an upstairs flat where they could feel safe from future floods. Although this move gave them peace of mind, they were finding it very difficult to get over the loss of the bungalow, into which they had invested so much effort and hope for the future.

People mourned not just the loss of economic value but also the memories and sentimental value associated with their belongings and property. Their responses showed the home to be more than just a material collection of static objects—instead, it was more like a living creation that had grown with its owners over the years by their work on it and their experiences of living there (Sims* et al., 2009). Recovering the home is therefore not as easy as just purchasing new things.

Understanding the impact that the strip-out has on residents is particularly important in a context where surveyors are increasingly questioning the need for such drastic remedial action. A surveyor who we interviewed during the project explained that many workers did not have sufficient experience to assess water damage properly, resulting in unnecessary major renovations being recommended that caused additional disruption to residents and excessive costs for insurance companies.

> "Given the amount of water that actually got inside these properties, they didn't need to be stripped out, depending on the type of water that was in there, how long it was

---

* Now Whittle.

in there for and the depth, to go in there and see properties completely stripped out, no floors, no joists, no plaster, no floors, no ceilings, was completely overkill. It cost the insurance companies probably double if not triple the amount of money; it caused policyholders to be out of their properties for way over two years. I go back to Hull even now and I'm still seeing people living in caravans. It didn't need to happen, and it's so infuriating when something like this happens." —Martin, interview

The issue of whether such extensive stripping-out was appropriate and necessary in Hull was beyond the scope of our research. However, this is clearly a vital topic for future investigation because, as we demonstrate in this chapter, many of the negative impacts associated with flood recovery stem not from the flood itself, but rather from the overwhelming disruption that results from such extensive rebuilding work. If this disruption could be reduced, then it seems likely that much of the costs (both economic and emotional) of recovery could be avoided.

20.4.2   THE WORK OF RECOVERY

With the strip-out completed, residents must continue with the task of rebuilding their homes. A huge amount of effort is involved in this process, with tasks ranging from chasing quotes and phoning insurance companies, to managing builders. Whittle et al. (2010) provide a list of the different kinds of recovery work that residents reported carrying out in their diaries and interviews, including endless phone calls, trying to find a house to rent, choosing new things for the house every weekend, project managing builders, managing paperwork (e.g., invoices and claims forms), cleaning (either after builders or trying to get rid of the smell), making choices all the time, keeping a personal record of all correspondence and phone calls, e-mailing companies to explain that they were not happy with the work, making arrangements to meet people, getting quotes, finding builders, posting and faxing things, driving across town to empty dehumidifier buckets, keeping an eye on the workmen, making tea for the builders, re-doing the jobs the builders had done wrong, remedying damage caused to the upstairs of house, driving back and forth between rented house and home to check on it, having to wait at home for deliveries (some of which never came), making a note of snagging issues and calling companies back to correct them, packing and unpacking boxes of belongings, contacting utility companies to redirect mail, phone, bills, etc., and sorting the garden after the home is completed. This work is all carried out alongside everyday tasks and activities such as going to work, taking the children to school, and going to the supermarket.

However, different kinds of work are involved, depending on the various ways in which residents become involved in the recovery process. Some residents took on the very active task of "project management."

Leanne was an active and confident lady in her early 60s who had recently retired. During her interview, which took place in a static mobile home on her driveway, the builders were hard at work and our conversation had a constant backdrop of banging, hammering, and the roar of power tools. It soon became apparent that Leanne had played a central role in the re-creation of her home. However, by placing herself in a project management role, she had created a lot of stress for herself:

> "You see, it's every day and you ring them and they are not in or the answer phone is on and they don't get back to you. You fax them copies of quotes or invoices or what have you, and you write on it, 'Please let me know if you have received this and if you are going to approve it or if it's this or if it's that, if it comes within whatever.' And they just don't answer you. So you don't know whether they've received the information or the fax or what. So you ring up and then you can't get through to them to find out either. And if you speak to anybody else, 'Well we are not dealing with your claim; you really need to speak to so-and-so.' And it's just like hitting a brick wall; you can imagine the frustration." —Leanne, interview

Listening to this account, it was clear that the project management task on which Leanne had embarked was physically, mentally, and emotionally stressful as she took on the responsibility of chasing and coordinating things that would otherwise have been done by the insurance company. However, by managing in this way, she was able to ensure that the repairs progressed quickly and that the work was done to the standard she had requested. Indeed, she felt that her drive and persistence had enabled her to be back in her home in time for Christmas 2007.

Other flooded residents took an alternative path to recovery and experienced different kinds of stresses as a result. Lucy was also retired. However, unlike Leanne, Lucy's husband decided to use their insurance company's builders. He felt that this strategy would involve less risk and would protect them from the kinds of stresses involved in organizing repairs and finding quotes and project managing tradesmen (in short, the kinds of stresses experienced by Leanne). However, his decision led to work—and stress—of an altogether different kind as, despite repeatedly trying to chase their insurers, the couple had to wait for months before any work was started on the house.

Sitting in her caravan for months with no work taking place made Lucy feel depressed and out of control—particularly as she had to watch all her neighbors (who had appointed their own builders or were with different insurers) make progress on their homes. By the time Christmas arrived and Leanne was back in her home, work on Lucy's house had not even started and she wrote in her diary about how low this made her feel:

> "Felt down and closed in again last night. If I could only believe repair work would start soon. I feel forgotten by the insurers. Seeing people in their homes settled back in hurts. God knows what it feels like to be homeless. Monday and New Year's Eve tomorrow, have to pin smile on and party at [relatives]. Thank God for family and friends." —Lucy, diary

> "Up early Thursday morning, no sign of the builders . . . I feel like giving up ever living in the house again. Went to launderette and did supermarket shop. Another cold night shut in the coffin, as I call the caravan."—Lucy, diary

There are two points to note here. First, Lucy's comments about the launderette are an important reminder that it is not just the flood work that residents must contend with. All the other routine tasks of daily living also continue during the recovery process and, in many cases, these become more difficult as a result of living in temporary accommodations. For example, Lucy had to take her laundry to the launderette because she had no washing machine, while Laura had to endure a much longer commute into work from her temporary rented house in East Hull, as she reported in her diary:

> "Don't believe it; all bridges going up AGAIN right on rush hour—what another start to the day. Living in East Hull and trying to get to City Centre or west near impossible! Big family do tonight (Silver Wedding Anniversary and 21st Birthday). A bit peeved as this party's at an indoor party room very near to our flooded home, so would have been handy but due to us living so far away, I have to drive so cannot join in a celebratory glass of wine or two!" —Laura, diary

Second, by comparing Lucy's case with that of Leanne, we can see that there are many different kinds of physical, mental, and emotional work involved in the recovery process, from the full-on involvement of project managing to the endless waiting game that is involved when other people are making decisions about your life. Even what appears as doing "nothing" requires work in the form of planning, chasing people, or battling feelings of frustration and despair. Add to this the difficulty of managing everyday life on top of the flood repairs and it is not hard to see why one of the most popular metaphors used by diarists to describe the recovery process is one of a "fight."

However, our research also indicates that this fight can become longer and harder than it needs to be as a result of the ways in which residents are treated following a flood. The next section describes why this is so.

20.4.3 WHEN COMPANIES BEHAVE BADLY

Floods are often said to result in mental health impacts such as stress and depression (Chilvers, 2008). However, it is important to be clear about why these problems might occur. Far from being due to some inherent "weakness" in the residents themselves (as is often implied in the way these problems are reported), our research shows that it is a result of the ways in which people are treated by the many different agencies that they come into contact with after a flood.

Laura's story provides a clear picture of how things could have been different had the recovery process been handled better. By May 2008, Laura and her husband had already experienced a string of problems with long delays and poor workmanship that led to their builders being sacked from the job. This issue was eventually resolved and progress began to be made on Laura's house. However, further problems occurred when the insurance company inexplicably failed to pay the rent on the temporary accommodation that they were living in while their home was being repaired. Initially at least, the issue of the unpaid rent appeared easy to resolve. However, this proved not to be the case, and the following month became incredibly stressful for Laura as she recorded in her diary:

> "Friday May 30th: While in the office, receive a call from the agent looking after rented house. Still no rent being paid . . . I call [letting agents] to explain that we are still in rented house and builders still repairing our house but they insist they cannot pay rent without loss adjuster's say-so. I call loss adjusters but he's on holiday and so I call

his office; no one there can help as his manger is on hols [holidays] as well! [Letting agents] also don't even have right loss adjuster as they mention [name] who was loss adjuster number 1! What a bunch of idiots—they do not have any correct details of our situation! I got really stressed by all these phone calls and got nowhere!" —Laura diary, Monday, June 2, 2008

"Flexi day off work so had a day planned to call Loss Adjuster, [letting agents], and [property agents] to try to sort out unpaid rent. Got really stressed by these people at loss adjustors who don't seem to be any help at all. [NameA], our loss adjustor, doesn't answer my calls and his manager isn't available to help. [Letting agents] who are employed to handle our rented property and storage don't have a clue and even still think [nameA] is our loss adjustor. USELESS PEOPLE! [nameA] was our first, [nameB] is our third one!... Awaited return calls all day... around 5 pm [letting agents] called to ask for [loss adjuster's] number to call him! AGAIN. IDIOTS don't have his contact details so why are they getting paid to handle my rented property and storage? USELESS! This day off work has been so stressful, the total feeling of being powerless—this day nearly sent me down the route of going to see my doctor again with need of help! These people have no empathy, they all should look at their work practice and how they ever get a job done I never know—I don't know how they sleep at night!" —Laura, diary, Monday, June 23, 2008

"Day off today after working on previous Saturday. I call insurance department and speak to them regarding my concerns. I get really upset and have trouble explaining without crying as he says he will call loss adjustor for his side of the story! This comment really upsets me as why should I lie? I insist for his address to post my six-page letter and all the copies of e-mails when [name] has said he will pay rent and storage and never has. I feel absolutely exhausted after this call and feel quite shaken." —Laura, diary

These diary extracts illustrate how the mishandling of the recovery process can have a huge impact on the stress levels of residents. Laura is a rational, intelligent person who is not "mentally ill." Instead, her stress results from the ways in which her recovery has been mismanaged by the various agencies involved. Given the difficulties she has experienced, it is hardly surprising that she contemplates going to her doctor for some help. Yet counseling or anti-depressants will not resolve the source of her frustration—she simply needs to know that her rent has been paid and that she can trust the companies that are supposed to be helping her.

It is important to point out that some diarists had good experiences with their insurers and builders, and we have not included Laura's example with the purpose of making specific claims about these industries. Instead, the key point to emphasize is the stress that results when a person's case is not handled professionally, leading to re-traumatizing effects for residents who might otherwise have coped well with the flood recovery process.

20.5 DISCUSSION AND CONCLUSION

The preceding discussion has three implications for our understanding and management of flood recovery:

1. As stated at the beginning of the chapter, it is important to realize that the recovery process is often more difficult to deal with than the flood event itself. In particular, the increasingly controversial practice of stripping out the entire downstairs of an affected house means that the depth of the flood itself makes little difference to the level of disruption experienced by the resident. The practice of the strip-out is, however, increasingly contentious within the damage management industry, where the relative efficacy of different drying methods is being debated. This subject was beyond the scope of our study and yet clearly there is potential for further investigation into the recovery of the built environment after future flooding events in order to ascertain the relative scientific merits of different forms of drying and reinstatement technology, and to explore the nature of the relationships that exist between builders, insurers, and the damage management industry.

2. The experiences reported here point to the existence of what we have termed a "recovery gap." This emerges during the longer process of recovery at the point where the legally defined contingency arrangements provided to the affected community by its local authority diminish and where the less-well-defined services provided by the private sector (e.g., insurance, builders, etc.) start (Whittle et al., 2010). From an emergency management perspective, once people have been warned and rescued (if necessary), and after the situation has been handed from the response agencies to the local authority, the formal responsibilities toward householders become more obscure, and it becomes much less clear as to who should support people—and how—over the following months and years. The challenge, then, is for government and stakeholders to find ways to fill this gap by ensuring that residents receive more effective support during this time.

3. If this gap is to be filled successfully, we must pay attention to the ways in which residents are treated by the various companies and agencies that they come into contact with after the flood. This is no easy task, due to the size and range of the organizations involved (for one diarist, we counted 15 different agencies as being involved in the recovery process). Laura's story reveals the problems that can occur when the original trauma of the flood is followed by the double-trauma of poor treatment by the different agencies involved. We suggest that one way around this problem is for companies to adopt an "ethic of care," whereby they see themselves not just as individuals doing a job, but as recovery workers with an active role to play in helping people get their lives and homes back on track (Whittle et al., 2010). To be effective, such an ethic would necessitate a certain amount of cultural change within the relevant sectors (e.g., building and insurance) and it would need endorsing by the relevant trade associations for this reason. The end result of such a process, however, could make a real positive difference to residents' experiences of flood recovery over the longer term.

REFERENCES

Cabinet Office/HM Treasury (2006). *Stern Review on the economics of climate change*. London: Cabinet Office/HM Treasury.

Cabinet Office (2008). *The Pitt Review: Lessons learned from the 2007 floods*. London: The Cabinet Office.

Chilvers, M. (2008). Survey shows flood toll a year on. Available at <http://news.bbc.co.uk/1/hi/england/south_yorkshire/7445944.stm> [Accessed March 30, 2010].

Coulthard, T., Frostick, L., Hardcastle, H., Jones, K., Rogers, D., Scott, M., and Bankoff, G. (2007). The June 2007 Floods in Hull: Final Report by the Independent Review Body. Kingston-upon-Hull, UK: Independent Review Body.

Defra (Department for Environment, Food and Rural Affairs) (2005). Making Space for Water: Developing a new Government Strategy for Flood and Coastal Erosion Risk Management in England: A Delivery Plan. London, UK: Department for Environment, Food and Rural Affairs.

Defra (Department for Environment, Food and Rural Affairs) (2009). Draft Flood and Water Management Act. London, UK: Department for Environment, Food and Rural Affairs.

Gordon, R. (2004). The social system as a site of disaster impact and resource for recovery. *Australian Journal of Emergency Management*, 19, 16–22.

Hulme, M. et al. (2002). Climate Change Scenarios for the United Kingdom: The UKCIP02 Report. UKCIP.

Kirschenbaum, A. (2002). Disaster preparedness: A conceptual and empirical reevaluation. *International Journal of Mass Emergencies and Disasters*, 20(1), 5–28.

Mort, M., Convery, I., Bailey, C., and Baxter, J. (2004). The Health and Social Consequences of the 2001 Foot and Mouth Disease Epidemic in North Cumbria. Available at <www.lancs.ac.uk/shm/dhr/research/healthandplace/fmdfinalreport.pdf> [Accessed April 1, 2010].

Sims, R., Medd, W., Mort, M., and Twigger-Ross, C. (2009). When a "home" becomes a "house": Care and caring in the flood recovery process. *Space and Culture*, 12(3), 303–316.

Tapsell, S., and Tunstall, S. (2001). The Health and Social Effects of the June 2000 Flooding in the North East Region. Report to the Environment Agency. Flood Hazard Research Centre, Middlesex University, pp. 1–143.

Tapsell, S., Burton, R., Oakes, S., and Parker, D. J. (2005). The Social Performance of Flood Warning Communications Technologies (No. TR W5C-016). Bristol: Environment Agency.

Thrush, D., Burningham, K., and Fielding, J. (2005). Vulnerability with Regard to Flood Warning and Flood Event: A Review of the Literature. R&D Report W5C-018/1. Bristol: Environment Agency.

Twigger-Ross, C. (2006). Managing the Social Aspects of Flooding: Synthesis Report. Environment Agency R&D Technical Report SC040033/SR6. Bristol: Environment Agency.

Werritty, A., Houston, D., Ball, T., Tavendale, A., and Black, A. (2007). Exploring the Social Impacts of Flood Risk and Flooding in Scotland. Scottish Executive.

Whittle, R., Medd, W., Deeming, H., Kashefi, E., Mort, M., Twigger Ross, C., Walker, G., and Watson, N. (2010). After the Rain—Learning the Lessons from Flood Recovery in Hull. Final Project Report for "Flood, Vulnerability and Urban Resilience: A Real-Time Study of Local Recovery Following the Floods of June 2007 in Hull." Lancaster, UK: Lancaster University.

21 Property-Level Flood Protection
Case Studies of Successful Schemes

Mary Dhonau and Jessica E. Lamond

CONTENTS

21.1 INTRODUCTION

Despite the millions of pounds spent by the Environment Agency each year on building and maintaining flood defenses, it is clear that not all properties in England

and Wales are protected or are planned to be protected by community-level flood defense. Many people and properties are flooded each year and in a major incident, the number of flood victims runs into thousands, sometimes tens of thousands. This is not likely to change as government policy in the United Kingdom and across Europe appears to be moving toward encouraging property owners to protect themselves (Ashley et al., 2007; Kelly and Garvin, 2007; Rooke, 2007).

One fact that is regularly overlooked by engineers and academics is the stress and trauma to those unwitting recipients of unwanted and uninvited floodwater. Flood victims are forced out of their homes for many months and sometimes for years, while they watch their homes become a building site as it is being restored. Figure 21.1 shows the devastation within a flooded home from which the owners were displaced to a caravan/mobile home for 2 years.

For those individuals who are frequently flooded or are at high risk from flood, one of the most important decisions they have to make is how to protect and arrange their property to minimize the damage and disruption caused the next time. It is perhaps surprising that, at the moment, most property owners do not take steps to protect themselves. This situation is not unique to the United Kingdom. Research has shown that acceptance is low in most flood-risk populations (Correia et al., 1998; BMRB, 2006; Grothmann and Reusswig, 2006; Thieken et al., 2006; Harries, 2007; Norwich Union, 2008). The reasons for this low acceptance have also been studied and are seen to include the high cost or perceived high cost of protection. However, there are also other reasons, such as getting the right information about products, not wanting to admit that the flooding will return, unwillingness to face the disruption of building work, and the sheer inertia that stops properties from being protected (Sims and Baumann, 1987; Grothmann and Reusswig, 2006; Norwich Union, 2008; Proverbs and Lamond, 2008).

An online survey conducted in 2010 by the National Flood Forum (NFF) of those who had been flooded also revealed that despite wanting to make their homes resilient to flooding, some flood victims had actively been discouraged by their loss

FIGURE 21.1 Home for 2 years while this was sorted out.

adjuster, surveyor, or builder. This highlights a general lack of knowledge across the industry as a whole.

One of the most valuable roles of the NFF is to support property owners who are planning to protect their properties by providing information about the types of protection available and to publicize the effectiveness or otherwise of property-level flood protection. The NFF has observed many examples of homeowners and businesses retrofitting protection measures, and discussed with and advised the owners. Some of these solutions are publicized on the website (NFF, 2006) but there are also case studies published elsewhere (e.g., Lamond et al., 2009) that demonstrate the uses of the wide range of different protection products available. This chapter briefly describes some different approaches to property-level flood protection and summarizes a selection of case studies that cover some of these approaches. It must be noted that the NFF does not endorse the use of any particular product or company that may be used in the case studies—they are used for illustration purposes only.

21.2 APPROACHES TO PROPERTY-LEVEL FLOOD PROTECTION

The first approach usually taken to property-level flood protection is to attempt to keep water out of buildings (often known as *resistance* or *dry-proofing*). Resistance is suitable in floods where the floodwater does not lurk about for too long, such as surface water flooding or overland runoff where the floodwater enters a property via the front door, leaves quickly by the back, causing mayhem en route. This kind of intervention is best used in floods of low velocity and depths below about a meter, as above that depth there is some risk to the structural stability of the buildings (Kelman and Spence, 2004; Zevenbergen et al., 2007). It can be difficult to keep water out for two main reasons. First, water can enter buildings in many different ways and they must all be protected. Second, water can seep through materials and so resistance may be particularly problematic in prolonged floods. However, many people do successfully waterproof their property and even if measures are not completely successful, the protection often buys time to evacuate the property and to move contents out of the way of floodwater. Further advice on the suitability of solutions can be found from the U.S. Army Corps of Engineers (USACE, 1998), Communities and Local Government (Bowker, Escarameia, and Tagg, 2007), and the Association of British Insurers (NFF/ABI, 2006), among others. However, it is always sensible to take the advice of a qualified surveyor when employing resistant measures.

Another approach is to allow water in and plan to recover quickly (often known as *resilience* or *wet-proofing*). Resilience can be employed in most flood risk property, although in many older and listed buildings it is essential that specialist advice is taken (Fidler, Wood, and Ridout, 2004). Resilience is often used as a backup to resistance. Once water is allowed into the building, the internal structures and contents become vulnerable, so additional resilient approaches include moving items to safety, protection with waterproof coatings, or the use of materials that are not damaged by water, absorb water slowly, or dry quickly.

The advantage of resilient measures is that they can prevent damage even when other measures fail, thus cutting down on the length of time people will have to be out of their homes. Some solutions do not require deployment, resulting in lower annoyance due to false alarms and lower risk of failure. The disadvantage is that the installation of resilient materials means changing the home on the inside, a process some property owners find difficult to accept as many people report that if such measures were adopted their home would no longer look like a home (Harries, 2008).

21.2.1 Resistant Solutions

Barriers to the entry of water can be set at a distance from the building such as enclosing walls, bunds, or temporary barriers. They can also be attached to the building openings such as door and window guards, or they can enclose the building such as veneer walling or rendering of walls. Many of the products are listed in the blue pages, a directory of flood protection products and services maintained by the NFF (2006); they can be categorized as follows.

21.2.1.1 Barriers at a Distance

Water can be kept at a distance from the property using barriers such as enclosing fences, walls, or specially designed flood barriers. This may not be possible for property that is not detached, unless several owners cooperate.

21.2.1.2 Barriers to Stop Water from Entering the Building through Openings

Closing the gaps where water can easily enter can be an effective defense against flooding, particularly if the flooding is likely to be shallow and of short duration. Sandbags are often used for this but they are not the most effective and can be difficult to get in an emergency. Some barriers have been awarded kite-mark status, which means that they have been tested against standards of strength and leakage rates by an independent body against the publically available standard for flood protection products. Products include

- Door panels
- Window panels
- Airbrick covers
- Automatic barriers
- Self-seal airbricks
- Absorbent cushions, also known as sandless sandbags
- Appliance vent and pet flap covers
- Sealants
- Flood-resistant doors
- Weephole covers

21.2.1.3 How to Stop Water from Coming through the Walls and Floors

Barriers may prevent water from entering the building through doors, windows, and other openings during a short-duration flood. However, most building materials will let water through slowly and so in a prolonged flood, water may come through walls and floors. Walls can be coated with protective coverings, covered with waterproof render or sealants, or temporarily wrapped in plastic. Permanent installations such as veneer walling and flood skirts are offered by a few suppliers.

Waterproof membranes can be installed under floors or they may be coated with waterproof materials. However, water pressure under floors can be considerable, and it is possible that floors may crack unless drain holes or drainage membranes are used. In severe floods, uplift may result from complete waterproofing and so it is always wise to consult an engineer. Products include

- Flood skirts
- Veneer walling
- Cementitious render
- Waterproof coatings
- Waterproof membranes
- Epoxy resin
- Cavity drainage systems
- Sump and pump systems

21.2.1.4   Products to Stop Water from Entering through the Toilet

During a flood, water will often enter buildings through plumbing systems, even when the remainder of the building openings are above flood level. Any sink or toilet installed on the ground floor may be vulnerable. This will add sewage to the mud and silt that is already part of most floodwater and can be very unpleasant. Products to prevent backflow include

- One-way valves
- Toilet stopper
- Toilet pan seals
- Inflatable pipe stopper
- Ubend stopper

21.2.1.5   Pumps to Remove Water That Seeps through

If water does enter the home for any reason, either through failure, overtopping, or seepage, then it may be necessary or desirable to pump it out. Sumps and pumps can also be part of the design of some protection systems to relieve pressure from vulnerable structures. Many types and sizes are available and of course they need some kind of power. During a flood, any pumping system will necessitate the use of a generator. If generators are used, they should not be operated in a closed environment.

21.2.1.6   Flood Alarms

Flood alarms are useful to give warning to evacuate premises or to deploy protection measures. The Environment Agency Floodline warnings will give advanced warning when flooding is likely in an area but personal flood alarms can be a useful additional safeguard as they can be targeted at a particular known danger zone. They can be particularly useful in areas where the threat is from flooding not covered by the EA warnings. Also available are telemetry systems that can be placed in a watercourse that does not have an EA flood warning. These systems can call individuals to alert them that the water is rising rapidly and a flood may be imminent.

21.2.2 RESILIENT SOLUTIONS

Resilient solutions are designed to protect contents, fixtures, and fittings once water has been allowed inside a building. The most commonly cited solutions are to move electric sockets above the water line and to use hard finishes such as tiles rather than carpets and wallpaper. This has led to a feeling that resilient homes may be aesthetically undesirable. However, solutions can be designed to suit the individual property and the preferences of the property owner. There is normally more than one way to protect each element and often the products eventually adopted are not specialized for flood protection. The following list shows the different approaches to resilient protection:

- Permanently remove items from the path of the flood.
- Remove or move items when it floods.
- Choose furnishings and fittings that water will not damage.
- Cover vulnerable items with permanent waterproof coatings.
- Cover vulnerable items with plastic in the event of flood.
- Use furnishings and fittings that are easy and cheap to replace and decide to sacrifice them.
- Adapt or convert your property to make it less susceptible to the damage caused by floodwater, for example, by raising the power sockets to dado level.
- Adapt the way you use your property through, for example, moving your kitchen to an upper floor.

Some helpful products are also listed here:

- Plastic skirting, architraves, and other timber replacement products
- Removable door hinges for internal and cupboard doors
- Varnishes and resins for treating wood products, making them impermeable to water
- Large plastic bags for covering items that cannot be removed
- Plastic or stainless steel kitchens
- Wall-mounted boilers

21.3 CASE STUDIES

The experience of residents who have installed flood protection products in general suggests that they have arrived by a process of research, expert advice, and trial and error at solutions that suit them and minimize the damage they suffer during flood events. The following seven case studies span a wide range of products and types of properties. All the property owners speak of the great peace of mind afforded by their protection, and some have undoubtedly saved themselves and their insurers a great deal of money.

21.3.1 Case Study 1: Resilience

A house in Oxford was flooded three times, in 2000, 2003, and 2007. After the first two floods, the house was restored like-for-like by the insurance company under the normal procedure. After the 2007 flood, the owner decided that enough was enough and undertook to plan a more resilient scheme. Figure 21.2 illustrates several features of the scheme that was funded mainly by the insurer where it came within the cost of like-for-like restoration cost and from other funds where available.

The previous oak flooring was replaced with an attractive stone and this was sealed to make it waterproof. The owner liked the previous finish of wooden paneling and did not want to replace it with tiles. After some deliberation, the paneling was replaced with simulated plastic wood with grilles inserted to allow the brickwork below to breathe. Sockets and appliances were raised above the flood level where possible, and a cheap kitchen with plastic legs was installed. Finally, a sump was installed but the planned pump fell outside the budget available and was not fitted. This system has yet to be tested in further flooding.

21.3.2 Case Study 2: Resistance

A bungalow newly built in 1996 flooded in the Easter 1998 floods to about two feet and caused great distress to the residents, who had no idea they were at risk. As one resident put it in trying to prevent the water from coming in, "We ran around like headless chickens, not really knowing what to do." Eventually, they gave up and evacuated the property. Later on, "The aftermath was terrible and I cried for days";

FIGURE 21.2 Resilient case study.

FIGURE 21.3   Installing door guards.

the insurance claim was on the order of £113,500. Several false alarms later, this resident purchased door guards at a cost of £5,000. The door guards have to be deployed
at the time of the flood; they are kept in the garage and take about 20 minutes to fit,
as shown in Figure 21.3.

In 2007 the area was flooded again and the floodguards protected the property. This
kind of product can give great peace of mind, protect against damage and insurance
claims, and at the time of a flood allow residents to evacuate to safety knowing that
they have done their best to protect their property. The panic involved in trying to construct some kind of defense from sandbags or plywood on short notice is avoided, and
the time can be used to move valuable or sentimental items in case of overtopping.

21.3.3   Case Study 3: Low-Cost Resilience

After the 2007 flood, a forward-thinking North Yorkshire resident campaigned for
community flood defenses to prevent himself and his neighbors from being flooded
again. Although the scheme came close to funding, changes in the Environment
Agency points system left the residents without hope of public help.

The next port of call was the insurance company, where a scheme for resilient
reinstatement of his property was met with a refusal from an insurer concerned that
the benefits of more expensive reinstatement could be reaped by a competitor if the
resident switched policies in the future.

Finally, the resident came up with a low-cost scheme that he hopes will minimize the damage if the worst were to happen again. He suggests the following low-
cost alternatives:

- Install the boiler or service meters on a wall or on a plinth off the floor.
- Change timber floors to concrete. If the cost precludes this, then look to fit airbrick covers.
- Fit all timber skirting boards, door frames, and architraves, and paint them to a finish on all sides, front and back.
- Look to tile all concrete floors with nonporous floor tiles with solid adhesive backing and nonporous grout.
- Do not paint over your hinges to your doors; leave them exposed so that they are easy to unscrew and remove quickly.
- Look to use a lime-based plaster or sand and cement render, and skim when replastering the walls.
- Consider using waterproof paints for all surfaces that you feel to be in the flood line.
- Fit non-return valves on all drainage pipes.
- Remove any floor-mounted kitchen appliances and either site on plinths or have integrated products.

21.3.4 CASE STUDY 4: SMART AIRBRICKS

This case study illustrates how even very shallow flooding can have a great impact. A disabled Gloucester resident was flooded twice in 2007; flooding was caused by inadequate drainage and only to about four inches—unfortunately, just above the height of the airbricks. The resident lived alone, had limited mobility, and therefore was unable to save any furnishings from the water once the floodwater had penetrated the house. Repairs cost £14,200. The attached neighbor also flooded at a cost of more than £33,000.

Smart airbricks were the solution for this disabled resident as they do not require deployment. These airbricks are installed permanently and operate as normal airbricks except during high water when internal balls are lifted by the water and close the airbrick, thus preventing water from entering the home. Fifteen airbricks were fitted by the resident and his neighbor at a total cost of £1,500. Next time the floodwater reached the building, the airbricks closed and water was prevented from entering the properties. The disabled resident said,

> "The bricks were delivered quickly. I asked my local builder which had repaired my house to fit them. He did within a couple of hours at a reasonable rate. I would recommend the SMART Airbricks to anyone who wants to sleep at night without worrying."

21.3.5 CASE STUDY 5: PROTECTING SHOPFRONTS

Many flood-affected areas are in the center of towns and naturally affect small businesses and retail premises. Shopfronts may require larger gates than doors on residential homes but many companies can supply such gates. However, the pharmacist featured in this case study had a standard-sized door guard that was deployed each night for 2 years, as shown in Figure 21.4.

FIGURE 21.4   The door guard in its regular place.

After a freak hailstorm, the benefits of his policy became clear. The hailstones drifted up to six feet high, blocking drains and causing flash clouding. Many other businesses suffered but the prudent pharmacist's storefront was protected by a simple door guard.

21.3.6   Case Study 6: Neighborhood Collaboration

It is often said that there is no point in protecting the openings in a semi-detached or terraced property if the water is going to come in from your unprotected neighbor's residence. In Case Study 3 above, two attached properties adopted the same solution at the same time and both benefited from the added protection. In this case study, a row of houses collaborated in building a flood defense wall around their houses that can be closed off in times of flood by flood gates. Figure 21.5 shows the wall, an attractive stone finish reinforced by concrete inside.

These properties had flooded in 1998 but failed to meet the Environment Agency criteria for flood defense funding. One resident instigated meetings and the result was that the residents funded their own community defense with advice from the Agency at a cost of £9,000 per property. In July 2007 at the height of a record-beating flood, the wall was eventually overtopped but it had allowed the residents seven hours to remove all vulnerable contents.

FIGURE 21.5 The protective wall.

21.3.7 Case Study 7: Complete Resilience

The homeowner displaced for 2 years in Figure 21.1 had also been flooded in the year 2000 and was determined to never again suffer such distress. Not only did she protect her home with door barriers but she also made a concerted effort to make the property resilient to flooding by using many different resilient approaches. In addition to the usual raising of the boiler, fuse box, and oil tank, and ceramic tiled floors, this resourceful resident installed demountable radiators, a raised fireplace, a wall-mounted sink, and even mounted the television on the wall.

Furniture was designed to be lightweight and easy to move, such as a folding table, lightweight chaise longue, plastic chairs, and an easily dismantled bookcase.

Finally, the immovables were designed to be flood resilient. Open-plan oak stairs will only require careful cleaning and have no void areas to trap moisture. The kitchen (see Figure 21.6) is both waterproof and beautiful. It is constructed from powder-coated and stainless steel with high-gloss solid acrylic doors and a tiled floor.

21.4 SUMMARY

Increased flood incidents and current UK government thinking are likely to result in a growing need for individuals to install their own property-level flood protection.

FIGURE 21.6   Floodproof kitchen.

The decision regarding whether to install measures and which to install is therefore a crucial one for residents at risk of flooding but is often neglected due to ignorance of appropriate measures; discouragement from insurers, builders, and loss adjusters; or the belief that others will act. Too often, the sandbag is the only solution that comes to mind, and experience shows that sandbags are not effective in preventing flooding.

Modern resistant solutions include door guards designed to seal against openings or slot into prefixed grooves, smart airbricks, self-closing barriers, tanking, and veneer walling. Modern resilient solutions include the use of state-of-the-art plastic replacement fittings, protective veneers, and fixtures designed to be light and easy to remove, such as removable door hinges.

The case studies in this chapter show how relatively straightforward these forms of property-level flood protection can be. With careful thought regarding the appropriate defense system, even very cheap solutions can be effective for some people. The majority of properties at risk of flooding in the United Kingdom could benefit from the installation of protection measures in terms of minimizing disruption during and after flooding and in increased peace of mind.

Many of these solutions are also cost effective, as the relatively inexpensive door guards and smart airbricks case studies illustrate. If deep or prolonged flooding renders resilience as the only solution, then the expense can be much more significant but damages prevented are likely to be commensurately large. Clearly, the costs and benefits of installing protection are different for different properties and property owners. Support and advice are available from the NFF, but in complicated scenarios the use of a qualified professional is recommended.

REFERENCES

Ashley, R., Blanksby, J., Chapman, J., and Zhou, J. (2007). Towards integrated approaches to reduce flood risk in urban areas. In Ashley, R., Garvin, S., Pasche, E., Vassilopoulos, A., and Zevenbergen, C. (Eds.), *Advances in Urban Flood Management*. London: Taylor & Francis.

BMRB (2006). Flooding: Is the Nation Prepared? Aspects of Social Research.

Bowker, P., Escarameia, P., and Tagg, A. (2007). Improving the Flood Performance of New Buildings—Flood Resilient Construction. London: Communities and Local Government.

Correia, F. N., Fordham, M., Saraiva, M. D. G., and Bernado, F. (1998). Flood hazard assessment and management: Interface with the public. *Water Resources Management*, 12, 209–227.

Fidler, J., Wood, C., and Ridout, B. (2004). Flooding and Historic Buildings. Technical advice note. In Wedd, K. (Ed.). London, UK: English Heritage.

Grothmann, T., and Reusswig, F. (2006). People at risk of flooding: Why some residents take precautionary action while others do not. *Natural Hazards*, 38, 101–120.

Harries, T. (2007). Householder responses to flood risk, the consequences of the search for ontological security. Middlesex, UK: Middlesex University.

Harries, T. (2008). Feeling secure or being secure? Why it can seem better not to protect yourself against a natural hazard. *Health, Risk and Society*, 10, 479–490.

Kelly, D. J., and Garvin, S. L. (2007). European flood strategies in support of resilient buildings. In Ashley, R., Garvin, S., Pasche, E., Vassilopoulos, A., and Zevenbergen, C. (Eds.), *Advances in Urban Flood Management*, London: Taylor & Francis.

Kelman, I., and Spence, R. (2004). An overview of flood actions on buildings. *Engineering Geology*, 73, 297–309.

Lamond, J., Dhonau, M., Rose, C., and Proverbs, D. (2009). Overcoming the barriers to installing property level flood protection — An overview of successful case studies. *Road Map Towards a Flood Resilient Urban Environment*. Paris: Urban Flood Management Cost Action Network C22.

NFF (National Flood Forum) (2006). National Flood Forum Home Page www.floodforum.org.uk.

NFF/ABI (National Flood Forum/Association of British Insurers) (2006). Flood Resilient Homes—What Homeowners Can Do to Reduce Flood Damage. London: Association of British Insurers.

Norwich Union (2008). Homeowners Fear Future Flooding but Fail to Take Measures to Protect Their Property. Norwich Union.

Proverbs, D., and Lamond, J. (2008). The Barriers to Resilient Reinstatement of Flood Damaged Homes. *4th International i-Rec Conference 2008. Building Resilience: Achieving Effective Post-Disaster Reconstruction*. Christchurch, New Zealand.

Rooke, D. (2007). The summer of storm. *Water and Environment Magazine*, 10, 8–9.

Sims, J. H., and Baumann, D. D. (1987). The adoption of residential flood mitigation measures. What price success? *Economic Geography*, 63, 259–273.

Thieken, A., H., Petrow, T., Kreibich, H., and Merz, B. (2006). Insurability and mitigation of flood losses in private households in Germany. *Risk Analysis*, 26, 1–13.

USACE (US Army Corps of Engineers) (1998). Flood Proofing Performance: Successes and Failures.

Zevenbergen, C., Gersonius, B., Puyan, N., and Van Herk, S. (2007). Economic feasibility and study of flood proofing domestic dwellings. In Ashley, R., Garvin, S., Pasche, E., Vassilopoulos, A., and Zevenbergen, C. (Eds.), *Advances in Urban Flood Management*. London: Taylor & Francis.

22 Improving Community Resilience

Education, Empowerment, or Encouragement?

Carly B. Rose, David G. Proverbs,
Ken I. Manktelow, and Colin A. Booth

CONTENTS

22.1 INTRODUCTION

Worldwide, it has been found that there are wide variations in the ways that people respond to probabilistic hazard forecasting: Some people may make extensive preparations, in line with advice from the relevant authorities, while others will take no action at all. Although the dangers arising from natural hazards are undeniably real, the element of uncertainty inherent in forecasting means the perceptions of the individuals in, for example, flood-risk areas, form an important aspect of human response. These perceptions may be influenced by factors such as familiarity with the hazard, the degree of "controllability" associated with the threat itself (Slovic, 2007), as well as personality types (Baumann and Sims, 1978) and belief systems (Lindell and Perry, 2000).

If resilience to natural hazards, such as flooding, is to increase in our communities, then it follows that more people will need to take appropriate steps. This, however, means changing existing behavior patterns, not simply providing information to aid in purchasing decisions. The study of behavior, and theories surrounding behavioral change, lie within the province of psychology, rather than flood risk management per se. An understanding of psychological theory can, therefore, shed light on the range of behaviors currently exhibited by at-risk communities, many of which may initially appear to be illogical or irrational responses to a hazard. This knowledge can, in turn, guide the development and appropriate targeting of intervention strategies such as education campaigns, community engagement initiatives, or, where appropriate, changes in policy and process by the decision-maker groups.

The term *coping strategies* will be used in this section to denote those actions affecting the ability to prevent, tolerate, avoid, or recover from flood impacts. Technical strategies for dealing with flooding are covered elsewhere in this volume; hence, this chapter focuses on the wider issues operating within the at-risk community, such as decision-making behaviors and the influence of intangible factors such as personality types and belief systems.

To place this discussion in context, the governance background surrounding flood risk in the United Kingdom will be outlined first.

22.2   UK SOCIO-POLITICAL CONTEXT

For many years, the predominant management tool for fluvial and coastal flooding in the United Kingdom was the construction of hard-engineered flood alleviation schemes; a well-known example is the Thames Barrier and the associated embankments. Completed in 1982, this system is designed to protect a large area of London from extreme tidal surges of the type that affected the east coast of England in 1953 (EA, 2008a). In England and Wales, these schemes were overseen by a succession of public-sector bodies, culminating in the present role of the Environment Agency, currently the lead agency in this respect. (In Scotland, the equivalent responsibility lies with the Scottish Environmental Protection Agency (SEPA) and in Northern Ireland, the Rivers Agency.)

To comply with the EU Floods Directive of 2007 (EU Commission, 2007), the UK government conducted a review of its long-term policy in this area, culminating in the document "Making Space for Water" (Defra, 2005). This clearly articulates the premise that floods cannot be prevented, but flood risk can be managed; it also emphasizes the need to adopt a more integrated and holistic approach, involving stakeholders at all levels. One of the stated aims of the strategy is for the public to become more aware of flood risk, and be empowered to take suitable action themselves where appropriate; for example, by incentivizing the uptake of property-level resilience and resistance measures via grant schemes (Defra, 2005). Essentially, this means that individuals will need to accept more of the responsibility for protecting homes and businesses from flood impacts.

This change in policy approach, although understood and accepted at governance level, did however represent a profound change for the residents of at-risk areas, in comparison with the somewhat paternalistic approach to "flood defense" that had previously been the accepted norm. Studies have shown, however, that the level of preparedness behaviors is positively correlated with the degree to which personal responsibility *is accepted*; conversely, where local councils or emergency management agencies are perceived to be responsible for the safety of residents, then preparedness actions are less likely to be adopted (Paton, 2003).

As an example, one form of preparedness that is freely available in the United Kingdom is a flood warning service covering many, but not all, flood-risk areas. Although the operating bodies—the Environment Agency and SEPA—make no charge for the service, take-up has been far from complete, as will now be seen.

22.3   FLOOD WARNING IN THE UNITED KINGDOM

In the summer of 2007, several areas of the United Kingdom experienced extreme flooding; as part of the post-event review, it was established that only 41% of eligible homes were registered for the free Flood Warnings Direct Service (Pitt, 2008). At that time, residents of England and Wales were required to opt in to receive flood warnings by telephone: one of the recommendations arising from the Pitt Review was that, instead, landline telephone numbers should automatically be included in the scheme unless people actively chose to opt out. A pilot study using this method demonstrated that only 2% of the sample declined the warning service (Environment Agency, 2008b); the new procedure was rolled out in February 2010, with an additional half a million telephone numbers being added to the warning system.

This change takes advantage of a common human behavior pattern, namely procrastination: in this example, the original procedure may have created an intention to register for warnings in the minds of the public, but this was not necessarily carried through into action. Reversing the approach, therefore, harnesses this tendency such that a failure to act (i.e., declining the service) now leads to a positive resilience outcome (more at-risk premises receiving warnings and therefore having time to deploy flood protection products or move possessions to safety). The psychological theory underpinning this is known as "inaction inertia," as discussed by Tykocinski and Pittman (1998).

There are other factors to consider in hazard warning systems—for example, the degree to which a warning institution is trusted can have a bearing on whether issued warnings will be believed and acted upon. Individuals may prefer to seek confirmation from other sources such as friends, relatives, or neighbors and if this is not forthcoming, they may disregard official alerts (Brown and Damery, 2002). Prior to 1996, flood warnings in England and Wales were issued by the National Rivers Authority (a predecessor body to the Environment Agency) but disseminated to the public via the police, who then cascaded the information. Local volunteer flood wardens were notified initially; they then contacted a designated list of residents with the news, either by telephone or by knocking on doors in their area (Harding and Parker, 1974). The new procedure, introduced by the Environment Agency in September 1996, made use of innovative communications technology to issue warnings by telephone, directly to households and businesses (EA, 2000). In addition to the prerecorded telephone warnings, an information and advice service for each area was provided; this, however, could only be accessed using the correct area warning code (similar to a personal identification number) via the telephone system. Thus, personal contact with local representatives, known to the at-risk community, was replaced by an efficient but depersonalized system, and it is interesting to speculate whether this procedural shift may have contributed to the perceptions of flooded residents in the following years, as in the following examples reported by Fielding et al. (2007):

"It was horrible. You panic if you get a phone call at two o'clock in the morning and you don't know what it's for." —Owner occupier, Southern Region

"The first time they phoned at night, I . . . got up to have a look obviously—well, what flood warning? There was no water in the road at all, not at any point . . . and the annoying thing is that you can't put the phone down or it will keep ringing because you haven't listened to the whole of the message. It rabbits on and on…and it isn't appropriate to what is happening out here. . . ." —Owner occupier, Thames Region

In more recent years, the approach has been modified somewhat, in that community flood warden schemes have again been established in many parts of England and Wales. The primary role of these volunteers is not to replace, but rather to enhance the automated messaging system by improving communication and community resilience (EA, 2009). These people are encouraged and supported in their roles by local authorities, as well as by the Environment Agency, in order to reduce the impacts of flooding; for example, identifying vulnerable people from within the community who may need extra help, and reporting any blocked drains and ditches to the appropriate authority (Leicester Leicestershire & Rutland Local Resilience Forum, n.d.; London Borough of Redbridge, n.d.; Cotswold District Council, n.d.). Other initiatives to promote community engagement are also being actively promoted: in Wales; for example, Flood Awareness Officers now work with residents on a one-to-one basis, advising on flood risks and assisting in the preparation of personal flood plans (EA, n.d.).

These changes have been prompted because education of the general public, by means of awareness campaigns alone, has been found to be insufficient to address issues such as flood risk. This matter is discussed in more depth in the next section.

22.4   AWARENESS-RAISING CAMPAIGNS

Across the world, many governments have invested significant resources in trying to raise awareness not only of flooding, but also of other natural hazards such as hurricanes, volcanic eruptions, and earthquakes. These campaigns typically hinge on the provision of information and advice designed to equip the at-risk population to take appropriate action. These initiatives are rarely as effective as might have been hoped; despite a long-running flood-awareness campaign within England and Wales, as detailed by Bonner (2006), a survey in 2007 found that only 60% of at-risk residents claimed to be aware that they lived in a flood-risk area, with a still smaller number (16%) having taken preventative action (Harries, 2007).

In this context, the public has often been seen as "irrational," with policy failures being attributed to public ignorance (Brown and Damery, 2002) or apathy (ABI/ NFF, 2004). This perception is, however, predicated on an intellectual worldview that people make decisions in accord with rational (i.e., logically explicable) processes; this is called the *Rational Actor Paradigm* (Jaeger et al., 2001). The consistency of the finding that awareness alone does *not* engender action, across hazards and across cultures, suggests that powerful additional factors are at work here (Paton,

Smith, and Johnston, 2000; McGee and Russell, 2003; McClure, 2006; Knocke and Kolivras, 2007).

One of these background influences is emotion; this may, not surprisingly, influence decision making with respect to flood risk, but it cannot be assumed that even a strong emotion such as fear will elicit a predictable response. Following the floods of summer 2007, it was found that 43% of those in the areas directly affected reported their fear of flooding had increased; nonetheless, 95% of the at-risk householders reported they had not taken any action to protect themselves from a future event (Norwich Union, 2008).

Difficulties in influencing the public do not affect the hazard warning community alone; an interesting commentary on the process of developing effective risk communication strategies is offered by Fischhoff (1995):

> Every year (or, perhaps, every day), some new industry or institution discovers that it, too, has a risk problem. It can, if it wishes, repeat the learning process that its predecessors have undergone. Or, it can attempt to short-circuit that process, and start with its product, namely the best available approaches to risk communication.

The sequence of strategy options identified by Fischhoff (1995) is shown in Table 22.1. A very similar process has indeed taken place in the provision of natural hazard preparedness advice; the final two stages of making the public "partners" in the process and adopting a portfolio of approaches are now underway in the United Kingdom (H M Government Cabinet Office, n.d.). Borrows (2007) recommended a partnership approach between flood risk professionals and those at risk, for example, by developing and using intermediaries from among the at-risk communities; this is now being addressed by the creation of community flood warden schemes. This type of approach can also help dispel the belief that flood protection is, or should be, the responsibility of the government, or local authorities, rather than householders (Brilly and Polic, 2005). The Environment Agency has acknowledged the importance of the social sciences in the context of its own work on flood risk science; it has recently described concepts such as "engaging with communities" and "building trust" as vital

---

TABLE 22.1
Developmental Stages in Risk Management

All we have to do is get the numbers right.

All we have to do is tell them the numbers.

All we have to do is explain what we mean by the numbers.

All we have to do is show them that they've accepted similar risks in the past.

All we have to do is show them that it's a good deal for them.

All we have to do is treat them nice.

All we have to do is make them partners.

All of the above.

Source: Adapted from Fischhoff, B. (1995). Risk perception and communication unplugged: Twenty years of process. *Risk Analysis,* 15(2), 137–145.

elements in its expanding research program (Henton, 2008). Similarly, in an Australian study, promoting community participation in flood warning programs from the outset was found to lead to the population taking ownership of the problem, rather than being passive recipients of instructions from the authorities (Dufty, 2008).

As detailed elsewhere in this book, accepting water ingress and enhancing the recovery process might be the optimum method for long-term flood resilience in the UK housing stock, but both aesthetic and emotional factors are found to impinge upon the decision-making process. Following the UK summer 2007 floods, for example, it was found that 46% of those affected "wanted their home put back exactly as it was before" and therefore chose not to make any changes to their property in the repair phase (Norwich Union, 2008). This may demonstrate a form of anxiety reduction, in that fitting any form of visible flood protection measures to the home would constitute an acknowledgment that normality may again be disrupted in the future. Proverbs and Lamond (2008) summarize this as follows. Individuals can achieve peace of mind by one of two methods: ignoring the risk of flooding or taking action against flood risk. How people make choices in risk contexts is discussed in the next section.

22.5   SOCIAL FACTORS AND BELIEF SYSTEMS IN DECISION MAKING

Because humans are social animals, decision-making processes are subject to the influence of society as a whole. There is the option to act as an individual or to conform to a group norm, and different individuals will be motivated to follow different choices. Homeowners may, for example, choose to buy and install flood-resilient products in an effort to protect their own property, but in so doing may attract opprobrium from neighbors, as such actions can be perceived as advertising that there is a flooding problem in that locality. This is believed to affect the salability of other homes in the area and thus creates an interesting dilemma, both for those wishing to conceal the truth as well as those wishing to protect their own homes and possessions (Garland, 2008).

Some individuals or groups within society may, however, be unable to take advantage of the resilience messages advocated in awareness-raising campaigns; additional measures may be required to reduce their vulnerability to impacts (Adeola, 2003). For example, those who have limited social networks may also find themselves disadvantaged in terms of emotional, as well as practical, support in the aftermath of a flood event (Green and Penning-Rowsell, 2004), and this can adversely affect their long-term recovery.

Factors such as sociocultural values, beliefs, or superstitions may also exert effects on different sectors of the at-risk population (Smith, 1996). A cultural misapprehension was noted in relation to the Easter 1998 flood event in Banbury, Oxfordshire; some recent immigrants to the United Kingdom expressed surprise, as they had not expected to be flooded in a developed country (Tapsell and Tunstall, 2008). People without previous flood experience have been found to envisage the consequences differently from people who had actually experienced severe flood losses: They underestimated the negative emotional consequences associated with a flood (Siegrist and

Gutscher, 2008). Such findings have implications for warning campaign development, in that these aspects should be addressed along with the more obvious tangible loss issues.

A key characteristic of climate change adaptation is the inherent uncertainty around the causes and, indeed, for some individuals and groups, doubts as to the existence of the problem itself. The roles of strength of belief in climate change and strength of belief in personal adaptive capacity have been examined in the context of adaptations to forestry management in Sweden (Blennow and Persson, 2009). It was found that the prerequisites for positive adaptation were strong beliefs in the hazard itself and the belief that, as an individual, a person has the power to do something about the hazard.

Another contributory factor in the decision-making process concerns personality types and traits—for example, the "locus of control" construct (Rotter, 1982). Where experience leads an individual to believe that he (or she) is responsible for the outcomes of his (or her) actions, that person is said to develop an "internal" locus of control. If forces external to the individual (such as "fate" or a religious entity) are perceived to be responsible for outcomes, however, the learning process is likely to result in the development of an "external" locus of control. For example, it has been identified that many health education programs seek to increase internality by encouraging patient responsibility for their own health care, as internals are more likely to engage in positive health behaviors (Wallston and Wallston, 1978).

In the field of hazard preparedness, a positive correlation between the locus of control score obtained and the purchase, or non-purchase, of flood insurance has been noted (Baumann and Sims, 1978). Similarly, McClure, Walkey, and Allen (1999) found that "internals" displayed preparedness for earthquakes, while "externals" tended toward passivity in the face of the same hazard. In two New Zealand studies, correlations were found between internality and the tendency to take mitigating action (Spittal et al., 2008), while externals were most likely to blame the government, or chance, for disasters (McClure, 2006). Adoption of best practices from other sectors could, therefore, be of benefit in developing resilience initiatives optimized for different sections of society, rather than a "one-size-fits-all" approach.

The complex interrelationships operating in this context have been examined in an effort to formulate a model of disaster preparedness and response. For example, Lindell and Perry (2000) provide such a model, based on findings from seismic hazard adjustments in the United States (Figure 22.1).

The authors acknowledge that past hazard experience is not directly represented within the figure; this is an important factor, however, in that information on vulnerability and resource efficacy is typically derived from this source. Further investigation and expansion of such models has also taken place; for example, Lindell and Hwang (2008) found factors affecting the basic causal chain in a multihazard environment included gender, age, income, and ethnicity.

Efforts to enhance resilience therefore need to encompass not only practical but also sociological and psychological adjustments if optimal improvement is to be achieved.

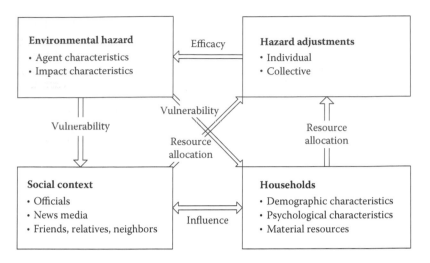

FIGURE 22.1   Interrelationships between environmental hazards, households, social context, and hazard adjustment. (Adapted from Lindell, M. K., and Perry, R. W. (2000). Household adjustment to earthquake hazard: A review of research. *Environment and Behavior*, 32, 461–501.)

22.6   CONCLUSION, SUMMARY, AND LOOKING FORWARD

Complex issues can be seen to arise in relation to flood preparedness: understanding and adapting to relevant sociological and psychological theory is required at the policy-maker level if resilience in communities is to be an attainable goal. Education of the at-risk population is only one element in the "toolbox" of techniques required; empowering and encouraging the public is also a valuable, indeed essential, requirement for success.

In the wider context of the challenges posed by climate change, it is hoped that these lessons will not need to be learned afresh in relation to each and every type of hazard; the human needs and responses that apply to flood risk may help to inform thinking on currently developing programs on heat wave, drought, and windstorm resilience.

REFERENCES

ABI and NFF (2004). Flood Protection: How Can We Provide Advice to Householders? Workshop Notes.

Adeola, F. O. (2003). Flood Hazard Vulnerability: A Study of Tropical Storm Allison (Tsa) Flood Impacts and Adaptation Modes in Louisiana. Boulder, CO, USA: Natural Hazards Research Center, University of Colorado.

Baumann, D. D., and Sims, J. H. (1978). Flood insurance: Some determinants of adoption. *Economic Geography*, 54(3), 189–196.

Blennow, K., and Persson, J. (2009). Climate change: Motivation for taking measures to adapt. *Global Environmental Change*, 19, 100–104.

Bonner, D. (2006). Environment Agency Flood Awareness Campaigns. [October 18–19, 2006] Available at <http://www.egmasa.es/europa/florispre/documentos%5cponenciasevent oespa%C3%B1a%5cencuentrodeespecialistas-18-19octubre2006%5c19-10-2006%5 cdavidbonner-floodawarenesscampaignstheexperiencesoftheenvironmentagency.pdf>.

Borrows, P. (2007). Should we rely on defences? In *5th Flood Management Conference*, 9–12 October 2007, Warrnambool, pp. 1–9.

Brilly, M., and Polic, M. (2005). Public perception of flood risks, flood forecasting and mitigation. *Natural Hazards and Earth System Sciences*, 5, 345–355.

Brown, J. D., and Damery, S. L. (2002). Managing flood risk in the UK: Towards an integration of social and technical perspectives. *Transactions of the Institute Of British Geographers*, 27(4), 412–426.

Cotswold District Council (no date). The Role of Community Flood Wardens. Available At <http://www.cotswold.gov.uk/nqcontent.cfm?a_id=9740>.

Defra (2004). Making Space for Water. London: Department For Environment, Food and Rural Affairs.

Defra (2005). Making Space for Water: Taking Forward a New Government Strategy for Flood and Coastal Erosion Risk Management in England. First Government Response to the Autumn 2004 Making Space for Water Consultation Exercise. London, UK: Department for Environment, Food and Rural Affairs.

Dufty, N. (2008). A new approach to community flood education. *The Australian Journal of Emergency Management*, 32(May 2), 4–8.

EA (Environment Agency) (2000). Flooding In Wales — October/November 2000. Available At <http://www.Environment-Agency.Gov.Uk/Static/Documents/Research/110701.Pdf>.

EA (Environment Agency) (2008a). Floodnews—United Action Is Our Best Flood Defence [October 31, 2008]. Available at <http://www.northlincs.gov.uk/nr/rdonlyres/5663e954-A272-4e20-94d4-B8047759388d/29682/floodnews_summer08.pdf>. London: Environment Agency.

EA (Environment Agency) (2008b). The Thames Barrier: A Description of Flooding Risks in London. Available at <http://www.environment-agency.gov.uk/homeandleisure/floods/105277.aspx>. London: Environment Agency.

EA (Environment Agency) (2009). Environment Agency Talks to Morpeth about Flood Options [Online]. Available at <http://www.environment-agency.gov.uk/news/110681.aspx?month=8&year=2009>. London: Environment Agency.

EA (Environment Agency) (no date). *Flood Awareness Wales*. London: Environment Agency. Available at <http://www.environment-agency.gov.uk/113810.aspx>.

EU Commission. Brussels: EU.

Fielding, J., Burningham, K., Thrush, D., and Catt, R. (2007). Effectiveness of Flood Warnings: Public Response to Flood Warning (R&D Technical Report SC020116). Bristol: Environment Agency.

Fischhoff, B. (1995). Risk Perception and Communication Unplugged: Twenty Years of Process. *Risk Analysis*, 15(2), 137–145.

Garland, P. (2008). Flooding — A Personal Perspective. In *Flood Repair Network Workshop* No. 4 08/05/08. Birmingham, Uk: Embassy House.

Green, C., and Penning-Rowsell, E. C. (2004). Flood insurance and government: 'Parasitic' and 'symbiotic' relations. *Geneva Papers on Risk and Insurance: Issues and Practice*, 29(3), 518–539.

H M Government Cabinet Office (n.d.). UK Resilience Home Page. Available at <http://www.cabinetoffice.gov.uk/ukresilience.aspx>.

Harding, D. M., and Parker, D. J. (1974). Flood hazard at Shrewsbury, United Kingdom. In White, G. F. (Ed.), *Natural Hazards*. London: Oxford University Press.

Harries, T. (2007). Householder Responses to Flood Risk; The Consequences of the Search for Ontological Security. Ph.D. thesis, Middlesex University.

Henton, T. (2008). The Value of Science to the Environment Agency: Flood Risk Science—Open Board Paper Ref Ea(08)45 Item 08. Available at <http://www.environment-agency. gov.uk/static/documents/utility/8science_2076173.pdf>.

Jaeger, C. C., Renn, O., Rosa, E., and Webler, T. (2001). *Risk, Uncertainty and Rational Action.* London: Earthscan Publications.

Knocke, E. T., and Kolivras, K. N. (2007). Flash flood awareness in southwest Virginia. *Risk Analysis*, 27(1), 155–169.

Leicester Leicestershire & Rutland Local Resilience Forum (n.d.). Community Flood Wardens. Available at <http://www.localresilienceforum.org.uk/help_your_community/ before_an_emergency/community_warden/>.

Lindell, M. K., and Hwang, S. N. (2008). Households' perceived personal risk and responses in a multihazard environment. *Risk Analysis*, 28(2), 539–536.

Lindell, M. K., and Perry, R. W. (2000). Household adjustment to earthquake hazard: A review of research. *Environment and Behavior*, 32, 461–501.

London Borough of Redbridge (n.d.). Flood Wardens. Available at: <http://www.redbridge. gov.uk/cms/parking_rubbish_and_streets/general_street_information/flooding/flood_ wardens.aspx>.

McClure, P. J. (2006). Guidelines for Encouraging Householders' Preparation for Earthquakes in New Zealand—Report for Building Research. Wellington, New Zealand: Victoria University of Wellington.

McClure, P. J., Walkey, F. H., and Allen, M. (1999). When earthquake damage is seen as preventable: attributions, locus of control and attitudes to risk. *International Journal of Applied Psychology: An International Review*, 48(2), 239–256.

McGee, T. K., and Russell, S. (2003). "Its just a natural way of life . . ." An investigation of wildfire preparedness in rural Australia. *Environmental Hazards*, 2003, 1–12.

Norwich Union (2008). Homeowners Fear Future Flooding but Fail to Take Measures to Protect Their Properties. Available at <http://www.norwichunion.com/press/stories/3981-homeowners-fear-future-flooding-but-fail-to-take-measures-to-protect-their-properties. htm> April 9, 2008.

Paton, D. (2003). Disaster preparedness: A socio-cognitive perspective. *Disaster Prevention and Management*, 12(3), 210–216.

Paton, D., Smith, L. M., and Johnston, D. (2000). Volcanic hazards: Risk perception and preparedness. *New Zealand Journal of Psychology*, 29(2), 86–91.

Pitt, M. (2008). The Pitt Review—Learning Lessons from the 2007 Floods. London: H M Government Cabinet Office.

Proverbs, D. G., and Lamond, J. E. (2008). The barriers to resilient reinstatement of flood damaged homes. In *4th International I-Rec Conference on Building Resilience: Achieving Effective Post-Disaster Reconstruction,* April 30 to May 1, 2008. Christchurch, New Zealand, pp. 1–13.

Rotter, J. (1982). *Generalised expectancies for internal versus external control of reinforcement. The development and application of social learning theory.* New York: Praeger Publishers.

Siegrist, M., and Gutscher, H. (2008). Natural hazards and motivation for mitigation behaviour: People cannot predict the affect evoked by a severe flood. *Risk Analysis*, 28(3), 771–778.

Slovic, P. (2007). *The Perception of Risk.* London: Earthscan.

Smith, B. (1996). Coping as a predictor of outcomes following the 1993 midwest flood. *Journal of Social Behavior and Personality*, 11(2), 225–239.

Spittal, M. J., Mcclure, P. J., Siegert, R. J., and Walkey, F. H. (2008). Predictors of two types of earthquake preparation—Survival activities and mitigation activities. *Environment and Behaviour*, 40, 798–817

Tapsell, S. M., and Tunstall, S. M. (2008). I wish I'd never heard of Banbury: The relationship between "place" and the health impacts from flooding. *Health & Place*, 14(2), 133–154.

Tykocinski, O., and Pittman, T. (1998). The consequences of doing nothing: Inaction inertia as avoidance of anticipated counterfactual regret. *Journal of Personality and Social Psychology*, 75(3), 607–616.

Wallston, B. S., and Wallston, K. A. (1978). Locus of control and health: A review of the literature. *Health Education Monographs*, 6, 107–117.

23 Financial Implications of Flooding and the Risk of Flooding on Households

Jessica E. Lamond

CONTENTS

23.1 INTRODUCTION

The cost of flood damage is considerable and in the short term can be devastating to homes and businesses without the benefit of insurance. The costs can be divided into long- and short-term impacts, tangible and intangible, and also into indirect and direct costs (2005). Examples of short-term direct costs include the expenditure necessary to cope with the immediate impact of water and bringing life back to normality, including physical repair costs, replacement of damaged or lost possessions, and alternative accommodations. Those fortunate to be insured for flood damage can be reassured that the majority of these costs will be paid by their insurer and they will not suffer financial hardship. However, they may be subject to longer-term costs or the risk that future floods will result in significant financial losses.

These longer-term costs are related to the risk that insurance coverage will not be renewed, will be subject to high excess payments, or will become unaffordable. This leads to the further concern that property without insurance for flooding will prove difficult or impossible to sell and there will be a reduction in the value of property at risk of flooding or recently flooded. From the perspective of the property owner, other long-term implications include the costs not covered by insurance; these might involve expenses due to increased travel costs, higher food costs in the absence of cooking facilities, increased cost of insurance, and the expenses of providing flood protection measures.

The anticipated loss of property value is backed up by economic theory, which predicts that property value will be impacted by known flood risk (Skantz and Strickland, 1987; Shilling, Sirmans, and Benjamin, 1989). This assumption is built into the analysis of the benefits of flood protection, often without any empirical evidence that the value loss is a real phenomenon (Chao, Floyd, and Holliday, 1998). In fact, empirical studies show a wide variation in observed impacts—from a positive uplift in recently flooded property to a loss of 30%. The differences can be explained by a variety of factors, including the insurance and risk disclosure regimes (Lamond, Proverbs, and Antwi, 2005).

This chapter discusses the theoretical responses of property markets to flooding and flood risk and then, using the UK example, shows how the actual response can manifest itself under the prevailing regime. Other long-term financial implications are discussed, and the chapter also considers how future changes in flood risk and risk regimes could have an impact on households.

23.2 THE THEORY OF PROPERTY VALUE LOSS DUE TO FLOOD RISK

In the aftermath of flooding, the shock felt by property owners naturally extends to the belief that a property at risk of such an event will not be sellable or will only sell at a very low price. Surveys of previously flooded households in the United Kingdom frequently cite the loss of property value as a real concern (EA/Defra, 2005; Harries, 2007; Thurston et al., 2008). This fear has also been stoked by negative media coverage (Cooper, 2004; Dolan, 2004; Jackson, 2005; Whittle, 2005).

Economic theory would predict that a possible discount of the price of residential property due to floodplain location might accrue from multiple sources stemming from the multiple impacts of flooding on property. Primary flood damage will reduce the habitable living space, making the property less desirable to live in. The discount a rational consumer will apply for this would be the expected cost to them of restoring the property to pre-flood condition over the lifetime of their ownership.

Indirect costs of flooding include evacuation costs in the expectation that residents may have to relocate while repairs are made. A rational discount will include the tangible expenses of relocation and extra travel expenses. It could also include the less tangible elements of disruption of the house as a primary social space giving access to convenient schooling and leisure activities.

Emotional well-being can suffer in the event of a flood. Health risks, stress impacts, loss of irreplaceable items due to flooding, and the fear of flooding constitute further disutilities of floodplain living. The value that homeowners place on these quality-of-life attributes could also lead to discount in floodplain property.

Long-term investment and financial security is a very serious issue in the purchase of homes. For many UK residents, the home is the most significant financial investment they will ever make. Many homeowners anticipate that their property will appreciate in value, allowing them to either move up the property ladder or realize capital for other purposes (e.g., retirement income). The expected long-term return on investment and short-term borrowing power may be seen to be reduced by the impact of climate change. Buyers may place a discount on the value of floodplain homes due to this perceived investment weakness.

To a rational buyer or seller, the value of a property at risk should be reduced by an amount equal to the expected lifetime losses due to flood risk. By this argument, if a property floods frequently and on each occasion has substantial repair costs, then the cost of continually restoring the property will quickly exceed the cost of buying an equivalent property outside the flood risk area, leaving aside any other flood losses. A rational individual will choose to live elsewhere, and the value of that property should reduce to nil or to the value of using that land for some other purpose not affected by the frequent flooding.

This theory that the loss of property value will be equivalent to expected losses due to flooding leads to property value loss being excluded from cost–benefit analyses of flood prevention schemes. Put simply, an individual can either choose to accept flood loss as damages while occupying the property or as loss of sale value if they sell up and move out. In theory, they will not be obliged to suffer both; they are substitutable losses. It can also be used to argue for the equity of charging floodplain residents for the protection offered by a new scheme as they will be seen to benefit by suffering less flood damage or by increased property value.

A similar argument can be applied to the cost of insurance. The purchase of flood insurance can be seen as a third choice in the avoidance of flood losses. The cost of insurance against flood damage, if based on the real risk of flooding, will represent a loss for the floodplain resident. The flood premium should be roughly equivalent to the expected damages plus some profit margin and management charges for the insurance company. Therefore, increased insurance costs are also substitutable for flood damage losses or property value loss and are, therefore, not counted in cost-benefit models for new flood protection schemes.

In a world of perfect market efficiency and fully symmetrical information between all parties—the homeowner, insurer, and potential floodplain resident—the financial consequences of flood risk can be easily understood and predicted. An efficient market in equilibrium will have already capitalized the expected flood loss into property value; a new flood will only change the value if it changes the future expected flood losses.

23.3   RESULTS FROM EMPIRICAL STUDIES

The theory described above is the economic view of the discount of property value due to flood risk. In reality, the experience of floodplain residents will often be very different from the theoretical model. One key theoretical assumption is the presence of information about flood risk that is equally available to all parties. This is not the case, and recent research reveals that market information issues ensure that the anticipated market loss often fails to materialize. Many buyers and even sellers are unaware of the flood risk to their property if there has been no recent flood event to call flood risk to mind (Lamond, Proverbs, and Hammond, 2009b). As flood risk is not a standard search within property transactions in the United Kingdom, the discovery is left to chance or to the background knowledge of the seller or his agents. Information, if sought from the best available sources, has improved over the past decade but is still not perfect and may be inaccurate. Under these conditions, it may be natural that there is scant empirical evidence of an existing discount for flood-prone property in the United Kingdom (Lamond and Proverbs, 2006; Lamond, 2008). This picture is consistent with other markets where floodplain regulation is light, such as in Canada (Schaeffer, 1990), Australia (Eves, 2002), and New Zealand (Montz, 1993). In these regimes, where risk is not highlighted during standard transactions, events that change market perception of risk such as a flood event may have an impact on property price toward or even exceeding the theoretically rational flood damage discount.

As an illustration, consider property in Bewdley, a frequently flooded area of the United Kingdom. Property in the floodplain showed a dip in realized price after the 2000 flood and again after a 2001 flood event but the price recovered (Lamond, Proverbs, and Hammond, 2009a). A similar pattern was seen in Australia by Lambley and Cordery (1997) who observed that some entrepreneurs may have made money as a result of anticipating the recovery.

Risk can also be communicated by a change in regulation or insurance regime, as illustrated by Skantz and Strickland (1987). Here the imposition of a different insurance regime has also been demonstrated to have an impact on the price of property in the floodplain. The risk has not changed but the costs or perceptions of risk have. Therefore, changes in insurance regimes can have direct and indirect effects on the operation of the market and in moving property prices.

These results demonstrate some differences between the economic model of property valuation in the presence of risk and the financial reality faced by the property owners. In a market such as the United Kingdom where flood risk is not capitalized into property value, a flood event may trigger financial hardship. In the short to medium term, the owner sees a devaluation of his property, which may mean that he lacks the necessary equity to move or to install protection measures. In the immediate aftermath of a flood, this discount may be much higher than the theoretical discount due to expected flood damage. The property devaluation may well be coupled with a much more expensive insurance premium, higher excesses, or withdrawal of insurance coverage. This represents either a real ongoing cost (of premium) or a potential large loss in the event of a future flood and may make it impossible to sell the property or to contemplate the prospect of having to fund the damages caused by any future flood.

23.4   INSURANCE COST AND AVAILABILITY

Where flood insurance is available, the cost of flood damage can be offset by purchasing insurance for buildings, contents, or both. This can enable occupation of property that might otherwise be seen as too risky to buy. In common with all insurance, the pooling of risk supports risk tolerance, and for a low-probability/high-cost event such as flooding, the role of insurance in managing risk can be critical to maintaining communities (ABI, 2005). Insurance regimes vary internationally, as detailed in Gaschen et al. (1998), such that insurance may not be available for all properties and affordability may also vary. To a private insurance market, not all risks are insurable at a rate that owners can afford and therefore those most at risk from flooding may find it impossible to obtain.

In the United States, flood insurance is managed under the National Flood Insurance Program (NFIP). This scheme is mandatory for flood victims who have previously claimed disaster relief and also for those requiring government-backed financing. Despite this, coverage is not high (Burby, 2001). Low perception of risk was responsible for the lack of insurance among the residents of North Dakota. According to a survey, half of the respondents cited conservative flood estimates, a belief in dikes and flood control devices, and a belief that the flood would not damage the home as very important. Shaw (2004) found that cost is a major factor for the 41% who make the decision not to fulfill the obligation to purchase flood coverage post flood.

In Germany, flood risk coverage is available as an extension to domestic insurance policies. Schwarze and Wagner (2004) found that it was included in only 3.5% of domestic insurance policies. The penetration in East Germany is higher due to the historic bundling into normal policy for East Germany. Flood experience is seen to boost demand for insurance but the availability of emergency relief and private donations are seen to weaken the incentive to insure and to implement preventative measures (Kreibich et al., 2005).

The UK system is different again as a result of an agreement between government and insurers entered into in the 1960s (Huber, 2004). Flood insurance in the United Kingdom is part of the standard household policy and in the past has been priced regardless of risk, effectively cross-subsidizing those at risk by the wider population. However, since 1998, the provision of universal subsidized flood insurance in the United Kingdom has been under threat and is currently only guaranteed until 2013 (ABI, 2008).

A move toward risk-based pricing for UK floodplain residents could mean that some high-risk residents could pay a lot more for their insurance, find their property uninsurable, or be effectively uninsured due to exclusion and excess conditions. Insurers could base their pricing on many factors, including scientifically estimated return probabilities, historic claims information, or a combination of the two. Risk-based pricing brings with it questions about the mortgagability and therefore salability of property that clearly goes beyond the immediate concern of paying premiums each month.

However, a recent questionnaire survey of UK residents at risk of flooding and recently flooded showed that, at present, the prevalence of risk-based pricing is not high (Lamond, Proverbs, and Hammond, 2009b). Most residents of floodplain property are finding it possible to obtain insurance at a competitive price despite some initial difficulties at renewal or in obtaining competitive quotes.

Just under half of the survey respondents described difficulties in obtaining insurance against flooding. For example, just 13% of respondents reported being refused a quote for insurance during the search for coverage. Only 3% of respondents reported being refused annual renewal of their policies due to flood risk. Having been flooded is more likely to cause difficulty in obtaining coverage than simply being at risk of flood.

One third of those previously flooded had experienced refusal to quote for insurance due to flood risk as compared to only 5% of those not flooded. Only one respondent reported being required to install resilient features as a condition of coverage, implying that insurers are not using the underwriting negotiation to encourage self-protection.

Despite the reported difficulties during the search for insurance, 93% of respondents reported having some kind of insurance. This level of coverage is typical in the United Kingdom. Gaschen et al. (1998) estimated that 95% of all UK households have buildings insurance. The reasons for not having coverage were also revealing; only one respondent (0.2% of respondents) cited flood risk as their reason for being unable to obtain buildings insurance, and only three respondents (0.7%) reported being unable to obtain contents insurance due to flood risk. However, four policyholders (1%) had to accept an excess of £2,500 or above. One respondent reported being asked to accept a £16,000 excess but had moved his policy as a result. Furthermore, 24 respondents (6%) suffered flood exclusion.

In summary, 10% of survey respondents at moderate or significant risk of flooding had no coverage for flood risk but the majority of these were covered for other risks. This is cause for concern but reassuringly far from a blanket ban on insurance in flood risk areas, as 90% of at-risk residents have achieved full insurance.

There was also evidence that some floodplain residents searched more widely for coverage than the non-floodplain population. The five largest companies used by non-floodplain residents represented 55% of policyholders. Within floodplain residents, they represented only 30%, with a wider range of smaller alternative companies supplying coverage. The survey also indicated that the cost of insurance was largely not based on flood risk. Median coverage was unrelated to risk category except for new residents but the variability of cost of coverage was higher among floodplain residents, reflecting their wider search strategies (Lamond, Proverbs, and Hammond, 2009b).

In comparison to other international insurance markets, the UK floodplain resident has been able to insure his property cheaply in the past and the evidence is that he will still be able to do so. Even the results from the Pitt survey (Pitt, 2008) showed little evidence of large-scale refusal of insurance. Fully risk-based premiums for these properties could amount to thousands of pounds annually and therefore the general insurance model is shielding property owners from the full economic implications of floodplain residency. The ABI Statement of Principles is due to expire in 2013 and it remains to be seen whether this will result in risk-based premiums for all. The danger in a competitive market where policy holders are shopping around, as the evidence shows they are currently doing, is that differential policies regarding risk-based premiums may result in adverse selection and the costs of future flooding being borne by the smaller insurers who are willing to accept high-risk policies.

23.5   THE COST OF MITIGATION MEASURES

It seems from the above analysis that the floodplain resident has neither an obligation nor a financial incentive via lower insurance premium to invest in flood mitigation. Indeed, in the short to medium term, a recently flooded resident may lack the means to borrow against his property value in order to fund protection schemes, and the insurer's policy of replacing like-with-like will also reduce the possible sources of funds. Property owners must also consider information costs (Camerer and Kunreuther, 1989; Kunreuther and Pauly, 2004); for a low-risk/high-impact event, the cost of obtaining information necessary to assess the correct avoidance response may well be too high. This applies not only to insurance purchases in markets where flood insurance is optional but also to the installation of protective measures. While some indication of risk is free of charge, the detailed information needed to properly assess flood risk and choose a sensible avoidance plan can be expensive and involve survey costs of £500 to £1,000 (Defra, 2008).

The perception of risk held by homeowners in the United Kingdom is therefore a key factor in their decision to invest—or not invest. The raising of awareness of flood risk among residents in the floodplain and their recruitment onto flood warnings schemes has been a target of Environment Agency policy (Hall et al., 2003). This is particularly important and difficult in areas that have not recently flooded because complacency may be generated by the absence of recent inundation. Conversely, in areas where flooding has frequently occurred, residents may have a very good idea about the likelihood of their own property suffering flood damage in a given event—maybe even better than the Environment Agency (Richardson, Reilly, and Jones, 2003).

Respondents to an insurance survey were asked what mitigation measures they had taken against flood damage (Lamond, Proverbs, and Hammond, 2009b). Just over half had registered for flood warnings and less than 10% had taken any of the other measures—namely, purchasing temporary barriers, installing permanent barriers, or installing resilient fixtures and fittings. This finding agrees with other research within at-risk populations (Burby, 2001; Brilly and Polic, 2005; Thieken et al., 2006) that invariably shows low levels of penetration of flood protection measures. Experiencing a flood had encouraged some residents to install resilient or permanent measures; 20% of previously flooded residents had taken other measures, whereas before the flood only 9% of these had taken other measures. The largest growth was in the installation of resilient fixtures and fittings by an additional 9%.

The average cost of installing protection for residential property has been estimated for the UK market at around £3,000 for a resistant system, but could cost more than £30,000 for resilience (ABI, 2006, 2009; Thurston et al., 2008). This could act as a deterrent to owners investigating further. Using evidence from those who have taken steps, the cost of flood protection was gathered for a study by collecting data from providers of flood protection in the United Kingdom (Lamond et al., 2009). Cost information was available for 22 case studies and was seen to be very variable, from the cost of a simple flood board to sophisticated veneer walling systems. The cost of solutions for domestic property varied from £50 for a single gate barrier to £30,000, with a median cost of £8,000 and a mean average of £10,142. Solutions that included resilient measures cost more, on average, than those without such measures. However, resilient measures can cost very little, especially those installed during restoration work. The cost of installing measures was often less than previous bills for restoration, implying that they may be beneficial for those who would be paying to restore the property. However, it has been observed that the system a property owner installs often fails, and further loss from flood damage and investment in alternative systems may be necessary.

Funding the mitigation measures can be problematic, particularly if recently flooded households have been subjected to increased expenses due to that flooding. No evidence is available to show that the installation of measures holds any benefit in terms of reduced insurance or increased property value to offset the cost. However, despite this, some residents are clearly prepared to fund such measures. Payment details were available for 26 of the protection case studies. The majority were entirely owner funded (i.e., 15 cases). Grants paid for five installations, five were paid for by a combination of owner and insurer, and one was wholly financed by the insurer. The cases where the insurer contributed were usually installed during restoration and included resistant and resilient features.

23.6 CONCLUSIONS

Long-term financial implications of flood risk and flooding are seen to be a complex phenomenon not predictable from economic theory and dependent on factors such as the insurance and regulatory regime within which the property market operates. Two major conclusions follow from these observations.

Governments and policy makers who accept the economic view that the removal of flood risk, for example, by the construction of flood defenses, can be viewed as a one-off gain for property owners for which they should be prepared to pay may cause unintended hardship for floodplain residents. The empirical evidence from real market transactions suggests that in many countries, flood risk is currently more widespread than the floodplain population and that populations inhabit the floodplain in ignorance. The move to any system in which the floodplain resident takes a greater part of the risk would need to take this into account and set in motion schemes to protect the financially vulnerable.

The expectation that floodplain residents will willingly embrace responsibility for their own protection because it is in their financial interest is ungrounded. Financially speaking, the floodplain resident is fairly well served by the status quo as long as they can obtain insurance at reasonable rates. It is for insurers, policy makers, and possibly researchers to demonstrate that floodplain residents will benefit from self-protection, either by setting in place incentives or penalties or by appealing to motives other than financial ones.

REFERENCES

ABI (Association of British Insurers) (2005). The Social Value of General Insurance. London: Association of British Insurers.

ABI (Association of British Insurers) (2006). Flood Resilient Homes—What Homeowners Can Do to Reduce Flood Damage. London: Association of British Insurers.

ABI (Association of British Insurers) (2008). Revised Statement of Principles on the Provision of Flood Insurance. London: Association of British Insurers.

ABI (Association of British Insurers) (2009). Resilient Reinstatement—The Cost of Flood Resilient Reinstatement of Domestic Properties. Research paper. London: Association of British Insurers.

Brilly, M., and Polic, M. (2005). Public perception of flood risks, flood forecasting and mitigation. *Natural Hazards and Earth System Sciences*, 5, 345–355.

Burby, R. J. (2001). Flood insurance and floodplain management: The US experience. *Global Environmental Change Part B: Environmental Hazards*, 3, 111–122.

Camerer, C. F., and Kunreuther, H. (1989). Decision processes for low probability events: Policy implications. *Journal of Policy Analysis and Management,* 8, 565–591.

Chao, P. T., Floyd, J. L., and Holliday, W. (1998). Empirical Studies of the Effect of Flood Risk on Housing Prices. IWR report. United States Army Corps of Engineers.

Cooper, K. (2004). House prices in flood areas could fall 80%. *The Sunday Times.* October 3, 2004. London.

Defra (Department for Environment, Food and Rural Affairs) (2008). Resilience Grants Pilot Projects. London, UK: Department for Environment, Food and Rural Affairs

Dolan, L. (2004). House prices set to plummet in flood-prone areas. *The Sunday Telegraph.* September 26, 2004. London.

EA/Defra (Environment Agency/Department of the Environment Food and Rural Affairs) (2005). The Appraisal of Human Related Intangible Impacts of Flooding. Technical report. Environment Agency/Department of the Environment Food and Rural Affairs.

Eves, C. (2002). The long-term impact of flooding on residential property values. *Property Management*, 20, 214–227.

Gaschen, S., Hausmann, P., Menzinger, I., and Schaad, W. (1998). Floods—An insurable risk? A market survey. Zurich: Swiss Re.

Hall, J. W., Meadowcroft, I. C., Sayers, P. B., and Bramley, M. E. (2003). Integrated flood risk management in England and Wales. *Natural Hazards Review*, 4, 126–135.

Harries, T. (2007). Householder Responses to Flood Risk, The Consequences of the Search for Ontological Security. Middlesex, UK: Middlesex University.

Huber, M. (2004). Insurability and regulatory reform: Is the English flood insurance regime able to adapt to climate change? *The Geneva Papers on Risk and Insurance*, 29, 169–182.

Jackson, P. (2005). Vox prop: Would you buy a house in a UK flood-risk area? *The Independent.* January 19, 2005. London.

Kreibich, H., Thieken, A., Petrow, T., Muller, M., and Merz, B. (2005). Flood loss reduction of private households due to building precautionary measures—Lessons learned from the Elbe flood in August 2002. *Natural Hazards and Earth System Sciences*, 5, 117–126.

Kunreuther, H., and Pauly, M. (2004). Neglecting disaster: Why don't people insure against large losses? *Journal of Risk and Uncertainty*, 28, 5–22.

Lambley, D., and Cordery, I. (1997). The effects of catastrophic flooding at Nyngan and some implications for emergency management. *Australian Journal of Emergency Management*, 12, 5–9.

Lamond, J. (2008). *The impact of flooding on the value of residential property in the UK.* School of Engineering and the Built Environment. Wolverhampton: University of Wolverhampton.

Lamond, J., Dhonau, M., Rose, C., and Proverbs, D. (2009a). Overcoming the barriers to installing property level flood protection—An overview of successful case studies. Road map towards a flood resilient urban environment. Paris: Urban Flood Management Cost Action Network C22.

Lamond, J., and Proverbs, D. (2006). Does the price impact of flooding fade away? *Structural Survey*, 24, 363–377.

Lamond, J., Proverbs, D., and Antwi, A. (2005). The effect of floods and floodplain designation on the value of property: An analysis of past studies. *2nd Probe Conference.* Glasgow, Scotland.

Lamond, J., Proverbs, D., and Hammond, F. (2009b). Accessibility of flood risk insurance in the UK—Confusion, competition and complacency. *Journal of Risk Research,* 12, 825–840.

Lamond, J., Proverbs, D., and Hammond, F. (2009). Flooding and property values. In Brown, S. (Ed.) *Findings in Built and Rural Environments (FiBRE)*. London: Royal Institution of Chartered Surveyors.

Montz, B. E. (1993). The hazard area disclosure in NZ, the impact on residential property values in 2 communities. *Applied Geography*, 13, 225–242.

Penning-Rowsell, E. C., Chatterton, J., and Wilson, T. (2005). *The Benefits of Flood and Coastal Risk Management, A Handbook of Assessment Techniques,* London: Middlesex University Press.

Pitt, M. (2008). The Pitt Review: Learning Lessons from the 2007 Floods. London: Cabinet Office.

Richardson, J., Reilly, J., and Jones, P. J. S. (2003). Community and public participation: Risk communication and improving decision making in flood and coastal defence. *Defra Flood and Coastal Management Conference*. Newcastle-under-Lyme, England: Keele University.

Schaeffer, K. A. (1990). The effect of floodplain designation regulation on residential property values: A case study in North York, Ontario. *Canadian Water Resources Journal*, 15, 1–14.

Schwarze, R., and Wagner, G. G. (2004). In the aftermath of Dresden: New directions in German flood insurance. *The Geneva Papers on Risk and Insurance*, 29, 154–168.

Shaw, M. M. (2004). Group flood insurance program and flood insurance purchase decisions. *International Journal of Mass Emergencies and Disasters*, 22, 59–75.

Shilling, J. D., Sirmans, C. F., and Benjamin, J. D. (1989). Flood insurance, wealth redistribution and urban property values. *Journal of Urban Economics*, 26, 43–53.

Skantz, T. R., and Strickland, T. H. (1987). House prices and a flood event: An empirical investigation of market efficiency. *Journal of Real Estate Research*, 2, 75–83.

Thieken, A., H., Petrow, T., Kreibich, H., and Merz, B. (2006). Insurability and mitigation of flood losses in private households in Germany. *Risk Analysis*, 26, 1–13.

Thurston, N., Finlinson, B., Breakspear, R., Williams, N., Shaw, J., and Chatterton, J. (2008). Developing the Evidence Base for Flood Resistance and Resilience. London: Environment Agency.

Whittle, J. (2005). Floods blight house prices. *News and Star*, January 13. Carlisle, United Kingdom.

24 Why Most "At-Risk" Homeowners Do Not Protect Their Homes from Flooding

Tim Harries

CONTENTS

24.1 INTRODUCTION

Despite the widespread and well-publicized availability of measures to retrofit properties so as to reduce flood risk (EA, 2010; NFF, 2010), take-up of these measures is low. Survey evidence reported in this chapter reveals that only 33% of people who have experienced a flood take steps to protect their homes from further flooding, and less than 8% of those do who have never been flooded. This chapter addresses the question of why this is so and asks what factors seem to encourage and discourage homeowners from taking steps that, at first glance, might seem obvious.

The chapter investigates the proposition that behavior in the face of risks such as flooding is influenced not only by the scientific evidence on the risk, but also by the way people perceive that risk and, indeed, choose to perceive it. Furthermore, it proposes and tests the hypothesis that responses to long-term flood risk might be influenced as much by people's responses to the available adaptation measures as by their evaluations of the extent of the risk. Behavior change strategies sometimes focus on people's understandings of the physical workings of risk processes and adaptation measures and try to plug the gaps in these understandings by providing more and better information (e.g., Bostrom et al., 1992; Atman et al., 1994; Siudak, 2001). However, although people's technical grasp of flood risk plays a part, this understanding, rather than simply reflecting the available scientific information, is influenced by other social and individual factors (Homan, 2001; Burningham, 2008), including the way in which people perceive the whole notion of adaptation.

To this end, the argument presented here draws on the author's analysis of survey data collected for the Department for Environment, Food and Rural Affairs (Defra) by Entec, Greenstreet Berman and John Chatterton Associates (Entec and Greenstreet Berman, 2008) ($n = 555$), and data collected by the author in semistructured interviews and focus groups conducted between 2005 and 2008 ($n = 60$).

24.2 SEMISTRUCTURED INTERVIEWS AND FOCUS GROUPS

The author (Harries) collected qualitative data between 2005 and 2008 in semistructured interviews and focus groups with 60 residents of areas in southeast England, Nottinghamshire, and Yorkshire. All were high-risk areas but some had been flooded recently while others had not. Furthermore, two of the recently flooded areas had benefited from publicly funded schemes that provided homeowners with free advice and free or subsidized flood protection. Participants were selected from across the range of types of household tenure, social grade, and household composition.

Interviews followed a loose structure that covered people's experiences of flooding and their experiences of, and attitudes toward, flood protection and resilience. These were recorded, fully transcribed, and then analyzed using thematic and discourse analysis (Fairclough, 2003).

The analysis of the qualitative data aimed to reveal the justifications and explanations that homeowners gave for their behavior (the thematic analysis) and also to examine the underlying representational structures and discourses that underpinned these rhetorical strategies (the discourse analysis). The next step was to use quantitative analysis of survey data to discern which, if any, of the incentives and objections identified in the qualitative analysis had a statistically significant impact on behavior.

24.3 THE SURVEY

The survey used for this purpose was collected for Defra in 2007 by the risk consultancy of Greenstreet Berman, using a questionnaire whose design was informed by the findings of the early elements of the above qualitative analysis. The full questionnaire is available in Entec and Greenstreet Berman (2008).

The sample population for the survey consisted of telephone numbers in postcode areas with a greater than 80% concentration of properties with a return period of 1:75 or higher according to the Environment Agency's National Flood Risk Assessment (NaFRA) 2006 Postcode Flood Likelihood Category Database. Individuals received telephone calls on weekdays between 9:00 am and 7:00 pm. Where calls were answered and people admitted to being previously aware of the flood risk, they were invited to take part in the research. Of the 6,000 telephone numbers called, half responded, a third of these claimed not to be aware of living in a flood risk area, and one in four of the remainder agreed to take part in the survey.

Of the 555 respondents, about half had no personal experience of flooding, a quarter had been flooded in outbuildings or gardens, and a quarter had been flooded in their homes. Less than 10% admitted to having taken any kind of property-level measure to reduce their exposure to flood risk.

A list of flood protection measures was read to respondents and they were asked, each in turn, to say whether they had ever used that measure. They were also asked to say whether they agreed or disagreed with a number of possible reasons FOR NOT putting in place protection measures and FOR putting them in place: for example, "My home is covered by insurance so I don't need to worry" and "My insurance premiums would go down or not go up so much." These statements were based on the themes that had emerged from the early qualitative interviews.

The answers to these questions were analyzed using two statistical methods designed for use with nominal data: the chi-squared test and multivariate logistic regression. The former was used to test the statistical significance of individual rhetorical positions and perceptions as predictors of behavior, while the latter was used to construct and test an integrative model of behavior (Bohrnstedt and Knoke, 1994).

24.4 PERCEPTIONS OF THE PROBABILITY OF FLOODING

The perceptual dimension with probably the greatest influence on behavior is that which concerns the probability of flooding. Although there is not always a link between probability perception and self-protection (Weinstein et al., 1991; Breakwell, 2007), the survey indicated that there is indeed such a link in the case of flood risk. Respondents who expected to be flooded within the next year were almost four times as likely as other respondents to have taken protection measures ($n = 515$, d.f. = 2, $p < 0.005$, $\chi^2 = 32.04$, OR = 3.73) and those that said they were likely to be flooded within the next 10 years were twice as likely to have done so ($n = 506$, d.f. = 2, $p < 0.05$, $\chi^2 = 8.52$, OR = 2.14).

It is worth mentioning some of the factors that influence homeowners' perceptions of the probability of flooding. In the United Kingdom, the Environment Agency provides residents with the estimated annual probability of flooding, expressed as a percentage, and also the estimated "return period." Although the provision of this type of data unquestionably has an influence on people's perceptions, numerous cognitive short-cuts, known as *heuristics*, cause their perceptions to differ from these scientifically produced figures. The most important of these are probably (1) *availability bias* (Tversky and Kahneman, 1973), the over-estimation of the probability of events that are more easily imagined; (2) *representativeness bias* (Nisbett and Ross,

1980), the assumption that recent patterns of events are representative of the future; and (3) *optimism bias* (Weinstein, 1980), people's tendency to believe that they are less vulnerable to a risk than others.

All of these heuristic effects are mediated by the experience of flood events. Experience of flooding creates images that are more vivid than those produced by abstract information about flood risk and thus increases availability bias; it also leads people to assume that flooding will happen again (representativeness bias). The influence of optimism bias is less clear. On the one hand, experience of a flood event will dent a person's belief that they are exempt from the risk. On the other hand, the vividness of the experience might encourage them to deny the risk event more fervently, as illustrated by the data from the semistructured interviews, which suggests that some of the respondents who had experienced flooding chose to represent flooding as a matter of "luck" in order to avoid the insecurity associated with a more probabilistic representation (Harries, 2008b). The survey analysis is unable to distinguish the mechanisms by which experience influences probability perception but did identify a correlation between risk perception and experience ($n = 494$, d.f. $= 4$, $p < 0.005$, $\chi^2 = 52.03$).

Despite the evident influence of the perception of flooding probability, it is striking that only 31% of those who expected to be flooded during the coming year had taken any action to protect themselves or their properties. The remainder of this chapter explores the reasons for this relatively moderate impact of risk perception by looking at the role played by a second important set of perceptions—those of the available adaptation measures. As indicated in Figure 24.1, adaptation perceptions mediate the influence of probability perception on behavior, particularly where homeowners are not very familiar with the measures available. As also shown in Figure 24.1, perceptions of adaptations can also have a feedback effect on the perception of flood probability, because optimism bias is more likely when people feel they are able to take control of a risk situation (Breakwell, 2007)—for example, when they believe that effective flood protection measures are available, or when the absence of any trustworthy measure makes denial the more attractive coping strategy.

In this chapter, discussion of the role played by perceptions of adaptation is presented in two sections, the first (see Section 24.5) of which looks at perceptions of financial and material costs and benefits, and the second of which considers perceptions of social and emotional issues (see Section 24.6).

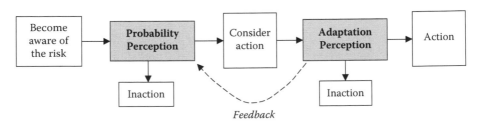

FIGURE 24.1 A linear representation of the risk response process.

24.5 PERCEPTIONS OF ADAPTATION: FINANCIAL AND MATERIAL ISSUES

Arguments about flood protection measures often focus on the material implications of taking or not taking such measures: their cost, their impact on property values, their benefits relative to household insurance, and their effectiveness in reducing costs. These are relatively comfortable arguments for people to use because they relate to one of the predominant discourses of the modern era—finance. However, the subsections that follow ask whether these arguments are genuine explanations of behavior or if, in fact, they primarily serve a rhetorical purpose.

24.5.1 THE PERCEIVED AFFORDABILITY OF ADAPTATION

Arguments about the first of these issues—cost—appear to be primarily rhetorical. Although 57% of survey respondents assumed that protection measures would be "too expensive" and cited this as a barrier to adaptation, there was no evidence of a correlation between perceptions of cost and the use of flood protection measures ($n = 519$, d.f. = 2, $p = 0.12$, $\chi^2 = 4.23$). The perception of adaptation measures as too costly was, however, correlated with the belief that adaptation is "not my responsibility" ($n = 527$, d.f. = 2, $p < 0.05$, $\chi^2 = 7.52$), and analysis of the in-depth interviews suggests that arguments about cost might be a proxy for concerns about the justice of asking individual homeowners to pay for their own protection.

Although cost will inevitably prevent some homeowners from implementing the more expensive protection measures, it is not a barrier to flood protection per se. Even the poorest of households are likely able to afford the £20 per house that it costs one respondent to make airbrick covers for himself and his neighbors, and rudimentary door boards made from a single piece of marine ply will also be relatively inexpensive. It seems probable, therefore, that arguments about cost are primarily a rhetorical strategy.

24.5.2 ANTICIPATED IMPACTS ON PROPERTY VALUES

A second perceptual issue that emerged from the semistructured interviews is the concern that protection measures would reduce the value of a property. These concerns are illustrated in the following excerpt from the semistructured interviews, in which a respondent describes the reactions he imagines prospective buyers of his own home would have if there were visible fittings for door boards:

> "Another thing that I'm concerned about, having doorboards fitted onto this house, is if we were to say put this house up for sale, you've got potential buyers coming and they're looking at these things and that's a straight put-off anyway, isn't it, you know. 'Well, why have you got flood guards up and that?'"—Julian, interviewee

The survey data indicate that these concerns might be a statistically significant deterrent to protective behavior. The 24% of owner-occupiers who said they hesitated to take adaptive measures in order to avoid revealing the flood risk to prospective buyers were less than a third as likely as others to have implemented such measures (n = 431, d.f. = 2, p < 0.05, χ^2 = 7.17).

This rationale is consistent with the behavior of the market. Negative impacts of flood events on property values in the United Kingdom are usually short-lived (Lamond et al., 2010), which suggests that, in the housing market, memories of flooding dissipate relatively quickly. As a result, homeowners have nothing to gain from the housing market and everything to lose if they use visible adaptation measures that remind people of the risk of flooding. Recent changes to the disclosure regime in the United Kingdom (DCLG, 2008) might yet impact the longer-term consequences of flooding on house prices, but in the meantime it is not unreasonable for homeowners concerned about the value of their property to decide not to use highly visible adaptation measures.

In contrast to the significance of the perception that protection might cause a fall in property values, there was no evidence that the promise of gains in property prices was a positive incentive for adaptation. Although 34% of owner-occupiers in the survey expressed the belief that taking adaptive measures would increase the value of their properties, this belief did not correlate with the implementation of protection measures (n = 423, d.f. = 2, p = 0.27, χ^2 = 2.59).

24.5.3 PERCEIVED COSTS AND BENEFITS OF INSURANCE

A third perception of the material impact of flood risk adaptations is the view that insurance represents an adequate alternative to physical protection. This is illustrated by two participants in the semistructured interviews who argue against the need for flood protection on the basis of the quality of their insurance:

> *Jean:* I'm fairly reassured because we've got an insurance and I've never sort of . . .
> *Paul:* Somehow covers almost everything. It's quite expensive but . . .
> *Jean:* I mean if I compare myself, you know, to [neighbor], he has no insurance and he would have to pay everything.
> *Paul:* And he's probably much more worried.
> *Jean:* He's probably ready to spend £700 or £800 to have a flood door because he has no insurance. I pay, erm, more than £700 a year probably on content and building I think, but I feel that I've protected myself like that.

The tendency toward less responsible behavior of those who are insulated from a financial risk by insurance is known in insurance literature as *moral hazard* (Grubel, 1971; Baker, 2002). Moral hazard also affects flood risk response. The availability of insurance encourages people to increase their exposure to flood-risk events (Cutter, 2006) and in the survey, the 27% of respondents who agreed with the statement, "My home is covered by insurance, so I don't need to worry [about adaptation measures]," were less than half as likely as other homeowners to have taken adaptive measures (n = 512, d.f. = 2, p < 0.05, χ^2 = 6.90, OR = 0.45).

On the other hand, the possibility that risk reduction measures will lead to a reduction in insurance premiums does not seem to influence behavior. Although the Association of British Insurers (ABI) claims that customers with property-level protection will normally be offered reductions in premiums (personal communication between the author and the ABI, 2008), this potential benefit was not spontaneously mentioned by any of the participants in the qualitative interviews, for whom only the threat of losing insurance coverage (or the promise of regaining it) appeared to be an incentive:

> *Respondent:* If [my insurer] were to ever say, "We're not insuring you anymore," then I'd be totally stuffed....
>
> *Interviewer:* Is that the factor in your thinking then about ... I don't know, about getting [door boards] on the doors for example? Would that make a difference?
>
> *Respondent:* It would if insurance wasn't available, yeah. Then I would have to think about doing something myself.
>
> *Interviewer:* How about if it were to bring the premiums down?
>
> *Respondent:* Yeah, that's a good point actually. Mmm.
>
> *Interviewer:* Have they said anything?
>
> *Respondent:* No, they haven't. That's a good point actually. I ought to ask them about that.

Furthermore, although 37% of survey respondents agreed, when prompted, that their insurance premiums would probably fall if they adopted flood protection measures, there was no correlation with adaptive behavior ($n = 507$, d.f. $= 2$, $p = 0.25$, $\chi^2 = 2.75$).

24.5.4   Perceived Reductions in Cost

Finally, the evidence does not suggest that financial savings provide an influential incentive for adaptive behavior. Although an annual flooding probability of 0.2% or greater is sufficient to make protection measures cost beneficial for the average home (Entec and Greenstreet Berman, 2008, p. 71), and although 57% of respondents claimed to believe that taking such measures would save them money, there was no correlation with behavior ($n = 508$, d.f. $= 2$, $p = 0.41$, $\chi^2 = 1.80$). Neither were financial benefits of protection measures mentioned spontaneously by most of the participants in the qualitative interviews and groups.

24.6   PERCEPTIONS OF ADAPTATIONS: SOCIAL AND EMOTIONAL ISSUES

Having discussed the impact of perceptions of the risk and of financial and material aspects of adaptive action, this next section of the chapter considers perceptions of the social and emotional impacts of flood protection measures. It focuses on three emotions that people sometimes anticipate will result if they implement household-level measures: anxiety, blame, and identity.

24.6.1   Anxiety

The first of these issues is anxiety. Unlike *fear*, which refers to something definite, the term *anxiety* describes the emotional response to a risk whose causes are insufficiently understood for people to know how to reestablish their sense of security

(Wilkinson, 2001). Anxiety can therefore be said to have an existential element and to challenge what the sociologist Anthony Giddens (1991) calls *ontological security*—people's faith in the continuity of their own identity and existence.

The potential for flood risk to cause anxiety and ontological insecurity is particularly acute when it affects people's homes. Home is more than just functional. It also has connotations of security and continuity and symbolizes, for many, the center of their lives and has been described as a "sacred spot" from which people can create a version of the universe that fits their desires (Cooper, 1976; Saunders, 1989; Smith, 1994; Mallett, 2004). As a result, invasion by a pollutant and forced relocation profoundly undermine people's feelings of essential security and mental health (Dupuis and Thorns, 1998; Ohl and Tapsell, 2000).

However, evidence on the relationship between anxiety and flood risk response is inconclusive (Breakwell, 2007). While in the short term anxiety can encourage protective behavior, in the longer term it is debilitating (Selye, 1956, as cited in Cox, 1978) and can lead to learned helplessness and cognitive impairment (Smith et al., 2003). Anxiety about flooding has been associated with a loss of confidence and indecision (Harries, 2008a), although it is unclear whether this is an explanatory relationship or whether, as indicated by Kallmen (2000), all three phenomena are the result of an anxious personality.

While clear evidence on the link between anxiety and self-protective behavior is lacking, the survey data suggest a fairly strong relationship between adaptive action and the desire to avoid anxiety. Those respondents who associated protection with feeling "safer" (77% of the total) were twice as likely to have implemented protective measures ($n = 508$, d.f. $= 1$, $p < 0.05$, $\chi^2 = 4.59$). On the other hand, this same desire leads people to argue against risk reduction measures if they believe that those measures will act as a reminder of the risk and therefore undermine the symbolic value of "home" as a place of safety (Harries, 2008b)—for example, highly visible measures and those that require deployment and, therefore, vigilance. Thus, one homeowner commented that having flood barriers for her doors would make her house feel less like a home and more "like a prison." Although people will become emotionally attached to almost any object in their possession (Thaler, 1980) and would probably become habituated to the presence of flood protection measures, this is not an effect that they anticipate when they consider the option of these forms of protection.

Any doubts about the reliability of protection measures exacerbate these anxieties because they add to the uncertainty. One respondent used an example from a different domain to explain this point:

Paul: I mean where I used to live, I had this veranda which had this problem of water coming in . . . And I was fixing and trying to do all these things, and then it . . . I mean, until it works, it's fine. But then all of a sudden, there is always, you know, something, that starts going wrong. That one little bit of this thing is not properly fixed or things, then you find dripping from one side. At that point, you panic more. I mean, probably you are . . . I got this impression, you feel safe but then when suddenly it's not working properly, erm, I got this impression that things then, erm, make you more anxious about and more worried as well. So unless you are absolutely sure that this [door board] here is going to work [. . .] because then you see all this water, and then you say, "Well, is this plastic here going to hold the pressure? Is it going to crack?" You know … there's always a risk—something to be worried about.

As the participant Paul points out, you will only stop feeling anxious when you are "absolutely sure" that a measure is going to protect you. This means that a protective measure that would be effective in nine out of ten floods might increase anxiety even if it reduced the mathematical probability of the ingress of water into a home:

Interviewer: You were just talking about preparing for holidays—potentially getting everything safe [from flooding] when you go away. … Is that something you'd actually want to do?

Paul: [… No] because then you will always have the day, where you say, "Well, this time I will not do anything." Then flood will come! [*Laughs*] Then, you know, at that point, there is no point, you know, because either you are 100% behind that; every time you do it in a very organized way, or you …

Interviewer: You mean every time you go away?

Paul: Yes, something like that, you know. During the winter, you know, because then there's no point if you do that nine times and then the tenth, you say, "Well, this winter I don't bother, I'm late and missing the plane." So you don't do it and then just at that time it happens; so at that point you think, "Well, I have done all that for nothing!"

Interviewer: So that would be worse than not doing it at all?

Paul: Yeah I think so.

24.6.2 BLAME AND REGRET

Paul's comments, above, illustrate a second psychosocial dimension of the perception of adaptation measures: the issue of anticipated blame and regret and the associated phenomenon of *blame-avoidance* (Jones and Berglas, 1978; Higgins, 1990). Taking adaptive measures increases the risk of attracting blame. If measures are never used or, worse, prove ineffective, those responsible for the decision to implement them expect to blame themselves and to feel themselves blamed by others. In the absence of clear norms of behavior, action attracts greater risk of blame than inaction (Fazio et al., 1982) and therefore poses a greater threat to perceived self-efficacy (Loomes and Sugden, 1982; Tykocinski and Pittman, 1998; van Dijk et al., 1999; Zeelenberg et al., 2002). As a result, acting against the norm is generally avoided. This is particularly the case in circumstances where protective behavior involves high levels of investment for uncertain gain (Baumeister, 1997), where failure of the measure might actually increase damage and thereby add an anthropogenic element to a risk that could otherwise have been represented as an "act of God" (Brown et al., 2005) and where adaptation implies an acceptance of responsibility and reduces an individual's ability to indulge in the rhetorical strategy of blaming others (Breakwell, 2007).

Blame shifting not only preserves self-image but also allows homeowners to hope for an eventual resolution of the risk situation. For example, a research participant on the verge of despair over the flooding in his area comforted himself with the argument that, "Sooner or later, one of the systems is going to realize what's going on. Some judge somewhere is going to notice this and is going to make them sort it out."

This is made possible only by the representation of the issue as being the responsibility of "the systems" rather than of the individual. Indeed, the use of the discourse of blame is evident throughout the qualitative interviews ("I blame the lock keepers"; "It's [name of water authority]; it's their drains; it's their sewer"; etc.).

However, although 19% of homeowners in the survey claimed that they were deterred from reducing the risk to their homes and families because it was "not my responsibility," this view did not correlate with adaptive behavior ($n = 524$, d.f. = 2, $p = 0.46$, $\chi^2 = 1.56$). This suggests that blame avoidance might be predominantly rhetorical and that although it influences the way in which issues and decisions are represented, it might have relatively little impact on practical decisions.

24.6.3 STIGMA AND SOCIAL IDENTITY

A third psychosocial consideration apparent from the semistructured interviews is the desire for the sense of security and belonging that results from shared social identity. The identity groups of greatest salience to the issue of flood risk usually consist of people in the same geographical vicinity who are exposed to similar risks. Indeed, as illustrated by the following quote, the reinforcement or formation of such groups is an example of what have been termed the "perceived benefits" of disaster (McMillen et al., 1997):

> [...] all the men just go out and you know they were discussing what was happening and getting the sandbags and stuff like that. [...] It's not often they all get together and are doing the same things, and helping each other so much.

According to social identity theory, groups maintain their sense of identity by emphasizing the positive characteristics of their own behaviors and the negative characteristics of the behaviors of outsiders (Abrams and Hogg, 1990). In situations where outsiders are urging cautious behavior, this can lead the *in-group* to shift toward the adoption in-group norms that involve increased risk taking (Hogg et al., 1990) and to the stigmatization of cautious behavior. To differentiate themselves from a perceived out-group that was urging precautionary behavior (i.e., the state), the participants of one focus group labeled anyone who took long-term flood protection measures as "nuts" and as having "more money than sense." In another case, a homeowner's decision to install a door board when his neighbors had not done so caused him to be accused of conjuring up an "apocalyptic image" of the future and of threatening the cohesion of the local community; hence the importance of involving entire social identity groups in resilience activities. When an entire in-group becomes engaged in adaptive behavior, it is assimilated into group identity and is no longer stigmatizing.

Implementing property-level measures in entire communities not only minimizes the problem of stigma, but also reduces the danger that self-protection will be seen as harmful to others. Individual adaptive measures are sometimes represented as divisive, with homeowners expressing concern that the use of protective measures will increase the depth of flooding for other, unprotected households, and that to pump water out of one home is to send it toward a neighbor. Introducing protection simultaneously to all local households alleviates this problem.

24.7   MODELING THE DRIVERS OF PROTECTIVE BEHAVIOR

In the above paragraphs, the use of the chi-square statistical test established the existence of significant correlations between behavior and four aspects of the adaptation perception. However, like all bivariate tests of association, the chi-square test is prone to yielding spurious associations and is unable to discriminate between direct associations and associations via intervening variables. For this reason, the final step in the analysis presented in this chapter is to apply multivariate logistic regression to the variables that have emerged as likely predictors of adaptive behavior—anticipated feelings of safety, anticipated reductions in disruption, beliefs about insurance, and concerns about selling the property. Using the *backward conditional* regression method recommended by Field (2005), nonsignificant variables were eliminated from the initial model until an optimum model was created. This final model (Table 24.1) contained only two of the initial predictor variables.

The model generated by this process indicates that the most important aspects of adaptation perception for the use of long-term protection against flooding are the belief that they will bring a feeling of safety and the perception that they offer no benefits not already provided by insurance. The nonsignificant value of the Hosmer and Lemeshow value suggests that this model provides a good fit with the survey data (see Field, 2005).

It is also worth noting that the first of these two variables, the expectation of an increase in feelings of safety, is correlated with perceived flood probability ($n = 512$, d.f. = 2, $p < 0.005$, $\chi^2 = 12.88$). This indicates that the impact of probability perception on behavior might be mediated by people's desire to feel safe in their homes—that is, that one of the reasons for the impact of probability perception on adaptation is the desire to eliminate the loss of a sense of security.

TABLE 24.1
Results of Logistic Regression onto the Dependent Variable Protection Measures

	n	B	S.E.	Wald	df	Sig.	Exp(B)	95.0% C.I. for Exp(B)	
								Lower	Upper
Protection measures would make me feel safer									
Agree	370	.742	.346	4.601	1	.032	2.100	1.066	4.137
My home is covered by insurance, so I don't need to worry									
Agree	129	–.858	.345	6.209	1	.013	.424	.216	.833
Constant		–2.082	.324	41.262	1	.000	.125		

$R^2 = 1.00$ (Hosmer and Lemeshow), 0.03 (Cox and Snell), 0.05 (Nagelkerke); $\chi^2 = 13.18$.

24.8 CONCLUSIONS

The analysis presented above identified a range of arguments and considerations that appear in homeowners' responses to the discourse of preemptive preparations against flooding. In talking about flood protection as a notion, they raise the problems of affordability and potentially negative impacts on property prices, display the tendency to shift the responsibility for flood risk onto others, and argue for the superiority of insurance to flood protection. Less obvious, yet visible on careful analysis, is evidence of a desire to minimize anxiety about flooding and to avoid being stigmatized or blamed if things go wrong.

However, bivariate analysis of the survey data suggested that only some of these factors were correlated with behavior, and the subsequent logistic regression analysis reduced the number of statistically significant predictors to just two: (1) the view that insurance is an adequate substitute for flood protection, a belief that is negatively correlated with protective behavior; and (2) the belief that flood protection will generate feelings of greater safety, a belief that is positively correlated with protective behavior. This analysis provides clear evidence for Billig's (1993) assertion that expressed attitudes and perceptions do not always have a direct relationship with behavior and that their use is sometimes purely rhetorical. It therefore reminds us of the dangers of taking at face value the reasons people give for implementing or not implementing flood risk reduction measures.

The findings of this study suggest that efforts to increase the take-up of flood protection measures among homeowners should concentrate on two aspects of people's perceptions of these measures: (1) the added advantages of having protection as well as insurance and (2) the potential of protection to provide a greater sense of safety. An emphasis on material benefits, the study suggests, is less likely to change behavior; and subsidies or price reductions are not necessarily essential components of any program to promote protection. However, flood protection products do need to be designed in a way that not only protects people's homes from damage, but that also makes them feel that they are protected. This means that homeowners need to have confidence that they will be effective—and evidence from the semistructured interviews indicates that this confidence only exists when people have either experienced the effectiveness of measures themselves, or when they have received reports of their effectiveness from someone they know and trust.

ACKNOWLEDGMENTS

The collection and analysis of qualitative data presented in this chapter was funded by an ESRC/Environment Agency CASE Studentship at the Flood Hazard Research Centre; an ESRC Placement Fellowship at the Department for Environment, Food and Rural Affairs (Defra); and an additional Defra grant. The collection of quantitative data was commissioned by Defra, who also supported its secondary analysis by the author. The writing of this chapter was made possible by an ESRC Postdoctoral Fellowship at the Geography Department of King's College, London.

REFERENCES

Abrams, D., and Hogg, M. (1990). *An Introduction to the Social Identity Approach*. Hemel Hempstead: Harvester Wheatsheaf.

Atman, C. J., Bostrom, A., Fischoff, B., and Morgan, M. G. (1994). Designing risk communications: Completing and correcting mental models of hazardous processes, Part 1. *Risk Analysis*, 14(5), 779–788.

Baker, T. (2002). Risk, insurance and the social construction of responsibility. In Baker, T., and Simon, J (Eds.), *Embracing Risk: The Changing Culture of Insurance and Responsibility*. Chicago: University of Chicago Press.

Baumeister, R. F. (1997). Esteem threat, self-regulatory breakdown, and emotional distress as factors in self-defeating behaviour. *Review of General Psychology*, 1(2), 145–174.

Billig, M. (1993). Studying the thinking society: Social representations, rhetoric and attitudes. In Breakwell, G. M., and Canter, D. (Eds.), *Empirical Approaches to Social Representations* (2nd edition). Oxford: Clarendon Press.

Bohrnstedt, G. W., and Knoke, D. (1984). *Statistics for Social Data Analysis* (3rd edition). Itasca, IL: F E Peacock Publishers.

Bostrom, A., Fischoff, B., and Morgan, G. M. (1992). Characterising mental models of hazardous processes: A method and its application to radon. *Journal of Social Issues*, 48, 4.

Breakwell, G. M. (2007). *The Psychology of Risk*. Cambridge: Cambridge University Press.

Brown, J. D., and Damery, S. L. (2002). Managing flood risk in the UK: Towards an integration of social and technical perspectives. *Transactions of the Institute of British Geographers*, 27, 412–426.

Brown, T. C., Petersen, G. L., Brodersen, M. R., Ford, V., and Bell, P. A. (2005). The judged seriousness of an environmental loss is a matter of what caused it. *Journal of Environmental Psychology*, 25, 13–21.

Burningham, K. (2008). A noisy road or noisy resident? A demonstration of the utility of social constructionism for analysing environmental problems. *The Sociological Review*, 46(3), 536–563.

Cooper, C. (1976). The house as symbol of the self. In Proshansky, H. H., Ittelson, W. H., and Rivlin, L. G. (Eds.), *Environmental Psychology: People and Their Settings* (2nd edition), pp. 435–448. New York: Holt, Rinehart and Winston.

Cox, T. (1978). *Stress*. Basingstoke: MacMillan.

Cutter, S. L. (2006). Moral hazard, social catastrophe: The changing face of vulnerability along the hurricane coasts. *The Annals of the American Academy of Political and Social Science*, 604(1), 102–112.

DCLG (Department for Communities and Local Government) (2008). Property Information Questionnaire – General Version. DCLG. Available at <http://www.communities.gov.uk/documents/housing/pdf/propertyquestionnairegeneral.pdf> [Accessed May 10, 2010].

Defra (Department for Environment Food and Rural Affairs) (2005). Making Space for Water—Taking Forward a New Government Strategy for Flood and Coastal Erosion Risk Management in England. London: Defra.

Dupuis, A., and Thorns, D. C. (1998). Home, home ownership and the search for ontological security. *The Sociological Review*, 46, 24–47.

EA (Environment Agency) (2010). Prepare Your Home or Business for Flooding. Environment Agency. Available at <http://www.environment-agency.gov.uk/homeandleisure/floods/31644.aspx> [Accessed March 29, 2010].

Entec and Greenstreet Berman (2008). Developing the Evidence Base for Flood Resilience. R&D Technical Report FD2607/TR. London: Defra.

Fairclough, N. (2003). *Analysing Discourse: Textual Analysis for Social Research*. London: Routledge.

Fazio, R. H., Sherman, S. J., and Herr P. M. (1982). The feature-positive effect in the self-perception process: Does not doing matter as much as doing? *Journal of Personality and Social Psychology*, 40, 404–411.

Field, A. (2005). *Discovering Statistics Using SPSS* (2nd edition). London: Sage.

Giddens, A. (1991). *Modernity and Self-Identity*. Cambridge: Polity Press.

Grothmann, T., and Reusswig, F. (2006). People at risk of flooding: Why some residents take precautionary action while others do not. *Natural Hazards*, 38, 1–2.

Grubel, H. G. (1971). Risk, uncertainty and moral hazard. *The Journal of Risk and Insurance*, 38(1), 99–106.

Harries, T. (2008a). Householder Responses to Flood Risk: The Consequences of the Search for Ontological Security. PhD thesis. Flood Hazard Research Centre, Middlesex University, Enfield, London.

Harries, T. (2008b). Feeling secure or being secure? Why it can seem better not to protect yourself against a natural hazard. *Health, Risk and Society*, 10, 5.

Higgins, R. L. (1990). Self-handicapping: Historical roots and contemporary branches. In Higgins, R. L., Snyder, C. R., and Berglas, S. (Eds.), *Self-Handicapping: The Paradox That Isn't*, pp. 1–31. New York: Plenum Press.

Hogg, M. A., Turner, J. C., and Davidson, B. (1990). Polarized norms and social frames of reference: A test of the self-categorization theory of group polarization. *Basic and Applied Social Psychology*, 90, 167–174.

Homan, J. (2001). A culturally sensitive approach to risk: 'Natural' hazard perception in Egypt and the UK. *Australian Journal of Emergency Management*, Winter, pp. 14–18.

Johnson, C., and Priest, S. (2008). Flood risk management in England: A changing landscape of risk responsibility? *International Journal of Water Resources Development*, 24(4), 513–525.

Jones, E., and Berglas, S. (1978). Control of attributions about the self through self-handicapping strategies: The appeal of alcohol and the role of underachievement. *Personality and Social Psychology Bulletin*, 4, 200–206.

Jones, E., Farina, A., Hastorf, A., Markus, H., Miller, D.T., and Scott, R. (1984). *Social Stigma: The Psychology of Marked Relationships*. New York: Freeman.

Kallmen, H. (2000). Manifest anxiety, general self-efficacy and locus of control as determinants of personal and general risk perception. *Journal of Risk Research*, 3, 111–120.

Kates, R.W. (1976). Experiencing the environment as hazard. In Wapner, S., Cohen, S., and Kaplan, B. (Eds.), *Experiencing the Environment*. New York: Plenum Press.

Kunreuther, H., and Slovic, P. (1986). Decision making in hazard and resource management. In Kates, R. W., and Burton, I. (Eds.) *Geography, Resources and Environment. Volume 2: Themes from the Work of Gilbert F. White*. Chicago: University of Chicago Press.

Lamond, J., Proverbs, D., and Hammond, F. (2010). The impact of flooding on the price of residential property: A transactional analysis of the UK market. *Housing Studies*, 25(3), 335–356.

Loomes, G., and Sugden, R. (1982). Regret theory: An alternative theory of rational choice under uncertainty. *Economic Journal*, 92, 805–824.

Mallett, S. (2004). Understanding home: A critical review of the literature. *The Sociological Review*, 52(1), 62–89.

McMillen, J. C., Smith, E. M., and Fisher, R. H. (1997). Perceived benefit and mental health after three types of disaster. *Journal of Consulting and Clinical Psychology*, 65(5), 733–739.

NFF (National Flood Forum) (2010). Flood Resilience. National Flood Forum. Available at <http://www.floodforum.org.uk/index.php?option=com_content&view=article&id=8&Itemid=4> [Accessed March 29, 2010].

Nisbett, R., and Ross, L. (1980). *Human Inference: Strategies and Shortcomings of Social Judgement.* London: Prentice-Hall.

Office for National Statistics (2001). *2001 census: Standard area statistics (England and Wales)* [computer file]: ESRC/JISC Census Programme, Census Dissemination Unit, MIMAS (University of Manchester) [online]. Available at <http://census.ac.uk> [Accessed August 2006].

Ohl, C. A., and Tapsell, S. M. (2000). Flooding and human health: The dangers posed are not always obvious. *British Medical Journal*, 321, 1167–1168.

Pitt, M. (2008). *Learning Lessons from the 2007 Floods*. London: Cabinet Office.

Risk & Policy Analysts Ltd., Flood Hazard Research Centre, EFTEC and CASPAR (2004). *The appraisal of human-related intangible impacts of flooding*. London: Defra.

Rose, N. (1999). *Powers of Freedom*. Cambridge: Cambridge University Press.

Saunders, P. (1989). The meaning of 'home' in contemporary English culture. *Housing Studies*, 4(3), 177–192.

Siegrist, M., and Gutscher, H. (2008). Natural hazards and motivation for mitigation behavior: People cannot predict the affect evoked by a severe flood. *Risk Analysis*, 28(3), 771–778.

Siudak, M. (2001). Role of education in reducing flash flood effects. In Grundfest, E., and Handmer, J. (Eds.), *Coping with Flash Floods*, pp. 15–18. London: Kluwer Academic Publishers.

Slovic, P. (2000). *The Perception of Risk*. London: Earthscan.

Smith, E. E., Nolen-Hoeksema, S., and Loftus, G. R. (2003). *Atkinson & Hilgard's Introduction to Psychology*. Belmont, CA: Wadsworth.

Smith, S. G. (1994). The essential qualities of the home. *Journal of Environmental Psychology*, 14(1), 31–46.

Thaler, R. (1980). Towards a positive theory of consumer choice. *Journal of Economic Behaviour and Organisation*, 1, 39–60.

Tunstall, S., Tapsell, S., Green, C., Floyd, P., and George, C. (2006). The health effects of flooding: Social research results from England and Wales. *Journal of Water and Health*, 4(3), 365–380.

Tversky, A., and Kahneman, D. (1973). Availability: A heuristic for judging frequency and probability. *Cognitive Psychology*, 5(2), 207–232.

Tykocinski, O., and Pittman, T. (1998). The consequences of doing nothing: Inaction inertia as avoidance of anticipated counterfactual regret. *Journal of Personality and Social Psychology*, 75(3), 607–616.

Van Dijk, W. W., Van der Pligt, J., and Zeelenberg, M. (1999). Effort invested in vain: The impact of effort on the intensity of disappointment and regret. *Motivation and Emotion*, 23, 203–220.

Walker, G., Burningham, K., Fielding, J., Smith, G., Thrush, D., and Fay, H. (2006). Addressing Environmental Inequalities: Flood Risk. Science report SC020061/SR1. Bristol: Environment Agency.

Weinstein, N. D. (1980). Unrealistic optimism about future life events. *Journal of Personality and Social Psychology*, 39, 806–820.

Weinstein, N. D., Sandman, P. M., and Roberts, N. E. (1991). Perceived perceptibility and self-protective behaviour. *Health Psychology*, 10, 25–33.

White, G. (1973). Natural hazards research. Reprinted in Kates, R. W., and Burton, I. (Eds.), *Geography, Resources and Environment. Volume 1: Selected Writings of Gilbert F. White*. Chicago: University of Chicago Press.

Wilkinson, I. (2001). *Anxiety in a Risk Society*. London: Routledge.

Zeelenberg, M., Van den Bos, K., Van Dijk, E., and Pieters, R. (2002). The inaction effect in the psychology of regret. *Journal of Personality and Social Psychology*, 82(3), 314–327.

25 Exploring the Effect of Perceptions of Social Responsibility on Community Resilience to Flooding

Aaron Mullins and Robby Soetanto

CONTENTS

25.1 INTRODUCTION

Human activity is having a large, detrimental effect on the environment, increasing climate change and thereby increasing the likelihood of severe flooding (IPCC, 2001). Furthermore, as climate change becomes an evermore serious threat, then flooding in the built environment will become evermore frequent and severe (McCarthy, 2007). Climate change is altering weather patterns all across the globe and creating changes that our global ecosystem is now struggling to cope with (Pitt, 2008). Our built environments have become increasingly merged with the natural environment, making both more susceptible to flooding. The aging physical infrastructure, rapid economic development, and growing populations all add to the vulnerability of our built environments to severe floods (Stewart and Bostrom, 2002).

Communities, organizations, and people in general are often ill-prepared to cope with flooding, with physical resilience measures proving to be largely ineffective and forecasts based on past events unable to accurately predict our ever-changing

343

world (Stewart and Bostrom, 2002). This has meant that society has become more vulnerable to the effects of flooding, and in 2007 there was widespread flooding in the United Kingdom that caused an enormous amount of damage as our fragile infrastructure was not able to cope with such extreme weather (Pitt, 2008).

This vulnerability within modern society is not limited to flooding events, as evidenced by the 2003 heat wave that caused a large loss of life throughout parts of Europe (Poumadère et al., 2005; Salagnac, 2007). Even the snowstorms that occurred in 2009 managed to cause chaos to transport networks and supply chains. Therefore, extreme weather events pose one of the biggest threats to UK society as climate change and the fragile infrastructure of our everyday lives combine to create this modern risk. To ensure the survival and well-being of individuals, it is of utmost importance that appropriate strategies are devised to improve the resilience of the community where these individuals live. This calls for a greater understanding of factors (e.g., drivers and barriers) influencing resilience and the interrelationships between key stakeholders of the community.

The research reported in this chapter explores perceptions of social responsibility as a way to enhance understanding of the decision-making process and interrelationships between three key community groups (policy makers, homeowners, and small businesses) in order to improve the resilience to flooding of the local community. The discussion suggests that a better understanding of social responsibility, the decision-making process, and interrelationships among members of the community will help joined-up thinking and optimize the selection of adaptation and mitigation strategies to flooding events.

25.2   THE FLOODING ISSUE

In 1953 an extreme flood in the Thames estuary and east coast region flooded 240,000 houses and killed more than 300 people. A tidal surge within the same area nowadays would cause damages of £80 to 100 billion to homes, businesses, and economic activity, affecting 1.25 million people (Parker and Penning-Rowsell, 2002). While expansion in particular locations may help accommodate the increasing population, it also increases a community's vulnerability to flooding, as there is more damage potential contained within smaller areas. Much of the land is already developed, or protected, forcing planning authorities to build close to, or actually within, tidal flood risk zones (Lonsdale et al., 2008). The impact of an extreme flooding event would also have an impact on a global scale, particularly in London where many business headquarters are located (Dawson et al., 2005). As the population continues to grow denser on floodplains across the United Kingdom, the vulnerability to extreme flooding events rises.

The UK floods in summer 2007 launched the largest rescue effort in Britain since World War II (Pitt, 2008). Despite being aware of flood warnings, many people did not expect the flooding to affect them and did not know what preventative steps to take or who to contact for help (Pitt, 2008). In the following year, many communities were still recovering from the floods and it is recognized that there are many lessons to be learned to improve the way we deal with flooding in the future (Pitt, 2008). Therefore, it is of utmost importance that measures to increase resilience to flooding are found.

25.3   UK RESILIENCE TO FLOODS

The severe flooding of 2007 came after the wettest May to July period ever recorded (Pitt, 2008). This indicates that the risks we face are increasing and we have not yet found a sufficient way to counter this risk. This is because although the government has been attempting to adapt to new risks, it has done so through the creation of new legislation and implementation of new civil protection measures, the majority of which have been built around an already stretched communication network and used already stretched resources. It should not fall to the formal organizations and institutions that are the functioning arm of the overburdened network to increase resilience to such events as they are too far embedded within the fragile infrastructure itself, adding frailties to the resilience measures themselves. These interdependent organizations have their place to increase resilience, but it may not be possible for them to achieve the kinds of results that could protect modern society to a sufficient level. Instead, it is the extended branches of the network, the communities themselves, that could make the greatest advances in creating resilience to flooding. This view was echoed by the Foresight Future Flooding report (Evans et al., 2004) and the Stern Review (Stern et al., 2006), which highlighted the importance of informing everyone about the risks posed by climate change and how it may affect their daily lives.

The uncertainty surrounding climate change, however, is mirrored in the uncertainty surrounding changes that will happen at the social and economic level over the course of time. This puts our man-made world at an increased risk of disaster, with more lives and livelihoods in danger of being swept away by levels of flooding that have never before been experienced. Therefore, the quest to protect our built environment from flooding has never been of such great importance and above all the forecasts and technologies of the modern age, it is still the people who remain the key to a successful defense. However, research has thus far neglected to fully investigate the impact of these findings within the built environment with which we are most familiar and is most salient to our needs—our own community.

25.4   MODERN COMMUNITIES: OVERRELIANCE
ON INTERCONNECTEDNESS

The majority of people in the United Kingdom live in urban areas that rely on an enormous amount of support from organizations to provide them with water, electricity, gas, communications, transport, and food, which are necessary elements of everyday life. The systems of this critical infrastructure rely on increasingly complex technology to provide them with greater interconnectedness. However, the networks that organizations use to support such a large amount of interdependencies are based on an outdated infrastructure that lacks the capacity to support our ever-more complicated lifestyles. Our societal infrastructure struggles to support us now, and the demands placed on this system of networks will only become greater over time (Pitt, 2008). This enormous amount of interconnectedness means that should an extreme flood take place, then these interdependencies leave communities vulnerable to the effects of flooding. Disasters often strike at the heart of the critical infrastructure and in a system where even the smallest of disturbances to the network can create enormous amounts of disruption to many people, disasters contain the potential to devastate our national infrastructure and thereby affect every aspect of modern life. This is a risk we are living with every day; it is important that society finds new ways to reduce its vulnerability and increase its resilience to flooding.

One of the main reasons why society is able to become more interconnected is through technological advancements in many industries; however, the 2007 floods also highlighted the danger of becoming reliant upon technology. In the Thames Region, the Regional Telemetry System partially failed, thus providing no data to the National Flood Forecasting System (NFFS) (Pitt, 2008). On one site, a failed river alarm resulted in 23% of all properties not receiving a flood warning in time (Pitt, 2008). A number of Environment Agency river-level gauges reached their recordable limit, were inundated by floodwater, or lost power, while others were inaccessible due to extreme flood conditions and therefore could not be read (Pitt, 2008). During the summer 2007 flood, 50% of the flood defenses that were tested by the floodwaters were overtopped (Pitt, 2008). These failings were found in technological resilience measures across the country and together they demonstrate why new, nontechnological resilience measures must be found. One of the main areas to emerge from the discussion of resilience of how this can be achieved is the idea of individuals being more socially responsible and accepting a greater level of individual responsibility for community resilience.

25.5 SOCIAL RESPONSIBILITY

Social responsibility has long been an important field of research for both academics and business practitioners, and continues to provide a valuable research area for those wishing to investigate modern societal issues (Gorte, 2005; Peterson and Jun, 2007). Personal responsibility for behavior is important to increase resilience, and understanding how people perceive themselves and each other in relation to a particular aspect may be a useful way of investigating that aspect itself. Therefore, exploring perceptions of social responsibility for flooding events will provide an excellent platform from which to investigate barriers and drivers to community resilience.

In 2009 the United Kingdom was hit by severe snowstorms that tested the resilience of many communities. The storms highlighted major discrepancies between what householders believed the Council was responsible for and what the Council believed it was responsible for. An example of this can be seen when, as the snowfall became heavier, the Council began prioritizing main roads, meeting what they believed to be their responsibility to the community. However, in doing so they left many residents isolated and feeling that the Council was not meeting its responsibility to the community. The resilience of many communities across the United Kingdom had been undermined by gaps in people's expectations of their own and other community groups' social responsibilities. These gaps are indicative of barriers to community resilience and are brought about by a lack of integration and joined-up decision making between residents, local businesses, and policy makers. Residents were not aware of the decisions being made by the Council or of resilience procedures that stated that grit bins (i.e., salt bins) would only be provided upon request.

The Council believed that it was attending to the needs of the whole community as resilience measures were in place; however, the community was not aware of these measures and believed the Council had failed them. In the eyes of the Council staff, residents had failed to meet their own expectations of social responsibility by failing to request grit (salt) and maintain their own resilience levels. This makes perceptions of social responsibility within and between community groups of vital importance to resilience research.

The emergency services and utility companies are responsible for many of the immediate impacts of flooding in the built environment, but the continued success-ful resilience of the community in the short to medium term relies on the groups that make up that community, such as the homeowners, SMEs (small and medium-sized enterprises), and policy makers. The Pitt Review (Pitt, 2008) supports the impor-tance of these three groups, highlighting that local government plays a central role in managing flood risk, with community groups, such as local flood groups and the National Flood Forum, helping to inform the public of the risks they face before, during, and after a flood event. The Environment Agency is forging stronger links within the community, conducting research and implementing action plans, such as the national project launched in 2008 that aims to record surface water flooding in order to produce a data set of the most vulnerable areas (Pitt, 2008). Businesses are beginning to understand the need for a business continuity plan, seeing it as a critical element of good business practice, gaining help from policy makers to increase their own level of resilience as well as better safeguarding the infrastructure that provides services to homeowners (Pitt, 2008). This highlights some of the interdependencies that the individuals within these three groups possess.

Communities are made up of individuals, each of whom can have an effect on their personal level of resilience to flooding, which in turn will have an effect on their community resilience. Thus, individuals have a responsibility to increase their own resilience and they can do so through the decisions they make about being aware of the risks faced by their community, accepting these risks, and engaging with the issue of flooding. Unfortunately, many people are unaware or are in denial about the risks they live with each day, and it is these counterproductive attitudes and flawed decision making that need to be changed in order to increase resilience. To instigate the necessary changes, researchers need to firstly understand how and why people reach the decisions they do about the risk of flooding, as well as understand how the interdependencies within the community can affect these decisions. These individu-als are not simply the residents within the community, but also heads of businesses and local policy makers, each of whom has a key role to play in increasing resilience. For example, why do local policy makers make the decision to build houses on flood-plains when they know that this decreases their community resilience to an extreme flooding event? Why do residents and businesses make the decision to occupy build-ings on floodplains when they know that this decreases their personal resilience to an extreme flooding event?

The example above indicates that there is a lack of individual and social respon-sibility being taken for actions that can affect personal and community resilience to flooding. We may live in a modern blame culture but there appears to be a lack of accountability for the tragedies that occur when the effects of disasters are increased because individuals have made poor decisions that have decreased their resilience to such events. Is it the fault of residents who choose to live there, or the fault of policy makers who choose to build there? Too often, floods are blamed on being an "Act of God" when in fact a clear pathway of poor decisions made over a long period of time has contributed to the final damage caused by flooding events. Furthermore, the overreliance on others that is fostered through our modern interdependent lifestyles can also contribute to attitudes, decisions, expectations, and behaviors that are detrimental to our resilience. It is time then for individuals to play a greater role in increasing both their personal and community resilience to ensure that in the future communities will be better protected against these events.

25.6   UNDERSTANDING INDIVIDUAL ROLES IN RESILIENCE

In the United States, personal responsibility is recognized by the Federal Emergency Management Agency (FEMA) as being the key to building a resilient community (Colten, Kates, and Laska, 2008). However, there are many views on how much of a threat climate change poses, indicated by some people suggesting that immediate action should be taken; others suggesting that the scientific evidence is unreliable; or given the uncertainty, nothing should be done until there is more reliable evidence; or simply not believing that climate change affects their lives in any way (Lorenzoni and Pidgeon, 2006). Given that modern society contains masses of interdependencies to function efficiently, it is reasonable to determine that it will require further collaboration and joined-up thinking between key community groups to efficiently increase community resilience. This need for integration is reflected in community resilience models that have stressed the importance of community participation and the ability to communicate community problems (Paton, 2007), as well as the need to integrate community stakeholders (Cutter et al., 2008). However, many existing models, while emphasizing that understanding interdependencies between community groups will be beneficial, also note that generic models of community resilience have thus far failed to specify the content of such interventions, knowledge that will be required to positively affect resilience factors (Paton, 2008).

This aspect is further emphasized by the need to integrate community groups within climate change education, as top-down information (i.e., policy makers telling people what should be done) does not work and bottom-up information (i.e., community groups integrating information together) is needed to improve risk communication and community resilience (Dufty, 2008). Therefore, while social responsibility has been highlighted as a potential key factor for effecting community resilience, it is yet to be explored in enough depth to provide contextual information toward understanding how and why these effects occur. However, what can be assumed is that in order for people to understand how and why they must be more socially responsible to increase their resilience to flooding, they must first understand what constitutes resilient behavior.

The Pitt Review (Pitt, 2008) uses the real-life example of a homeowner who was flooded in 2000 and then again in 2007; but having adopted a number of resilience measures after the first flood, the homeowner had reached a level of resilience where he was able to return to normal very quickly. This homeowner made the decision to increase his individual resilience to flooding, which in turn has increased the resilience level of his community and placed less of a strain on resources and infrastructure. Unfortunately, the overall take-up of resilience measures is low, even for simple, low-cost measures (Pitt, 2008). Some of these practical resilient measures may mean additional costs, but will reduce flood damages in the future (Soetanto et al., 2008). However, many residents simply refused to accept that their properties might flood again, and it is this lack of responsibility to themselves and their community that undermines current resilience measures. Norwich Union found that 46% of people did not believe that it was their responsibility to take resilience measures, stating that this responsibility lay instead with local authorities and the government (Pitt, 2008). These kinds of perceptions create barriers to resilience, with each community group believing that the other is responsible for taking resilience measures.

The Pitt Review (Pitt, 2008) also provides information about farmers in Upton-upon-Severn who used their equipment to minimize flood damage, displaying a high level of social responsibility. It is important to identify the level of social responsibility an individual must possess in order to make the decision to engage in resilience-promoting behavior, and what social and psychological barriers lie in the way of achieving this. The Pitt Review (Pitt, 2008) calls for a greater degree of personal resilience and a community consisting of a greater number of socially responsible individuals would have a higher resilience to flooding due to their combined resilience levels. These individuals would understand their role within the community, rather than believing that it is someone else's responsibility and being overly reliant on other community groups. In turn, the better prepared an individual, business, or local authority is, the less they will be affected by the flood and the more time and resources they will have to fulfill the roles that do require them to help others within the community.

Therefore, it is important to understand how the three key community groups perceive their own level of responsibility, and what they perceive to be the responsibility of others, in order to highlight where barriers to resilience are being formed. If we understand communities as being a complex system of interdependencies, the resilience of that community is determined by the system's ability to absorb disturbance, self-organize, and the capacity to learn and adapt. Therefore, it is the attitudes, perceptions, decision making, and behaviors that members of a community adopt or display prior to a flooding event that can determine the ability of that community to absorb the disturbance. Furthermore, these aspects may also determine their motivation and ability for self-organization during the event and how much they are willing to learn from the event in order to change their perceptions and behaviors so that resilience can be increased in the future. Therefore, research needs to fully investigate what current perceptions of social responsibility exist within the three community groups and how their interrelationships may affect their own resilience levels, as well as that of their community. It is only when we know what the current perceptions of social responsibility are within and between community groups that we can take the necessary action to overcome any barriers to community resilience that may exist.

25.7 CONCLUSION

This chapter has highlighted that flooding represents a serious threat to UK communities. To better protect ourselves, we need to understand the barriers to and drivers for resilience at the community level. Community resilience is largely affected by the decision making of its key community groups, and therefore a better understanding of factors affecting the decision-making process is required. This chapter has highlighted the potential of social responsibility to affect decision making. However, there are a number of considerations to be taken into account by researchers in these areas.

It is important to understand that many social and psychological factors may not be distinct from each other and may influence and affect each other, as well as the overall decision-making process. This can be seen where a better understanding of perceived social roles and responsibilities would provide a context for exploring perceptions of flooding risk. We can take one key community group, homeowners, as an example: If an individual did not believe that the risk of flooding was great, then he may not engage in any resilience-enhancing actions. However, simply stating that there is a linear relationship between perceptions of risk and engagement in resilience-enhancing measures does not provide a full enough picture to inform future resilience measures. Instead, understanding how that individual perceives the level of involvement that homeowners currently have with these issues and the responsibilities they have as a community group in relation to these issues requires further exploration. What resilience-enhancing actions do homeowners believe they should be engaging in, or are able to engage in? Do homeowners feel that they have certain responsibilities that they should meet? How do the perceptions of the role of homeowners change before and during flooding events? This example and the questions it raises also extend to the other two community groups of SMEs and policy makers. This then raises questions of how do these three key community groups view each other's roles and responsibilities, and are there any gaps between the expectations of others and an understanding of one's own role? These gaps would be potential barriers to increasing community resilience.

Future research should also consider the effects that each stage of the decision-making process will have on the other stages; for example, community-level decision making may be improved through increasing acceptance of other views and integrating agencies within the community in the decision-making process, such as homeowners, SMEs, and policy makers. Each person is a decision maker in their own right; however, it is often joint decisions and a joint effort that are required. This integration may be able to promote trust, increase access to reliable information and communication of risk, as well as aid in identifying individual roles regarding resilience to flooding events. Policy makers could work with SMEs and residents to remove any "policy obstacles," thereby demonstrating how a greater understanding of social responsibility can highlight flaws in current resilience policies. It will also allow policy makers to see exactly where perceptions of social responsibility differ within the community, representing issues that need to be addressed so that SMEs and residents become more aware and more accepting of risk and better understand their own roles in increasing community resilience. These are but a few examples of the broader considerations needed to make this process work, attempt to counter the many failings of previous measures, and change patterns of resilience-reducing coping strategies and behaviors within the community by promoting engagement with the issue. One of the first tasks faced by researchers, however, is establishing a common framework for measuring and monitoring social responsibility within the community. Such a framework will provide a platform for integration and joined-up thinking between key community groups. There is much work to be done in this field but what can be concluded is that the role of perceptions of social responsibility is extremely important when trying to protect our built environments from flooding disasters.

REFERENCES

Colten, C. E., Kates, R. W., and Laska, S. B. (2008). Community Resilience: Lessons from New Orleans and Hurricane Katrina. CARRI Research Report 3, Community and Regional Resilience Initiative, Oak Ridge, TN: Oak Ridge National Laboratory.

Cutter, S. L., Barnes, L., Berry, M., Burton, C., Evans, E., Tate, E., and Webb, J. (2008). A place-based model for understanding community resilience to natural disasters. *Global Environmental Change*, 18, 598–605.

Dawson, R. J., Hall, J. W., Bates, P. D., and Nicholls, R. J. (2005). Quantified analysis of the probability of flooding in the Thames Estuary under imaginable worst case sea-level rise scenarios. *Water Resources Development*, 21(4), 577–591.

Dufty, N. (2008). *A New Approach to Flood Education*. Parramatta: Molino Stewart Pty Ltd:.

Evans, E. P., Ashley, R., Hall, J. W., Penning-Rowsell, E. P., Sayers, P.B., Thorne, C. R., and Watkinson, A. R. (2004). *Foresight future flooding, scientific summary: Volume 2: Managing future risks.* London: Office of Science and Technology.

Gorte, J. (2005). Corporate social responsibility: Close to victory. *The Journal of Investing*, 14(3) 140–141.

IPCC (Intergovernmental Panel on Climate Change) (2001). *Climate change 2001. Impacts, adaptation, and vulnerability.* Cambridge: Cambridge University Press.

Lonsdale, K., Downing, T., Nicholls, R., Parker, D., Vafeidis, A., Dawson, R., and Hall, J. (2008) Plausible responses to the threat of rapid sea-level rise in the Thames Estuary. *Climatic Change,* 91(1/2), 145–169.

Lorenzoni, I., and Pidgeon, N. (2006). Public views on climate change: European and USA perspectives. *Climatic Change*, 77(1–2), 73–95.

McCarthy, M. (2007). Extreme conditions: What's happening to our weather? *The Independent*, August 28, 2007.

Parker, D. J., and Penning-Rowsell, E. C. (2002). *The Case for Flood Protection for London and the Thames Gateway*. Flood Hazard Research Centre, prepared for the Environment Agency.

Paton, D. (2007). Measuring and Monitoring Resilience in Auckland. GNS Science Report 2007/18.

Paton, D. (2008). Modelling Societal Resilience to Pandemic Hazards in Auckland. GNS Science Report, 2008/13.

Penning-Rowsell, E. C., Chatterton, J. B., Wilson, T., and Potter, E. (2002). Autumn 2000 Floods in England and Wales, Assessment of National Economic and Financial Losses, Draft Final Report to the Environment Agency. March 2002. Middlesex University, Enfield: Flood Hazards Research Centre.

Peterson, R. T., and Jun, M. (2007). Perceptions on Social Responsibility: The Entrepreneurial Vision. Business Society OnlineFirst, published on September 18, 2007 as doi:10.1177/0007650307305758. Available at <http://bas.sagepub.com/cgi/rapidpdf/0007650307305758v1> {Accessed March 2, 2010].

Pitt, M. (2008). The Pitt Review—Learning Lessons from the 2007 Floods. London: Cabinet Office.

Poumadère, M., Mays, C., Le Mer, S., and Blong, R. (2005). The 2003 heat wave in France: Dangerous climate change here and now. *Risk Analysis*, 25, 1483–1494.

Salagnac, J. L. (2007). Lessons from the 2003 heat wave: A French perspective. *Building Research and Information*, 35(4), 450–457.

Soetanto, R., Proverbs, D. G., Lamond, J., and Samwinga, V. (2008). Residential properties in England and Wales: An evaluation of repair strategies towards attaining flood resilience. In *Hazards and the Built Environment: Attaining Built-In Resilience*, Bosher, L. (*Ed.*). London: Taylor & Francis.

Stern, N., Peters, S., Bakhshi, V., Bowen, A., Cameron, C., Catovsky, S., Crane, D., Cruickshank, S., Dietz, S., Edmonson, N., Garbett, S.-L., Hamid, L., Hoffman, G., Ingram, D., Jones, B., Patmore, N., Radcliffe, H., Sathiyarajah, R., Stock, M., Taylor, C., Vernon, T., Wanjie, H., and Zenghelis, D. (2006). *Stern Review: The Economics of Climate Change*. London: HM Treasury.

Stewart, T. R., and Bostrom, A. (2002). *Extreme event decision making: Workshop report*. Albany, New York: University at Albany, June.

26 Lessons for the Future

Felix N. Hammond, Colin A. Booth, Jessica E. Lamond, and David G. Proverbs

CONTENTS

26.1 INTRODUCTION

Most societies are confronted with concerns about climate change and its potential impact. While the debate surrounding the major cause(s) of global warming remains unsettled, its reality and effects on human well-being and long-term survival are undisputed. An immediate and easily observable repercussion of climate change in comparatively recent times is the significant seasonal shifts in the Earth's hydrological cycles, as attributed by the available scientific evidence. As a consequence of global warming, oceans are becoming warmer, sea levels are rising, and ice caps are melting, releasing a greater volume of water into the cycles and varying its balance and patterns. This shift has resulted in changes in the Earth's water circulation and availability.* The sensitivity of hydrological cycles to global temperature variability has differed between regions and, in some cases, between countries. Whereas in some countries this has led to water shortages, other countries and regions have endured rises in the magnitude and frequency of floods. Focusing attention on floods, the resulting changes to the hydrological cycle is compounded, in most countries by the inadequacies of the sewerage and drainage systems and increased urban development. Extreme flood events have followed intense tidal waves (tsunamis), storms, and exceptionally heavy rainfalls that have invariably inundated the existing sewerage and drainage systems.

26.2 AN OVERVIEW OF THE LESSONS LEARNED

Flood risk management will no doubt benefit from this collection of chapters from researchers, practitioners, and policy makers on various aspects of flooding. The twenty-five chapters in this book have examined four key themes pertaining to flooding: impacts (Chapters 2 through 6), recovery and repair (Chapters 7 through 11),

---

* Intergovernmental Panel on Climate Change, Working Group I, Climate change, the IPCC scientific assessment (Cambridge University Press, Cambridge, 1990).

mitigation and adaptation (Chapters 12 through 19), and community perspectives (Chapters 20 through 25). All have been examined within the context of the built environment.

The first chapter set the scene for the ensuing chapters, highlighting the ongoing uncertainty in climate change science, causal factors, exposure and vulnerability to flood events, flood defenses, and management. The ability to forestall, cope with or adapt ably, or even to offer advance warning and improve preparedness against future flood events depends on the accuracy of flood forecasting. It has always been difficult to be precise about future climatic or weather events. Following a detailed examination of existing flood forecasting techniques, with emphasis on meteorological ensemble prediction system products, novel propositions to reduce predictive uncertainty and improve reliability of flood forecasting were proffered (Chapter 2).

The essence of flood forecasting is to warn and prepare. Thus, the availability of refined forecasting models can do very little to help in adapting, coping, or mitigating the hazards of flood unless they are complemented by a system that makes use of the forecasting data to adequately warn and prepare for the management of the actual incident. Current flood warning systems in the United Kingdom have proved effective in preventing loss of life and damage. However, in the drive to continuously improve, the aptness of the current deterministic warning system—where flood warnings are issued when it is certain that a location will flood—has been questioned. A move toward a probabilistic warning system is favored, which will make it possible to warn on the basis of risk rather than certainty, leaving adequate opportunity for preparation for the management of the incident (Chapter 3). The critical fact about a flood episode is its impacts. The impacts of flooding can be wide ranging—economic, social, health, and political. Analysis of the UK 2007 flooding episode, using Barnsley Metropolitan Borough as a case study, affected a total of forty-eight separate localities (Chapter 4). In this particular case, all 352 properties identified within the Environmental Agency Local Flood Warning Plan were inundated with profound impacts on both the council and the community. This, as would be expected, led to a severe disruption of normal life, including transportation and utility services. The health impacts of flooding persist long after the flood event is over (Chapter 5). The obvious health consequences of floods are the deaths and injuries often highlighted in the media. The less obvious and immediate health consequences (such as infections, chemical hazards, and mental health problems) that become obvious after the frenzy tend to fall on the blind side of the media. Yet, these may be more considerable than the immediate health issues. For example, it was shown that during the 12-month post-flood era following the Bristol flood of 1968, the mortality rate rose by 50% in the flooded part of the city compared to no appreciable increase in the nonflooded area. The economic and financial damage caused by flooding can also be huge (Chapter 13)—not only for individuals but also for government, industry, and particularly for SMEs (Chapter 19).

When flooding occurs, the affected must, as much as possible, recover and attempt to put right whatever has been damaged. Recovering and repairing the damage caused by floods requires money or other resources. Depending on the scale of damage, the amount required may exceed the affordability limits of flooded households or organizations. In countries with a sufficiently well-developed insurance market, many will depend on their insurers to provide the money needed for the recovery and repairs due. Flood insurance is well developed in the United Kingdom and is currently widely available at affordable prices (Chapter 7). However, because the insurance model requires that the insured show adequate investment in risk management and resilience, making insurance widely available in the future, particularly in the United Kingdom, will partly depend on the government's long-term funding commitments to flood defense. Beyond the economics, there are real practical and technical issues involved in flood repair and recovery. Flooded buildings require thorough drying to prevent rot, mold, and degradation of the building fabric. This may be based on a Restoration Industry Association drying standard (Chapter 8). Reinstatement is a crucial practical post-flood action (Chapter 9). A more concerning issue pertaining to recovery are the disruptive effects that flooding tends to have on those living in the affected property (Chapter 20). This is worsened by any gaps that may emerge between publicly provided contingency measures and assistance obtained through the private sector, such as insurance companies and so on. This is leading to a growing need for individuals to install their own property-level flood protection, such as door guards designed to seal against opening or slot into prefixed grooves, smart airbricks, self-closing barriers, tanking, and veneer walling (Chapter 21). Rehabilitation involves the management of the emotional and physical turmoil of the victims as well as reinstating their damaged properties. It is not enough just to rehabilitate the affected building. In the heat of the moment, insurance procurement managers and suppliers shift attention to cost minimization and volume at the expense of soft but vital issues such as customer care, empathy, and competent project management, particularly in the case of small claims. Such soft issues are difficult to overlook with big claimants.

Putting the victim's damaged property back to the exact state in which it would have been had the flood not occurred is a very difficult task and there cannot be pinpoint accuracies in such matters. This means that it is possible that the repair undertaken by the insurance company may fall short or surpass its pre-flood state. But how far short is acceptable? This is broadly a subjective question that can only be addressed or negotiated by the very parties involved. But where the parties are of disproportionate bargaining strengths, some legal guidance may help in objectivizing the quality of the workmanship, materials, and processes (Chapter 10). The situation can also be capitalized on to enhance the flood performance of the affected building (Chapter 11). This may entail a large investment in structural protection. Different standards may be required for different categories of properties based on their propensity for flooding. Properties that are likely to experience repeat flooding within, say, 25 years may need to be repaired or reinstated to a much higher resilience standard than those that are unlikely to experience repeat flooding.

On a wide scale, government-level funding may be used to develop flood defenses for the benefits of individual properties on floodplains. The question that needs addressing, however, is the following: Who should ultimately pay for the costs? Should these be funded from the general tax funds, or should the costs be transferred to the beneficiary property owners? (Chapter 23). Government funding could also be employed to ensure that individual properties are flood resilient or resistant to keep floodwater out of the property (Chapter 18) if performance standards are acceptable.

Affordability and possible negative impacts on property prices remain the major obstacles to owners providing property-level pre-emptive protection (Chapter 24). However, the desire to minimize anxiety about flooding has also played a significant role in discouraging the provision of property-level protection.

Managing flooding in a sustainable way is the way forward to effectively handle the issues involved in flooding (Chapter 12). This can be achieved through land use planning regulation (Chapter 16), through the creative design of urban spaces (Chapter 14), and through incorporating sustainable drainage in urban settings (Chapter 15). That said, there are political expediencies that may make this a difficult endeavor. Many of the political actions needed in this direction do not necessarily involve public expenditure. They entail, however, a political will to resist the lobbying of interest groups such as property developers and ecology lobbies to block spatial planning policies designed to reduce vulnerability through stricter and more flood-conscious building regulations. Flood management may involve adaptation in order to better cope with floods (Chapter 13). Interestingly, enterprises tend to adopt a minimalism strategy, concentrating more on technical issues to the neglect of comprehensive risk management. One of the fundamental causes of flooding is the ease with which existing drainage systems are inundated by surface water and heavy rains (Chapter 15). The UK's aging sewer network continues to accentuate flood risks in that country (Chapter 6). The way forward here lies with sustainable drainage systems, which have been proven to perform better than traditional piped systems (Chapter 15) and resilient critical infrastructure (Chapter 17).

An important aspect of flooding that can hardly be overlooked is the resilience of the community itself—its capacity to pull together and provide immediate and long-term assistance and help to the affected. Educating and empowering communities to develop resilience for prospective flooding is invaluable for effective and efficient flood management (Chapter 22). However, the success of this approach may be frustrated by other social and psychological issues (Chapter 25).

26.3   CONCLUSION

Flooding is an inevitable phenomenon that seems to have intensified in recent times. There are governmental-level and private-sector initiatives designed to lower the risk of flooding, and cope with it or its impacts. Some of these measures have worked satisfactorily, while others have not. Within this volume, best practice and novel alternative suggestions for helping to lower the risk of flooding, cope with it, or mitigate its damaging consequences have been offered both from theoretical and practical perspectives. Further, questions have been raised that identify areas where further development and research are needed. Therefore, policy and practice initiatives may greatly benefit from these contributions.

Index